Grid Integration of Solar Photovoltaic Systems

Grid Integration of Solar Photovoltaic Systems

Majid Jamil
Professor and Head
Department of Electrical Engineering
Jamia Millia Islamia
New Delhi, India

M. Rizwan
Assistant Professor
Department of Electrical Engineering
Delhi Technological University
Delhi, India

D. P. Kothari
Dean (Research & Development)
J. D. College of Engineering & Management, Nagpur, India
Former Director i/c, IIT Delhi, India
Former VC, VIT Vellore, India
Former Principal, VRCE, Nagpur, India

CRC Press is an imprint of the
Taylor & Francis Group, an **informa** business

CRC Press
Taylor & Francis Group
6000 Broken Sound Parkway NW, Suite 300
Boca Raton, FL 33487-2742

Printed on acid-free paper

International Standard Book Number-13: 978-1-4987-9832-7 (Hardback)

Visit the Taylor & Francis Web site at
http://www.taylorandfrancis.com

and the CRC Press Web site at
http://www.crcpress.com

Contents

Preface

It has been widely accepted that power generation from fossil fuels has to be switched over to a renewable and clean source of energy as quickly as possible. In fact, several initiatives focused on power generation from solar and wind sources have been designed and implemented during the past two decades. The sun is an abundant source of solar energy that is intermittent, eco-friendly, and freely available throughout the world. It can be used as thermal energy as well it can be converted directly to power through solar photovoltaic (PV) technology.

Recently, the penetration of solar PV installations has increased substantially due to several global initiatives undertaken by academia and industries, but it is yet to be considered one of the mainstream renewable energy technologies due to the high cost of power generation and challenges like the efficiency of the solar cell, storage of solar energy, integration of solar to grid, solar panel protection, converter cost, and efficiency. To make the solar PV system commercially viable, the unit generation cost of electricity needs to be reduced, which, in turn, calls for the development of a low-cost, highly efficient power conversion system or scheme to deliver the required electrical power. Hence, it is always critical to study and design the most appropriate solar power system and assess its performance to ensure maximum power capture from solar modules along with perfecting power quality, reliability, and efficiency. Again, the size of the distributed PV plant varies from a few kW to several MW, for which the type and configuration of the inverter also vary. Therefore, the inverter has to be properly selected, as the design and performance of the overall system depend mainly on the inverter. Researchers, engineers, technicians, and PV installers also need to be properly trained. In view of this, there is a need for a book that may be useful for researchers, engineers, students of electrical and mechanical engineering, and even technicians who work in the field and face the day-to-day problems related to design, installation, fault detection, etc.

The objective of writing this book is to prepare a text for PV system users covering all possible aspects of solar PV technologies to provide a platform to disseminate knowledge covering the fundamentals as well as a wide scope for research. The book is written in simple language so that persons who do not have knowledge of this field may easily understand and become familiar with the concept of PV technology. Many pictures, circuits, and graphs on various topics covered in the book are also provided. Practical examples are provided in this book so that a solar PV practitioner may be able to design, install, and analyze the performance of the system. Possible faults and their remedies are also presented in easy-to-understand terms.

The book is divided into 15 chapters. In Chapter 1, exposure to renewable energy sources is presented. Initiative taken by the Indian government to make solar energy feasible and economical are discussed. In Chapter 2, the basics of electrical engineering are included so that even a user who does not have a basic knowledge of electrical engineering may be able to understand a PV system. Predicting the availability of energy before deciding on the size of the power plant is very important; therefore, in Chapter 3, measurement and estimation of solar irradiance are included. In Chapter 4, solar cells, modules, and arrays are discussed. MPPT techniques play an important role in the design of solar PV systems, and a lot of research is underway in this field, so various techniques are discussed in depth in Chapter 5. Solar PV systems require specific power electronic converters to convert the power generated into useful power that can be directly interconnected with the utility grid and/or can be used for specific consumer applications locally. The roles of these power converters are critical, particularly when used in one of the most expensive energy-generating sources such as solar PV. Because the power converter/inverter is the interface of the distributed power system, any premature failure will make the entire system defunct; thus, this component needs to be selected carefully. These power converters need to be selected and designed optimally in order to provide maximum energy efficiency, ensuring reliability and safety of the overall solar PV system for different applications. In Chapter 6, different aspects of converters are discussed. Because the power generated from PV systems fluctuates, a form of backup is required. Therefore, energy storage for PV applications is explored in Chapter 7. Mounting structures and the balance of the system are described in Chapters 8 and 9, respectively. Assessing the site for PV applications is described in Chapter 10. Challenges and possible solutions to the integration of solar photovoltaic systems to the electrical grid are discussed in Chapter 11. Protection is an important component of any reliable system and is discussed in Chapter 12. The success of any technology depends on its cost effectiveness. The economics of generating power from a PV system is presented in Chapter 13. The performance of grid-connected PV systems is described in Chapter 14. The design of the inductor and isolation transformer is discussed in Chapter 15.

About the Author

Dr. Majid Jamil is a professor in the Department of Electrical Engineering, Jamia Millia Islamia (A Central University) in New Delhi, India. Presently he is also Head of the Department. Dr Jamil has more than 25 years of research and teaching experience. He has also served as an Assistant Professor at BITS Pilani Dubai Campus during 2003–2006. Dr Jamil has published more than 90 research papers in internationally refereed journals and conferences. Dr Jamil has guided 10 Ph Ds and 20 M. Tech. dissertations. Presently he is supervising 6 PhDs. He has Edited 2 books and has written three book chapters also. Dr Jamil has received grant of more than Rs 2.0 crores from AICTE and DST, Govt of India for research projects. He has also received the best paper award in a conference at University of California, Berkeley, USA in 2009. Prof. Jamil is fellow of institutes of Engineers and IETE, Senior member of IEEE, and life member of ISTE and ICTP. His areas of interest are power systems, energy, energy auditing and intelligent techniques.

Dr. M. Rizwan did his post-doctoral research at Virginia Polytechnic Institute and State University, USA. He received Ph.D. degree in Power Engineering from Jamia Millia Islamia, New Delhi. He is Assistant Professor at Department of Electrical Engineering, Delhi Technological University, Delhi since 2010. He has more than 15 years of teaching and research experience. He has published and presented more than 65 research papers in reputed international and national journals and conference proceedings. Dr. Rizwan has successfully completed three research projects in the area of renewable energy systems. His area of interest includes power system engineering, renewable energy systems particularly solar photovoltaic, building energy management, smart grid and soft computing applications in power systems. He is the recipient of Raman Fellowships for Post-Doctoral Research for Indian Scholars in USA for the Year 2016–17. Also he was selected for UGC research award for the period of 2014–2016. He is Sr. Member IEEE, Life Member, ISTE, Life Member SSI, Member International Association of Engineers (IAENG), and many other reputed societies. He is also associated with many journals including *IEEE Transactions, International Journal of Electrical Power and Energy Systems* (Elsevier), *Renewable and Sustainable Energy Reviews* (Elsevier), *Solar Energy* (Elsevier), and *International Journal of Sustainable Energy* (Taylor and Francis) in different capacities.

Dr. D. P. Kothari is presently the Dean (Research & Development) J. D. College of Engineering & Management, Nagpur, India. He obtained his BE (Electrical) in 1967, ME (Power Systems) in 1969, and PhD in 1975 from BITS, Pillani, Rajasthan, India. From 1969 to 1977, he was involved in teaching and developing several courses at BITS Pillani. Earlier, Dr. Kothari served as vice chancellor, VIT, Vellore; director in charge deputy director (Administration), and as head of the Centre of Energy Studies at the Indian Institute of Technology, Delhi; and as principal, VRCE, Nagpur. He was a visiting professor at the Royal Melbourne Institute of Technology, Melbourne, Australia, from 1982–1983 and 1989. He was also an NSF Fellow at Purdue University, Indiana, in 1992.

Dr. Kothari, who is a recipient of the Most Active Researcher Award, has published and presented 812 research papers in various national and international journals and conferences, guided 48 PhD scholars and 68 MTech students, and authored 49 books in various allied areas. He has delivered several keynote addresses and invited lectures at both national and international conferences. He has also delivered 42 video lectures on YouTube with a maximum of 40,000 hits!

Dr. Kothari is a Fellow of the National Academy of Engineering (FNAE), Fellow of the Indian National Academy of Science (FNASc), Fellow of Institution of Engineers (FIE), Fellow IEEE, and Hon. Fellow ISTE.

His many awards include the National Khosla Award for Lifetime Achievements in Engineering (2005) from IIT, Roorkee, and the University Grants Commission (UGC), government of India, has bestowed the UGC National Swami Pranavandana Saraswati Award (2005) in the field of education for his outstanding scholarly contributions.

He is also the recipient of the Lifetime Achievement Award (2009) conferred by the World Management Congress, New Delhi, for his contribution to the areas of educational planning and administration. Recently he received an Excellent Academic Award at IIT Guwahati by NPSC-2014. He has received six Lifetime Achievement Awards from various other agencies as well.

Acknowledgment

It is a great pleasure to acknowledge the assistance and contributions of many individuals to this effort. First we would like to thank Aastha Sharma, Shikha Garg, Todd Perry, and the other team members at CRC Press/Taylor & Francis Group, as well as Christina Nyren at Diacritech for editing and layout, for enabling us to publish this book. We would like to express our gratitude to all those who provided support, offered comments, allowed us to quote their remarks, and assisted in the editing, proofreading, and design of this book.

We would also like to extend our thank to Prof. Yogesh Singh Vice Chancellor, DTU; Prof. Madhusudan Singh, Head, Department of Electrical Engineering, DTU; Jamia Administration (Prof. Moinuddin, Prof. Parmod Kumar, Prof. Mini S Thomas, Prof. Munna Khan, Prof. Anwar Shehzad Siddiquie, Prof. Sirajuddin Ahmad, Prof. Mumtaz Ahmad Khan, Prof. Shabana Mehfuz, Dr. Sheeraz Kirmani, Dr. Ahmad Shariq Anees, and Dr. K. M. Rafi); and our Ph.D. students (Mohd. Ehtesham, Shamshad Ali, Priyanka Chaudhary, Astitva Kumar, Tausif Ahmad, Mohini Yadav, Arjun Balyan and Tanushree Bhattachargee).

We wish to thank the authorities of JD College of Engineering & Management, Nagpur, India, for their constant help & cooperation in writing this book, without which this book would not have seen the light of the day.

Above all, we gratefully acknowledge the support, encouragement, and patience of our families. It was a long and difficult journey for them.

Last but not least, we beg forgiveness of all those who have been with us over the course of the years and whose names we have failed to mention.

Majid Jamil
M. Rizwan
D. P. Kothari

1

Exposure to Renewable Energy Sources

1.1 Conventional Energy Sources

To consider a country a developed nation, the per capita energy consumption needs to be considered, which is an important key factor. The maximum generation of energy depends on the needs of a country, and this need can only be fulfilled when everyone gets a sufficient amount of energy in the form of electricity, transport, agriculture etc. In order to satisfy these requirements, today the world is mainly dependent on conventional resources such as coal, oil, natural gas, etc. India imports about 75% crude oil at present, and a steep increase in this figure is expected in the near future because of India's growing economy and rapid development. 18% of the world's population resides in India, but uses only 6% of world's primary energy in spite of being the third largest in power generation. Nearly 240 million people live in India without electricity. Developing countries like India and China face the problem of increasing per capita electricity consumption, and for that they will have to increase electricity generation. They still need to provide electricity to the entire population. Developed countries like the United States and European countries will need to produce more electricity if they want to maintain their current quality of life.

Most electricity is produced through the combustion of coal, oil, and natural gas. The combustion of these fossil fuels increases the amount of carbon dioxide in the atmosphere and adds several other polluting gases. Carbon dioxide is a strong greenhouse gas responsible for rising global temperatures. The second aspect of using fossil fuels is that their proven reserves are not going to last very long, even for a century, as a power generating source. This calls for finding an alternative source for producing electricity.

The planning of electricity production is influenced by two major factors. One important factor is the basic raw material, which could be any of the fossil fuels, and all three of them are known to increase global warming and environmental pollution. Second are the matters that have relevance to the economics of electrical power. There are huge power losses in terms of generation, transmission, and distribution. It is a challenging task to reduce the losses or identify an alternative method like a smart grid, micro grid, etc., so that the losses may be reduced and the cost may also come down. People want access to electricity but do not want to increase global warming, and both the general public and those who plan and establish the power plants are convinced that fossil fuels are responsible to a very great extent for the addition of greenhouse gases. In addition, the poisonous gases and particulate matter they add to the atmosphere are a proven danger to the respiratory systems of human and animals, as well as damaging to plants. To avoid global warming and the deleterious effects of these pollutants, efforts have been directed toward nonconventional or renewable energy sources (RES), but they have proved to be uneconomical as of now.

1.1.1 Energy Resources and Their Potential

1.1.1.1 Oil

The world has been divided into three groups based on oil and its use. Some countries are rich in oil, and another group is highly industrialized with a rich agriculture. The representatives of the first group are mainly the Middle Eastern or Arabian countries. The OECD (Organisation for Economic Co-operation and Development) represents the other group. There is a third group, which neither has oil nor is developed to the extent of the OECD group. The question has always been lurking if the oil-producing countries will keep on supplying oil to the world until the supply runs dry or they decide that they should keep the oil for their own use or sell to only those countries that approve their policies.

The present situation of supplying oil to the developed countries started because the Arab countries were resource poor in terms of food. They got food in return for oil—not as a barter agreement but as a gesture. Gradually the Arab countries increased their economic strength and started other businesses and started asserting their preferences in the politics of the world. A kind of hidden difference has arisen between the developed countries and the Arabs. The present crisis of terrorism and American actions in the Middle East are exposing the differences even further, and as the oil taps are exhausted, the differences will become more apparent. In the early days, Arabs were happy to get food in return for oil, and this tended to maintain a balance. The balance has been tilted in favor of oil suppliers, however, as can be seen, between 1950 and 1973 the exchange rate between food and oil was *one bushel of food for one barrel of crude oil.* This increased in accordance with exchange rates: *two bushels per barrel in 1974, five bushels per barrel between 1975 and 1998, six bushels in 1999, and now ten bushels or more per barrel of crude oil.* It is, of course, a fact that oil is also needed for agriculture and transport of food, but the Arab countries have found food suppliers outside of the OECD. They are also interested in building seawater *greenhouses.* This technology, although new and untried, offers cheap desalination and efficient food production. Successful implementation of this technology will bring Arab countries closer to self-sufficiency and away from the OECD. This is going to be a hallmark in terms of politics if this occurs.

1.1.1.2 Natural Gas

The technologies for natural gas-fired electricity generation have improved during the last decade and a half. It is expected that the use of natural gas in electricity production will increase by 87% by 2020. The most reliable estimates suggest that in the developed countries natural gas will support 30% of electricity generation, whereas the developing countries will derive 17% of their electricity from natural gas. Use of natural gas in electricity generation in combined-cycle gas turbines has been developed technically, and many countries are intent on using it. Further, the gas produces less carbon dioxide than both coal and oil while producing the same amount of electricity. It is cleaner fuel and is being used for public transport, such as in Delhi, for the specific purpose of keeping the environment clean.

The consumption of gas for electricity is already high in countries of the former Soviet Union with almost 33% of consumption occurring in that region. By 2025 it is expected that up to 63% of electricity will be generated from gas in the countries of the former Soviet Union. The East European countries were producing 9% of their electricity from gas in 2001, and they expect to import more gas from Russia to increase electricity generation up to 50%.

Western Europe consumed 413 billion cubic meters of gas for electricity generation in 2000 and is likely to increase it to 670 billion cubic meters by 2025. With a decreasing share of nuclear energy, the region is likely to have the gas share in electricity generation increase from 17% in 2001 to 38% in 2025. Western European countries have experienced fluctuating natural gas use, starting in 1973 when the oil crisis took place. In 1975 the European Union restricted the use of gas for power production. It was at 5% in 1981 and remained at that level during the 1980s. In the early years of the 1990s the region started getting gas from Russia, North Africa, and newly discovered fields in the North Sea, and thus the share of gas in the electricity market increased, and this pattern is being maintained.

Asia and Africa are not yet big users of gas. Japan consumes almost one fourth of all the gas consumed by Asian countries, and it imports all its gas as a liquefied product (LNG). India will turn out to be a big gas consumer, and it generates about 13.6% of its electricity from gas.

The United States imports most of its gas from Canada and draws a good amount through pipelines from the Alaskan North Sea. The electricity share of gas is likely to increase from 18% in 2001 to 24% in 2025. In recent years all the newly constructed power plants in the United States, accounting for 141,000 MW, are gas based. Imports from overseas will likely continue to increase. The United States imported 4.8 million metric tons of gas in 2002, which was 4% of the world's consumption. The import was doubled in 2003 to about 46 million metric tons in 2010. Thus, the United States is importing gas in large quantites, next to oil, as energy sources. Canada is planning to increase electricity generation by gas from 3% to 11% by 2025.

Natural gas prices are on the increase. Since 1999, prices have more than doubled. In 2000 gas was quoted for $2.55 per million BTU, but in 2003 the price rose to $6.31 per million BTU. Three countries that are likely to benefit from international gas trading are Russia, Iran, and Qatar, which hold gas deposits of 31%, 15%, and 9%, respectively.

1.1.1.3 Coal

Coal is the oldest fuel used for electricity generation. It produces a larger quantity of carbon dioxide in comparison with oil and gas, along with many other air pollutants, which makes it unpopular in developed countries. It produces sulfur dioxide, nitrous oxide, mercury, and particulate matter in addition to carbon dioxide. 64% of total coal production is used for electricity generation. The production stood at the equivalent of 94.5 EJ (Exa Joule) in 2001. It is expected that production will increase to 138 EJ by 2025. However, coal's share in electricity generation is going to reduce according to all estimates. It was responsible for 34% of electricity in 2001 and for 40% of electricity in 2005, but its share in terms of power will reduce to 31% by 2025. Europe was producing 20% of its electricity through combustion of coal, but it will produce only 12% of electricity in this manner by 2025. In 2001 the United States consumed 40% of all the coal, whereas China and India together used 27%. Both China and India produced 72% of their electricity by using coal. China is going to increase its electricity production from coal to 73% by 2025. On the other hand, India's share of coal for producing electricity is going to drop to 63%. The coal share of U.S. electricity is predicted to remain roughly at 50% by 2025.

In Eastern Europe and the former Soviet Union, the contribution of coal to electricity was 27% in 2001. But with the availability of gas from Russia, this share is going to reduce 6% by 2025. The burning of coal has made Eastern Europe the most polluted region in the world. Europe, in general is moving away from using coal for electricity. Most coal mines there were government subsidized. The European Union has adopted a policy of reducing or removing such subsidies, and only three countries—the UK, Germany, and Spain—continue to produce hard coal. The world deposits of coal are described in Table 1.1.

TABLE 1.1

Coal Deposits in the World

Region	Deposit (Gt)	Percent
Asia Pacific	292.5	29.7
North America	257.8	26.2
Former Soviet Union	230	23.4
Europe	125.4	12.7
Africa	55.4	5.6
South & Central America	21.8	2.2
Middle East	1.7	0.2
World	984.6	100

A typical 600-MW coal-fired power plant will consume 8,193 tons of coal every day, which may have to be brought from coal mines over a distance, which could be 1,000 km on average. The transport of such a huge mass of coal will require regular train movements every day, and several such plants are situated throughout the countries of the world. This requires an efficient rail system with good haulage capacity, but the transport itself will burn a great deal of fuel and add carbon dioxide and other pollutants. When coal is burned, it typically produces 17,900 tons of nitrous oxide, 8,200 tons of sulfur dioxide, 4,207,200 tons of carbon dioxide, and 70 kg of mercury. Additionally, the plants emit fly ash and radioactive material. The modern design of the plant is such that the plant and its surroundings appear clean, but all the gases and ash that come out of chimney affect areas 50 km to 100 km from the plant. Studies conducted in the United States between 1975 and 1985 concluded that an individual living within 1 km of a coal-fired power plant receives a radiation dose that is 1% to 5% above the average natural background level. In contrast, the typical increase from living at the boundary of a nuclear power plant is about 0.3% to 1% above the natural background level. There is no regulation of radiation from coal-fired plants. Mercury is a much bigger health issue that has not been addressed yet. Mercury is a toxic, persistent, biocumulative pollutant that affects human neurological development. Infants and growing children are most affected by mercury in the air. The technology for capturing mercury from flue gas is not yet developed, and it will be difficult to separate mercury from flue gas at a concentration of 1 ppb (part per billion).

1.1.1.4 Hydropower

It was the free kinetic energy of flowing water that provided energy for the Industrial Revolution 150 years ago. Bladed wheels were placed in the path of flowing water and water falling from a great height. The motion of the water turbines was first used to run the machines and then to run the electricity generators. Later, hydropower was fully dedicated to electricity generation because the water turbines could be made in such a large size and the technology of building dams was developed, which provided a source of water year round. There are several dams now in countries like the United States, China, Brazil, Venezuela, Sudan, Pakistan, and India. In the early 1900s Tata built a hydroelectricity plant in Western Ghats in India. A giant leap was taken when the Bhakra–Nangal hydro project was completed after independence. Several dams were built afterward.

India has the potential for about 84,000 MW of electricity through the hydro route. Sources exploited so far account for 15,000 MW. Further development of hydropower has been restricted mainly due to general public resistance. The dams create a large water body by displacing the population and destroying the habitats of several life forms. Further, the depth of the water reservoir behind the dam keeps reducing as silt deposits. Dredging operations for removing the silt are difficult and costly. Small hydro plants are preferred over large ones. The small plants produce power in terms of a few kW, which could be sufficient for a small region. Rivers flowing through hilly areas, such as in the Indian Himalayan regions, are most suitable for small hydro plants.

The global consumption in 2000 was 2,726 TWh/year. Although Canada, Brazil, the United States, and China were the leaders, together consuming 43% of hydroelectricity, Norway produced almost 99% and Brazil 87% of their electricity demands from this source. Industrialized countries consume 65% of global hydroelectricity. By 2020 hydroelectricity consumption is forecast to increase to 4,000 TWh/year, with developing countries contributing 800 and industrialized countries 600 TWh/year. The gap in hydroelectricity generation between industrialized and developing countries will narrow, whereas the gap between per capita consumption will further increase by 2020. Central Asia, South America–Caribbean, Sub-Saharan Africa, and South Asia were the regions that saw extensive construction of hydroelectricity generation facilities recently. It is of concern to note that whereas 5,738 TWh/year hydro capacity was installed in 1997, the electricity obtained from this source was only 2,600 TWh/year, which accounts for a 55% loss of capacity.

1.1.1.5 Nuclear Energy

Nuclear energy, contrary to many beliefs, offers a source of energy that is much cleaner. It is free of the greenhouse effect and does not throw particulate matter into the atmosphere. The bulk of nuclear fuel is much smaller than any of the fossil fuels to produce the same amount of electricity, and its transportation causes much less atmospheric pollution. The cost of a nuclear power plant is comparable to the cost of any thermal power plant and is less than that of a renewable-energy-source power plant. However, the problem with the nuclear source is the lack of public acceptance because there is a general perception that these plants are the source of nuclear radiation, which is a killer and through the nuclear waste of spent fuel, the risk is going to perpetuate.

The public perception has generally stemmed from the United States' policy of discarding the nuclear pathway for electricity in the light of large-scale public sentiment against nuclear power, despite the fact that it was the United States from which countries like Japan and France drew encouragement to depend upon nuclear electricity. Then in the back of the public mind there exists an image that relates nuclear energy with the atomic bomb, which devastated the two cities of Japan and ended the Second World War on a melancholy note. The American people have largely been fed the idea that terrorists may get hold of nuclear material and use it against any group of people. Therefore they have rejected the very idea of using nuclear energy.

On the threshold of economic and industrial revival after the Second World War, the Japanese were determined not to be left behind regarding this technology. Japan's science and technology agency was given a budget of $1.88 billion to start. The first commercial power plant began operation in 1970. Today 41 commercial reactors in Japan supply 33,000 MW of electrical power, which is equal to about 27% of the country's power requirements. Japan bought its first gas-cooled, graphite-moderated nuclear reactor from the UK. When Westinghouse and General Electric of the United States offered pressurized water reactor (PWR) and boiling water reactor (BWR) designs, the Japanese were the first to import them. Then Westinghouse and General Electric started manufacturing them under license.

Although initially done for the prestige, Japan now finds that nuclear energy is a must because it imports 80% of its energy in form of Middle Eastern oil and Chinese coal. Japan is the fourth largest per capita consumer of electricity in the world.

Perhaps a lack of energy sources was the reason that prompted the French government to establish their Atomic Energy Commission (CEA) immediately after the war in 1945. Unfortunately, they concentrated more on the development of bombs than on nuclear electricity to which they reverted only after exploding the test nuclear device in 1960. By that time one PWR and three gas-graphite commercial nuclear reactors were also under construction. The *Electricité de France* (EDF) and CEA had to face a great deal of resistance from the public, bureaucrats, and politicians. The path to nuclear energy became clear after French President Giscard d'Estaing declared in 1977 that nuclear electricity was the policy of the government. Several plants were established in the late 1970s. Nuclear power accounts for 75% of French electrical output. 54 reactors in France produce 63,000 MW of power, which provides 4,931 kWh of electricity per person. This compares very favorably with 2,265 kWh/person in the United States and 1,450 kWh/person in Japan. France has developed facilities for reprocessing spent nuclear fuels, which are used by Japan and many European countries.

Germany, Norway, and Finland have decided to scale down their nuclear electricity operation. Germany believes that it can rely upon renewable energy sources. A politician in Germany remarked that if the country could not produce all the energy it needed, it would buy it from France.

India started activity in this field early in 1948 when their Atomic Energy Commission was established under the chairmanship of Homi Bhaba and leadership of Prime Minister Jawahar Lal Nehru. The commission, now named Bhaba Atomic Research Centre, has carried out developmental work and research in the area of nuclear science and engineering. A General Electric–built nuclear power plant started operating at Tarapur in 1969. It currently produces 2,700 MW of electrical power in 14 plants. This is expected to increase to 20,000 MW by 2020. The problem is that the basic mineral containing uranium is available in very small quantities in India, although a possible material could be thorium, which is more readily available. Success will depend upon efficiently using thorium as a nuclear fuel. Meanwhile, India has to depend upon other sources for fuel, and agreement with the United States has become a political issue. Still it is desirable that the government should boost a nuclear program for electricity. Research should be supported in laboratories and universities.

India's favor for nuclear electricity offers the double advantage of providing business and not burning fossil fuels, which is the cause of the greenhouse effect and climate change. By not producing carbon dioxide, sulfur dioxide, nitrous oxide, and particulate matter we could keep our environment that much cleaner.

At present, India produces about 1,650 MW of electricity from its 17 reactors, whose combined capacity is 4,120 MW. These reactors are not being worked to full capacity because of an inadequate and erratic fuel supply. India still depends upon coal power, which is highly polluting in terms of greenhouse gases, air-polluting gases, and particulate matter. The contribution of coal to Indian electricity is to the extent of 55%, and it will increase in the future if options for RES such as solar, biomass, wind, hydro, etc., and nuclear electricity do not become available. Highly ambitious estimates place the contribution from the nuclear source at 40,000 MW by 2020, but careful estimates by the BARC (Bhabha Atomic Research Centre) limit nuclear electricity to 10,000 MW even if adequate fuel supplies are ensured. The government is planning to add electricity-generating stations of larger capacities even without the help of any other country. Examples are the ongoing work of two units of 220 MW each at Rawatbhata, one of 200 MW at Kaiga, and two units of 1,000 MW each at Koondankulam. To keep the size of the unit small, around 220 MW will be better to prevent massive power shortages in case of failure, although the economy of scale may inspire larger size units.

The corporate sector in India has started planning to move to nuclear electricity. Reliance Power has already started negotiating with leading international companies in the field of nuclear power. The company is trying to form strategic alliances. Tata Power, another energy giant in the private sector, plans to either go alone or in a joint venture after the nuclear electricity sector is opened to private investment. Tata Power plans an initial investment of Rs. 12,000 crores. Bharat Heavy Electricals (BHEL), a leading manufacturer of power equipment in the country, is planning to invest Rs. 1 lakh crores in the next five years to expand facilities in the area of nuclear power. Larsen and Toubro (L&T), a manufacturing giant, also plans to enter the manufacturing business of nuclear equipment. About 40 companies, including Videocon and Sajjan Jindal Group, are in talks with foreign companies to collaborate on a planned investment of 2 lakh crore rupees. Delegations from NSG (Nuclear Supplier Group) countries have started arriving in India to explore the nuclear electricity market. The waiver from NSG has brought all-around cheer in the share market, and many companies have boosted stocks. However, all these plans are contingent upon the successful completion of an Indo–American nuclear deal.

Power generation in India in different regions of the country is shown in Figure 1.1, and a pictorial representation of India's installed capacity is shown in Figure 1.2.

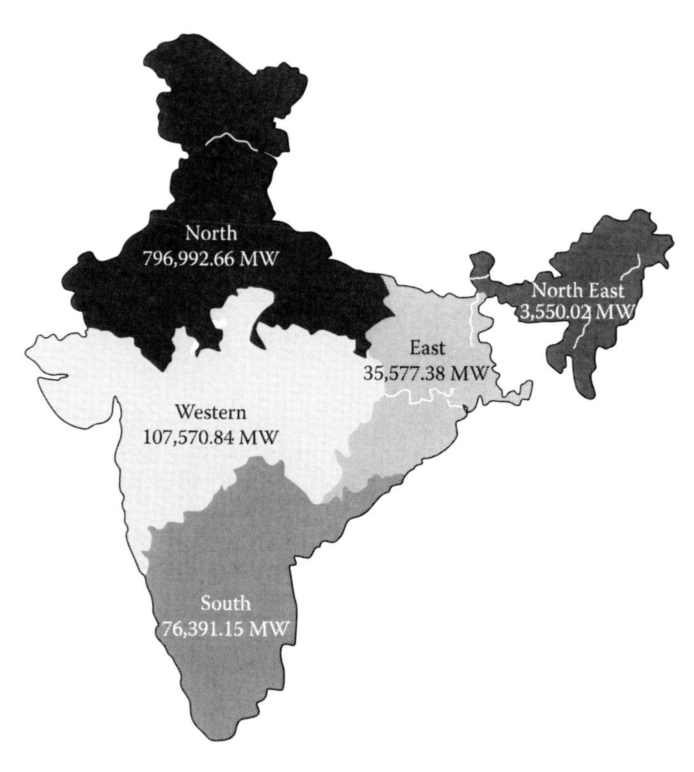

FIGURE 1.1
Regional electricity board of India with their installed generation capacity.

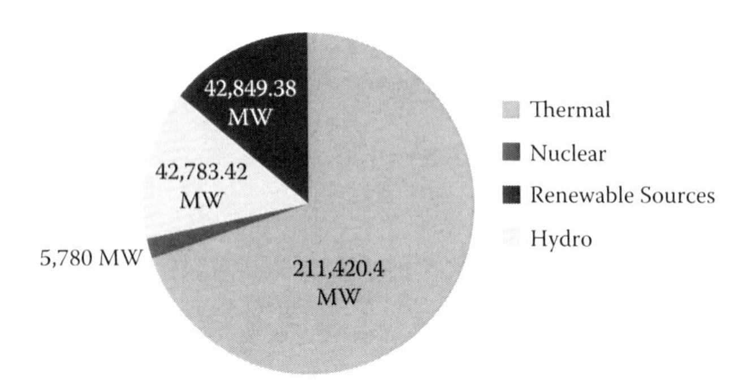

FIGURE 1.2
Pictorial representation of India's installed capacity.

1.2 Need for Renewable Energy Sources (RES)

Energy is most essential for the economic, social, and industrial development of a country. It is expected that in coming years the energy consumption will be many times more than at present. India is fast-growing country, expanding its industrial growth on one hand, and on the other hand there is a huge shortage of power in the country. Most of the power India was producing came from conventional sources until 2000; then, realizing the adverse effect of power generation from these conventional sources, India started generating power from RES. During the past decade a lot of investments have been made, and at present more than 14% of total installed power in India is generated from RES.

The role of RES has increased considerably in current years due to increasing energy demand with minimum ecological impact. RES are environmentally friendly, clean, safe, and sustainable sources of power. RES can complement the present gap of supply and demand and at the same time provide neat and clean energy. India has a huge potential of RES. Solar, wind, and bio all have excellent potential to generate power. The share of RES and total installed capacity of the Indian grid at different five-year plans is shown in Table 1.2.

Figure 1.3 shows the growth of RES during different five-year-plans. During the 11th and 12th plans, growth from RES increased many times as compared to previous plans. Figure 1.4 presents growth of total installed capacity and growth of RES during different plans.

The renewable energy sector is now a well-known sector in India with significant installed grid-quality renewable power, which has reached 45 GW as of June 2016. A large number of decentralized renewable energy systems such as biogas plants, solar water heating systems, biomass gasifiers, Solar Photovoltaic (SPV) systems, etc., have been promoted under various schemes of the Ministry of New and Renewable Energy (MNRE).

The government of India has already initiated many plans to provide power to remote areas of India. Solar photovoltaic is a free source and available easily everywhere in India and is being groomed for several government schemes like Jawaharlal Nehru National Solar Mission (JNNSM), state government policies, NSM (National Solar Mission), etc. With these schemes, the government of India has two main aims: first to provide power to all and second to generate power with fewer carbon contents emitted into the environment.

TABLE 1.2

Share of Renewable Energy Sources (RES) and Total Installed Capacity of Indian Grid at Different Plans

Plan Period	Share of RES (MW)	Total Installed Capacity (MW)
6th Plan (1980–1985)	0	42,584.72
7th Plan (1985–1990)	18.14	63,636.34
2nd Annual Plan (1990–1992)	31.88	69,065.19
8th Plan (1992–1997)	902.01	85,797.37
9th Plan (1997–2002)	1628.39	103,410.04
10th Plan (2002–2007)	7760.6	132,329.21
11th Plan (2007–2012)	24,503.45	199,877.03
12th Plan (2012–April 30, 2016)	43,086.82	302,833.2

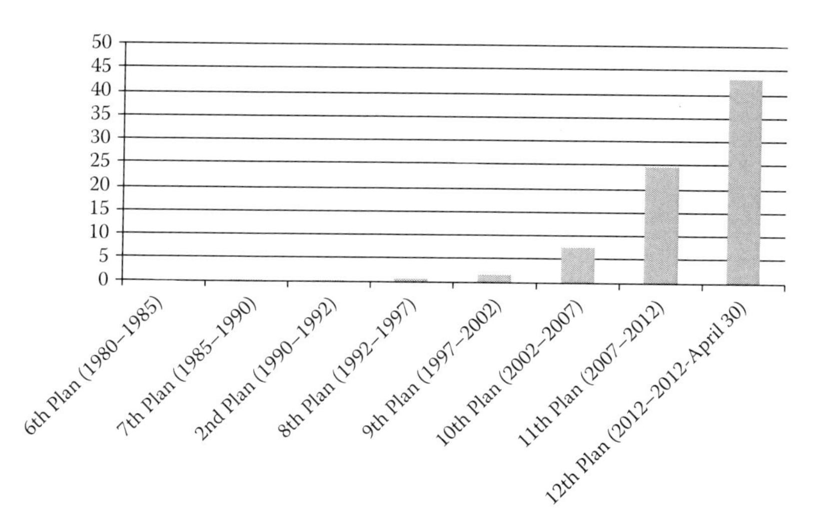

FIGURE 1.3
Growth of renewable energy sources (RES) during different plans.

With successful implementation of these schemes in India, the current status is shown in Figure 1.5.

Figure 1.5 shows the installed capacity of renewable energy in India through May 2016. The total renewable energy used to generate power is about 45 GW, and this figure is expected to rise fourfold until the end of 2022. Based on the initiatives of the government of India and technology development, the projected generation until 2022 is shown in Figure 1.6, and the percentage shares each renewable energy source is expected to have in India in 2022 is shown in Figure 1.7.

Region-wise data projected by the year 2022 are shown in Figure 1.8. The figure shows that the main focuses for solar power are in the Northern and Western regions due to ample amounts of sunlight. In addition, the figure illustrates the growth and development in the field of renewable energies and enhances the dependency of power on renewable sources, particularly on solar.

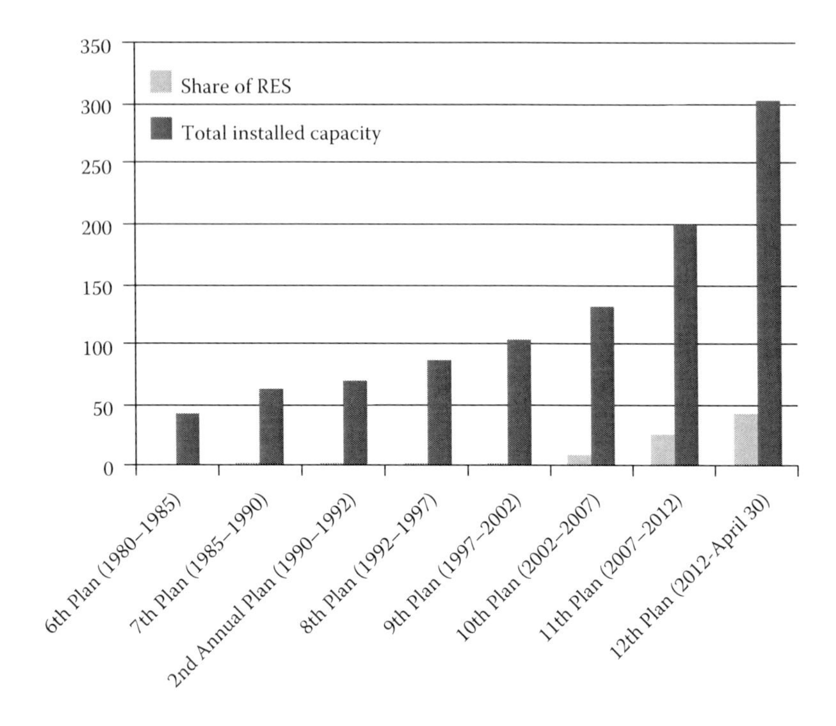

FIGURE 1.4
Growth of total installed capacity and growth of renewable energy sources (RES) during different plans.

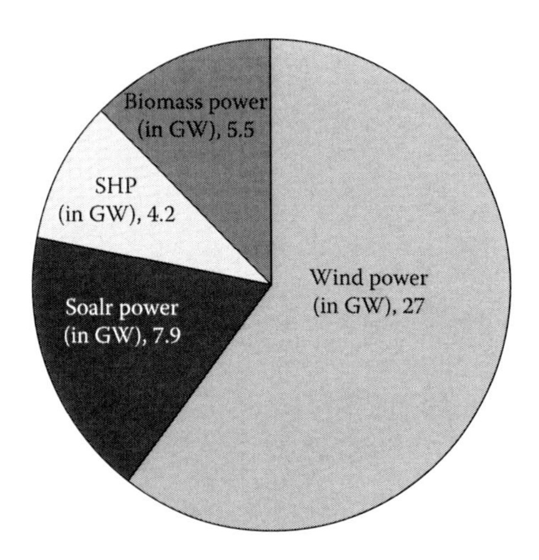

FIGURE 1.5
India's renewable energy installed capacity until May 2016 (45 GW).

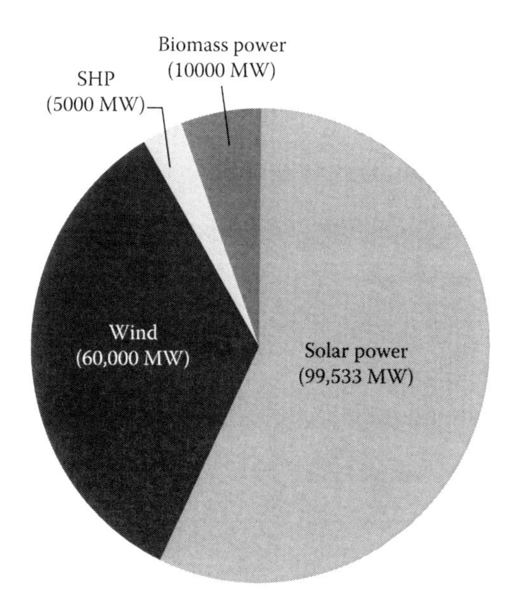

FIGURE 1.6
Projected renewable energy in India by the year 2022.

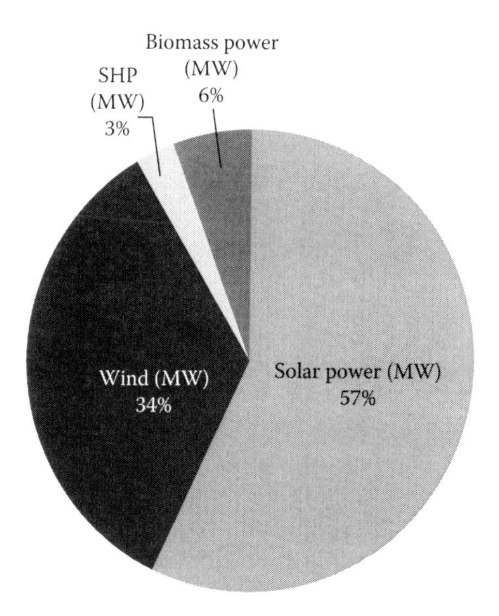

FIGURE 1.7
Renewable energy shares to be achieved by the year 2022 in India.

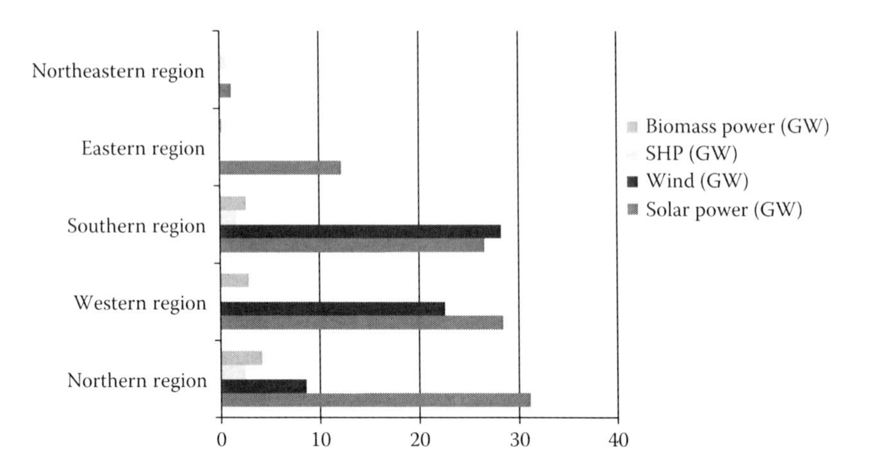

FIGURE 1.8
Region-wise distribution of renewable energy by the year 2022 in India.

1.3 Potential Renewable Energy Sources (RES) for Power Generation

1.3.1 Solar Energy

India has 300 clear days in a year, which is considered quite good for solar power generation. India is receiving on average solar energy of about 5 to 7 kWh/m^2 in most parts of the country. Realizing the huge potential of solar, wind, and biofuels, the government of India has established a separate ministry—the Ministry of New and Renewable Energy (MNRE)—which has initiated a number of programs to make these resources affordable for power generation. New schemes to harness the vast solar energy potential in the country have been announced, and subsidies have been provided to install solar power plants. The Jawaharlal Nehru National Solar Mission was launched on January 11, 2010, by the prime minister. The mission has set the ambitious target of deploying 20,000 MW of grid-connected solar power by 2022, which is aimed at reducing the cost of solar power generation in the country through (i) long-term policy; (ii) large-scale deployment goals; (iii) aggressive research and development (R&D); and (iv) domestic production of critical raw materials, components, and products to achieve grid tariff parity by 2022. The mission will create an enabling policy framework to achieve this objective and make India a global leader in solar energy.

Implementation of a scheme for the development of solar parks and ultra-mega solar power projects in the country commencing from 2014–2015 and onwards (i.e., from 2014–2015 to 2018–2019) has been announced. The scheme aims to provide a huge impetus to solar energy generation by acting as a flagship demonstration facility to encourage project developers and investors, prompting additional projects of a similar nature and triggering economies of scale for cost reductions, technical improvements, and large-scale reductions in greenhouse gas (GHG) emissions. It would enable states to bring in significant investment from project developers, meet its solar renewable purchase obligation (RPO) mandate, and provide employment opportunities to the local population. The state will also reduce its carbon footprint by avoiding emissions equivalent to the solar park's installed capacity and generation. Further, the state will avoid procuring expensive fossil fuels to power conventional power plants

1.3.2 Wind Energy

India has achieved great lead in terms of generating power from wind. India has the number-four position in the world, with wind power–installed capacity of 27 GW as of June 2016. The government is encouraging the participation of private partners in installation of wind-based power plants in the country and providing a friendly environment for these plants.

Wind in India is influenced by the strong southwest summer monsoon season, which starts in May to June, when cool and humid air moves toward the land, and the weaker northeast winter monsoon season, which starts in October, when cool and dry air moves toward the ocean. From March to August, wind is uniformly strong over the whole Indian coast, except the eastern coast. Wind speeds from November to March are relatively weak, although higher winds are available during part of this time on the Tamil Nadu coastline. These two monsoon winds make the generation potential in Tamil Nadu much higher than in other states. The potential for wind power generation for grid interaction has been estimated at about 45,000 MW on a macro level based on data collected from 10 states considering only 1% land availability.

1.3.3 Biomass Energy

Biomass has always been an important energy source for the country, considering the benefits it offers. It is renewable, widely available, carbon neutral, and has the potential to provide significant employment in the rural areas. Biomass is also capable of providing firm energy. About 32% of the total primary energy use in the country is still derived from biomass, and more than 70% of the country's population depends upon it for its energy needs. The MNRE has realized the potential and role of biomass energy in the Indian context and hence has initiated a number of programs to promote efficient technologies for its use in various sectors of the economy to ensure derivation of maximum benefits. Biomass power generation in India is an industry that attracts investments of over Rs. 600 crores every year, generating more than 5,000 million units of electricity and yearly employment of more than 10 million man-days in the rural areas. Bagasse-based cogeneration in sugar mills and biomass power generation have been taken up under a biomass power and cogeneration program.

This program has been implemented with the main objective of promoting technologies for optimum use of the country's biomass resources for grid power generation. Biomass materials used for power generation include bagasse, rice husk, straw, cotton stalks, coconut shells, soya husks, deoiled cakes, coffee waste, jute wastes, groundnut shells, sawdust, etc.

The current availability of biomass in India is estimated at about 500 million metric tons per year. Studies sponsored by the ministry have estimated surplus biomass availability at about 120 to 150 million metric tons per annum, covering agricultural and forestry residues, corresponding to a potential of about 18,000 MW. An additional 7,000 MW of power could be generated through bagasse-based cogeneration in the country's 550 sugar mills, if these sugar mills were to adopt technically and economically optimal levels of cogeneration to extract power from the bagasse produced by them.

States that taken leading positions in the implementation of cogeneration projects are Andhra Pradesh, Tamil Nadu, Karnataka, and Uttar Pradesh. The states that are leading projects of biomass power are Karnataka, Andhra Pradesh, Maharashtra, Tamil Nadu, and Chhattisgarh.

1.3.4 Small Hydropower Plants

The Small Hydro Power (SHP) Programme is one of the thrust areas of power generation from renewable in the MNRE. It has been recognized that small hydropower projects can play a critical role in improving the overall energy scenario of the country—in particular for remote and inaccessible areas. The ministry is encouraging development of small hydro projects in both the public and private sectors. Equal attention is being paid to grid-interactive and decentralized projects.

The ministry's aim is that the SHP installed capacity should be about 7,000 MW by the end of the 12th Plan. The focus of the SHP program is to lower the cost of equipment, increase its reliability, and set up projects in areas that give the maximum advantage in terms of capacity utilization. The MNRE has been vested with the responsibility of developing SHP projects with up to 25-MW station capacities. The estimated **potential** for power generation in the country from such plants is about 20,000 MW. Most of the potential is in the Himalayan states as river-based projects and in other states on irrigation canals. The SHP program is now essentially private investment driven. Projects are normally economically viable, and the private sector is showing lot of interest in investing in SHP projects. The viability of these projects improves with increase in the project capacity. The ministry's aim is that at least 50% of the potential in the country is harnessed in the next 10 years.

1.3.4.1 Hydropower Project Classification

Hydropower projects are generally categorized in two segments: small and large hydro. In India, hydro projects up to 25-MW station capacities have been categorized as SHP projects. Whereas the Ministry of Power is responsible for large hydro projects, the mandate for the small hydropower projects (up to 25 MW) is given to the MNRE. Small hydropower projects are further classified as presented in Table 1.3.

1.3.5 Geothermal Energy and Its Potential in India

Geothermal energy is one of the potential alternative sources of energy that has been successfully meeting both industrial and domestic energy requirements in many parts of the world over the last few decades. Geothermal comes from two Greek words: *geo*, which means "earth," and *therme*, which means "heat." Thus, geothermal energy is the heat from the earth. It is a clean and sustainable source of energy. Resources of geothermal energy range from the moderate-to-low temperatures of hot spring systems to hot rock found a few miles beneath the earth's surface, and down even deeper to the extremely high temperatures of molten rocks. Below the earth's crust, there is a layer of hot and molten rocks called magma. Heat is continually produced there, mostly from the decay of naturally radioactive materials such as uranium and potassium. Heat flows outward from the earth's

TABLE 1.3

Classification of Small Hydro Power Plants

Class	Station Capacity in kW
Micro Hydro	Up to 100
Mini Hydro	101 to 2,000
Small Hydro	2,001 to 25,000

interior. Normally, the crust of the earth insulates us from the interior heat. The mantle is semi-molten, the outer core is liquid, and the inner core is solid. It is interesting to mention here that the amount of heat within 10,000 meters of earth's surface is 50,000 times more energy than all the oil and natural gas resources in the world. In fact, geothermal energy is one of the oldest natural sources of heat and dates back to the Roman times, when the heat from the earth was used instead of fire to heat rooms and/or warm water for baths. Presently, it is being used as a source to produce electricity, mainly along plate margins.

India has huge potential to become a leading contributor in generating ecofriendly and cost-effective geothermal power. Around 6.5 percent of electricity generation in the world could be provided with the help of geothermal energy, and India would have to play a bigger role in this direction in the coming years. But power generation through geothermal resources is still in nascent stages in India. The Geological Survey of India has identified about 340 geothermal hot springs in the country. Most of them are in the low surface temperature range from 37°C to 90°C, which is suitable for direct heat applications. These springs are grouped into seven geothermal provinces: Himalayan (Puga, Chhumathang), Sahara Valley, Cambay Basin, Son-Narmada-Tapi (SONATA) lineament belt, West Coast, Godavari basin, and Mahanadi basin. Some of the prominent geothermal resources include Puga Valley and Chhumathang in Jammu and Kashmir, Manikaran in Himachal Pradesh, Jalgaon in Maharashtra, and Tapovan in Uttarakhand. A new location of geothermal power energy has also been found in Tattapani in Chhattisgarh. In addition, Gujarat is set to tap geothermal electricity through resources that are available in Cambay between Narmada and Tapi River. Puga, which is located at a distance of about 180 km from Leh in the Ladakh region of Jammu and Kashmir across the great Himalayan range, is considered to be a good potential source of geothermal energy. In Puga Valley, hot spring temperatures vary from 30°C to 84°C (boiling point) and discharge up to 300 liters/minute. A total of 34 boreholes ranging in depths from 28.5 m to 384.7 m have been drilled in Puga Valley. Thermal manifestations come in the form of hot springs, hot pools, sulfur condensates, and borax evaporates with an aerial extent of 4 km. The hottest thermal spring shows a temperature of 84°C, and the maximum discharge from a single spring is 5 liters/second.

1.3.6 Wave Energy

Oceans cover 70% of the earth's surface and represent an enormous amount of energy in the form of waves, tidal currents, marine currents, and thermal gradients. The energy potential of our seas and oceans well exceeds our present energy needs. India has a long coastline with estuaries and gulfs where tides are strong enough to move turbines for electrical power generation. A variety of different technologies are currently under development throughout the world to harness this energy in all its forms, including waves (40,000 MW), tides (9000 MW), and thermal gradients (180,000 MW). Deployment is currently limited, but the sector has the potential to grow, fueling economic growth, reducing the carbon footprint, and creating jobs not only along the coasts but also inland along its supply chains.

As the government of India steps up its effort to reach the objectives to complete its renewable energy and climate change objectives post 2022, it is opportune to explore all possible avenues to stimulate innovation, create economic growth and new jobs, and reduce our carbon footprint. Given the long-term energy need through this abundant source, action needs to be taken now on R&D upfront in order to ensure that the ocean energy sector can play a meaningful part in achieving our objectives in coming decades. The MNRE looks over the horizon at a promising new technology and considers the various options available to support its development. Over 100 different ocean energy technologies are currently under development in more than 30 countries. Most types of technologies are currently at the demonstration stage or the initial stage of commercialization.

1.3.7 Tidal Energy

Tidal power is the only form of energy that derives directly from the relative motions of the earth–moon system, and to a lesser extent from the earth–sun system. The tidal forces produced by the moon and sun, in combination with the Earth's rotation, are responsible for the generation of the tides.

Tidal power, also called tidal energy, is a form of hydropower that converts the energy of tides into electricity or other useful forms of power. Although not yet widely used, tidal power has potential for future electricity generation. Tides are more predictable than wind energy and solar power. Among sources of renewable energy, tidal power has traditionally suffered from relatively high cost and limited availability of sites with sufficiently high tidal ranges or flow velocities, thus constricting its total availability. However, many recent technological developments and improvements, both in design (e.g., dynamic tidal power, tidal lagoons) and turbine technology (e.g., new axial turbines, crossflow turbines), indicate that the total availability of tidal power may be much higher than previously assumed and that economic and environmental costs may be brought down to competitive levels.

Because the Earth's tides are caused by forces due to gravitational interaction with the moon and the sun and the earth's rotation, tidal power is practically inexhaustible and classified as a renewable energy source. The first tidal power station was the Rance tidal power plant built over a period of six years from 1960 to 1966 at La Rance, France. It has 240 MW installed capacity. RES installed capacity as in 2016 is presented in Figure 1.9. The estimated and cumulative achievement of renewable energy sources as on April 2016 is presented in Table 1.4.

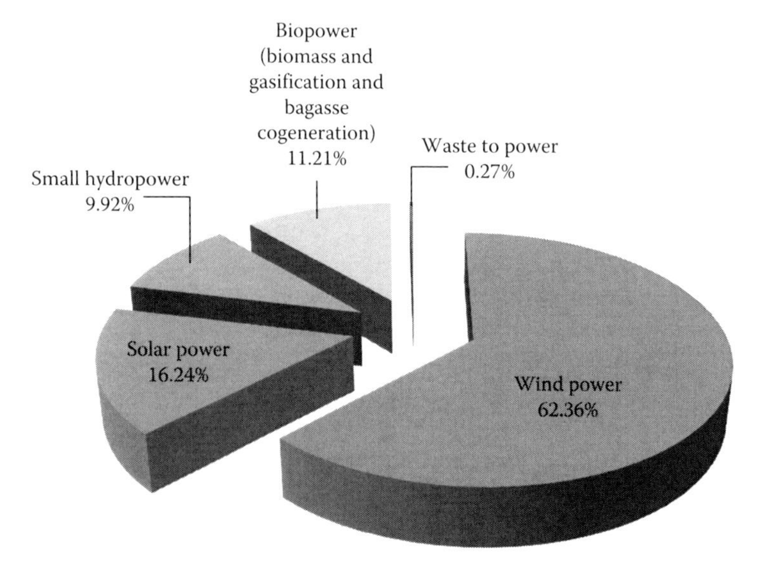

FIGURE 1.9
Pictorial representation of India's renewable energy sources (RES) installed capacity.

TABLE 1.4

Estimated and Cumulative Achievement of Renewable Energy Sources (RES) Connected to the Indian Grid

Sector	FY 2016–2017		Cumulative Achievements
	Target	**Achievement**	**(as of April 30, 2016)**
I. GRID-INTERACTIVE POWER (CAPACITIES IN MW)			
Wind Power	4,000.00	0.45	26,867.11
Solar Power	12,000.00	235.00	6,997.85
Small Hydropower	250.00	1.50	4,275.45
Bio Power (Biomass & Gasification and Bagasse Cogeneration)	400.00	0.00	4,831.33
Waste to Power	10.00	0.00	115.08
Total	**16,660.00**	**23,695.05**	**43,086.82**

- A British tidal energy company, Atlantis Resources, is expected to set up a tidal power plant with the capacity to generate over 250 MW in the Gulf of Kutch or Khambhat.
- India's first attempt to harness tidal power for generating electricity would be in the form of a 3-MW plant proposed at the Durgaduani creek in the Sundarbans delta of West Bengal.
- The Gulf of Kutch and Gulf of Cambay in Gujarat and Ganga delta in the Sundarbans, the world's largest mangrove forest, are the three sites identified as potential areas for tidal power generation.

1.3.8 Off-Grid Renewable Power

Two significant items affect the renewable energy priorities of India as compared to priorities of other nations. First, renewable energy supplies provide power to those rural populations that are inaccessible through conventional transmission grids, such as hills, islands, and far-flung areas away from a generation center—perhaps the remoter areas can only get electricity through renewable sources. Second, RE, being a distributed mode of power generation, is accessible to 1.25 billion to generate their own power in their backyards. Table 1.5 summarizes off-grid/distributed RE systems. Although these achievements look impressive, there is a huge requirement in terms of deployment for such off-grid/distributed/decentralized systems.

1.4 Government Initiatives for Solar Photovoltaic Systems

Solar PV has one of the highest capital costs of all RES, but it has the lowest operational cost, due to the very low maintenance and repair needs. For solar energy to become a widely used renewable source of energy, it is imperative that the capital costs are reduced significantly.

TABLE 1.5

Off-Grid and Captive Power Renewable Energy Systems

	FY 2016–2017		Cumulative
Sector	Target	Achievement	Achievement as of April 30, 2016
OFF-GRID/CAPTIVE POWER (CAPACITIES IN MW_{EQ})			
Waste to Energy	15.00	0.00	160.16
Biomass (Non-Bagasse) Cogeneration	60.00	0.00	651.91
Biomass	2.00	0.00	18.15
Gasifiers-Rural-Industrial	8.00	0.00	164.24
Aero-Generators/Hybrid Systems	0.30	0.00	2.69
SPV Systems	100.00	0.00	313.88
Water Mills/Micro Hydro	1.00	0.00	18.71
Total	**186.30**	**0.00**	**1,329.74**
III. OTHER RENEWABLE ENERGY SYSTEMS			
Family Biogas Plants (in Lakhs)	1.10	0.00	48.55

Solar PV is fast-changing industry, given the pace of technological and policy changes. India, where most regions enjoy nearly 300 sunny days a year, is an ideal market for solar power companies. Currently, India has around 60 companies assembling and supplying solar photovoltaic systems, 9 companies manufacturing solar cells, and 19 companies manufacturing photovoltaic modules or panels, according to an Indian Semiconductor Association study. However, spurred by factors like an increased demand for clean power, an energy-starved industry, and the falling cost of solar power generation, companies in this space are coming up with a noteworthy number of domestic projects. It has also helped that the government is lending support to such projects through state electricity boards with subsidies. PV installations in India today almost entirely comprise small-capacity applications. They are most visible in lighting applications (street lighting and home lighting systems) in the cities and towns and in small electrification systems and solar lanterns in rural areas. PV has also begun to be deployed to a small degree in powering water pump sets.

The central Indian government has recently approved 12 proposals under the Special Incentives Package Scheme (SIPS). Put together, these proposals could bring in about Rs. 76,573 crore of investment to the domestic solar power sector. Under SIPS, the government provides an incentive of 20 percent of the capital expenditure during the first 10 years to a unit located in a special economic zone (SEZ). Units based outside of the SEZ get a 25% incentive. Incentives could be in the form of capital subsidies or equity participation.

1.5 Future of Solar Energy in India

The Indian government has announced the National Solar Mission, outlining ambitious long-term plans to attain an installed solar power generation capacity of 20,000

MW by the year 2020, which would be increased to 100,000 MW by the year 2030 and further to 200,000 MW by the year 2050. To unfold in three phases, it aims to achieve parity with coal-based thermal power generation by 2030. The mission envisages an investment of Rs. 91,684 crore over the next 30 years. This will include an interest subsidy to the tune of Rs. 7,300 crores. The plan also aims to reduce the cost of solar power generation by 2017 to 2020 in order to make solar power competitive with power generated from fossil fuels.

The mission underlines the government's intention to give a boost to solar energy and is a purposeful step by India toward climate change mitigation. An analysis done by Greenpeace shows that the Jawaharlal Nehru National Solar Mission plan could ensure an annual reduction of 434 million tons of CO_2 emissions every year by 2050 based on the assumption that solar power will replace that provided by fossil fuels.

Questions

1. What is energy security? Explain the importance of various nonconventional energy sources in the present context of global warming.
2. Write your views on energy options for India, keeping the socioeconomic environment considerations in mind.
3. Discuss the global and Indian scenarios of solar photovoltaic systems.

2

Electrical Fundamentals

2.1 SI Units

The system of units used in engineering and science are usually abbreviated to SI units and are based on the metric system. The basic units in the SI system are listed with their symbols in Table 2.1.

TABLE 2.1

Basic SI Units

Quantity	Unit
Length	meter, m
Mass	kilogram, kg
Time	second, s
Electric Current	ampere, A
Thermodynamic Temperature	kelvin, K
Luminous Intensity	candela, cd
Amount of Substance	mole, mol

2.1.1 Charge

The unit of charge is the coulomb (C) where 1 coulomb is 1 ampere second (1 coulomb = 6.24×10^{18} electrons). The coulomb is defined as the quantity of electricity that flows through a given point in an electric circuit when a current of 1 ampere is maintained for 1 second. Thus, charge in coulombs is

$$Q = It$$

where I is the current in amperes and t is the time in seconds.

2.1.2 Power

Power is defined as the rate of doing work, or transferring energy. Power is measured in watts

$$P = W / t$$

where W is the work done or energy transferred in joules and t is the time in seconds. Thus, energy in joules is

$$W = P \times t$$

2.2 Electrical Potential and E.M.F.

The unit of electrical potential is the volt (V), where 1 volt is 1 joule per coulomb. One volt is defined as the difference in potential between two points in a conductor, which, when carrying a current of 1 ampere, dissipates a power of 1 watt:

$$\text{Volts} = \frac{\text{watts}}{\text{amperes}} = \frac{\text{joules/second}}{\text{amperes}} = \frac{\text{joules}}{\text{ampereseconds}} = \frac{\text{joules}}{\text{coulombs}}$$

A change in electric potential between two points in an electric circuit is called a potential difference. The electromotive force (emf) provided by a source of energy such as a battery or a generator is measured in volts.

2.3 Resistance and Conductance

The unit of electric resistance is the ohm (Ω) where 1 ohm is 1 volt per ampere. It is defined as the resistance between two points in a conductor when a constant electric potential of 1 volt applied at the two points produces a current flow of 1 ampere in the conductor. Thus, resistance is measured in ohms

$$R = V / I$$

where V is the potential difference across the two points in volts and I is the current flowing between the two points in amperes.

The reciprocal of resistance is called conductance and is measured in siemens (S). Thus, conductance in siemens is

$$G = 1 / R$$

where R is the resistance in ohms.

2.4 Ohm's Law

Ohm's law states that the current I flowing in a circuit is directly proportional to the applied voltage V and is inversely proportional to the resistance R, provided the temperature remains constant. Thus,

$$I = \frac{V}{R} \text{ or } V = IR \text{ or } R = \frac{V}{I}$$

2.5 Power and Energy

Power, P, in an electric circuit is given by the product of potential difference, V, and current, I. The unit of power is the watt, W. Hence

$$P = V \times I \text{ watts}$$

From Ohm's law, $V = IR$
Substituting for V in the earlier equation gives
$P = (IR) \times I$
that is,

$$P = I^2 R \text{ watts}$$

Also from Ohm's law, $I = V/R$.
Substituting for I in the equation earlier gives:

$$P = V \times \frac{V}{R}$$

$$\text{i.e.,} \quad P = \frac{V^2}{R} \text{ watts}$$

These are thus three possible formulae that may be used for calculating power.

$$\text{Electrical energy} = \text{power} \times \text{time}$$

The unit of electrical energy is kilowatt-hours (kWh).

2.6 Series Circuits

Three resistors—R_1, R_2, and R_3—connected end to end (i.e., in series) with a battery source of V volts.

Because the circuit is closed, a current I will flow, and the p.d. across each resistor may be determined from the voltmeter reading V_1, V_2, and V_3.

In series circuit

a. The current I is the same in all parts of the circuit.
b. The sum of the voltages V_1, V_2, and V_3 is equal to the total applied voltage, V, that is,

$$V = V_1 + V_2 + V_3$$

From Ohm's law:
$V_1 = IR_1$, $V_2 = IR_2$, $V_3 = IR_3$, and $V = IR$
where R is the total circuit resistance.

Because $V = V_1 + V_2 + V_3$
then $IR = IR_1 + IR_2 + IR_3$
Dividing throughout by I gives:

$$R = R_1 + R_2 + R_3$$

Thus for a series circuit, the total resistance is obtained by adding together the values of the separate resistances.

2.7 Parallel Networks

Three resistors, R_1, R_2, and R_3, connected across each other (i.e., in parallel), across a battery source of V volts.

In parallel circuit

a. The sum of the currents I_1, I_2, and I_3 is equal to the total circuit current, I (i.e., $I = I_1 + I_2 + I_3$).
b. The source p.d., V volts, is the same across each of the resistors.

From Ohm's law:

$$I_1 = \frac{V}{R_1},\ I_2 = \frac{V}{R_2},\ I_3 = \frac{V}{R_3},\ \text{and } I = \frac{V}{R}$$

where R is the total circuit resistance.
Because $I = I_1 + I_2 + I_3$
then

$$\frac{V}{R} = \frac{V}{R_1} + \frac{V}{R_2} + \frac{V}{R_3}$$

dividing throughout by V gives:

$$\frac{1}{R} = \frac{1}{R_1} + \frac{1}{R_2} + \frac{1}{R_3}$$

This equation must be used when finding the total resistance R of a parallel circuit. For the special case of two resistors in parallel:

$$\frac{1}{R} = \frac{1}{R_1} + \frac{1}{R_2} = \frac{R_2 + R_1}{R_1 R_2}$$

hence:

$$R = \frac{R_1 R_2}{R_1 + R_2} \quad \left(\text{i.e., } \frac{\text{product}}{\text{sum}}\right)$$

3

Measurement and Estimation of Solar Irradiance

3.1 The Structure of the Sun

The sun is a sphere of intensely hot gaseous matter with a diameter of 1.39×10^9 m and is, on average, 1.5×10^{11} m from the earth. The sun has an effective blackbody temperature of 5777 K. The temperature in the central interior region is variously estimated at 8×10^6 to 40×10^6 K, and the density is estimated to be about 100 times that of water. The sun is 333,400 times more massive than the earth and contains 99.86% of the mass of the entire solar system. It consists of 78% hydrogen, 20% helium, and 2% of other elements. It is estimated that 90% of the energy is generated in the region of 0 to 0.23 R (where R is the radius of the sun), which contains 40% of the mass of the sun, and the density is about 105 kg/m^3. At a distance 0.7 R from the center, the temperature drops to about 130,000 K and the density drops to 70 kg/m^3; the zone from 0.7 to 1.0 R is known as the convective zone, where the temperature drops to about 6000 K and the density to about 10 to 5 kg/m^3.

3.2 The Solar Irradiance Spectrum

The solar irradiance spectrum is mainly concentrated in the wavelength range of 0.25 – 3.0 µm. The given solar irradiance spectrum is divided into the different bands like UV, Visible and IR band on the basis of wavelength. About 48% of the total energy lies in the range 0.38 – 0.78 µm (Visible band), 46% energy lies in the band of 0.78 – 3.0 µm (IR) and only 6% of the energy lies in the range of 0.2 – 0.38 µm i.e. UV region.

The extraterrestrial solar irradiance (I_{sc}) is 1367 W/m^2. However, the solar irradiance is varying wrt variation in the earth-sun distance there is a variation of ±3 percent in the extraterrestrial radiation flux and the same can be calculated using the equation given below.

$$I_{on} = I_{sc}\left(1 + 0.033\cos\left(\frac{360n}{365}\right)\right) \tag{3.1}$$

where I_{on} is the extraterrestrial radiation measured on the plane normal to the radiation on the nth day of the year and I_{sc} is the solar constant.

3.2.1 Solar Constant and Solar irradiance

The solar constant is the amount of solar irradiance passing through a unit area at the top of the atmosphere perpendicular to the direction of the irradiance at the mean earth-sun distance. Its value is 1367 W/m^2.

Solar irradiance is the energy per unit area received from the sun in the form of electromagnetic radiation.

3.2.2 Depletion of Solar Radiation by the Atmosphere

The earth is surrounded by an atmosphere containing various gases, dust and other suspended particles, water vapor and clouds of various types. The solar irradiance during its passage in the atmosphere gets partly absorbed, scattered and reflected in different wavelength bands selectively. Irradiance gets absorbed in water vapor, Ozone, CO_2, O_2 in certain wavelengths. Radiation gets scattered by molecules of different gases and small dust particles known as Rayleigh scattering where the intensity is inversely proportional to the fourth power of wavelength of light. If the size of the particles are larger than the wavelength of light then Mie Scattering will takes place. There will be a reflection of radiation due to clouds, particles of larger size and other material in the atmosphere. Considerable amount of solar irradiance also gets absorbed by clouds which are of several types. Some irradiance gets reflected back in the atmosphere due to reflection from the ground, from the clouds, and scattering. This fraction of irradiance reflected back is called albedo of the ground and on an average the albedo is 0.3. The solar irradiance which reaches on the earth surface is called direct solar irradiance or beam irradiance. The irradiance which gets reflected, absorbed or scattered is not completely lost in the atmosphere and comes back on the surface of the earth in the short wavelength region and called diffused solar irradiance. The sum of the diffuse and direct irradiance on the surface of the earth is called global or total solar irradiance.

3.2.3 Factors Affecting the Availability of Solar Energy on a Collector Surface

Geographic location
Site location of collector
Collector orientation and tilt
Time of day
Time of year
Atmospheric conditions
Type of collector

3.3 Radiation Instruments

Pyranometer
Pyrheliometer
Pyrgeometer
Net radiometer
Sunshine recorder

These instruments are classified as first class, second class, or third class depending on their sensitivity, stability, and accuracy.

3.3.1 Solar Irradiance Components

Direct radiation: Direct transmission of solar radiation to the earth's surface

Diffuse solar radiation: Scattered by molecules and aerosols upon entering the earth's atmosphere

Global solar radiation = Direct irradiance + diffuse solar irradiance

Concentrators use direct radiation plus a small portion of scattered radiation.

Flat-plate collectors use direct and diffuse solar radiation and also reflected radiation.

3.3.2 Instruments Used

A pyranometer is used to measure global solar irradiance on a horizontal surface.

A pyranometer with shading ring is used to measure diffuse solar irradiance.

Direct radiation from sun is measured by a pyrheliometer.

3.3.2.1 Detectors for Measuring Radiation

Calorimetric sensors measure the radiant energy incident on a high-conductivity metal coated with a nonselective black paint of high absorptance. Thermomechanical sensors measure radiant flux through the bending of a bimetallic strip. These sensors consist of two dissimilar metallic wires with their ends connected.

Photoelectric sensors such as photovoltaic instruments are most numerous in the field of solar radiation measurement. A photovoltaic device is made of a semiconducting material such as silicon.

3.3.3 Measuring Diffuse Radiation

The same instrument used to measure total or global radiation is used to measure diffuse radiation. A suitable device (disc or shadow ring) is used to prevent direct solar radiation from reaching the receiver (pyranometer).

The following factors affect the device's accuracy:

Multiple reflections within the glass cover affect the accuracy of the measurement. In calculating the correction factor, it is assumed that the sky is isotropic. A part of the circumsolar radiation is also prevented from reaching the receiver by the shading device. The dimensions of the receivers are not adequately standardized.

3.4 Why Solar Energy Estimation?

The data on hourly global solar irradiance are essential for many applications, including solar power generation. Such data are also equally important in the design of energy-efficient buildings. Information on global solar irradiance received at any station is useful not only to the locality where the irradiance data are collected, but also for the wider world community. The potential of solar energy is enormous and could be utilized for many applications, including power generation. Such potential provides the opportunity to generate power on a distributed basis in a country like India where millions of people living in remote areas are unable to access the grid supply. In many countries, including India, hourly data on global solar radiation measurements are not available, even for those stations where measurements have already been done. Because accurate measurement involves cost and manpower, this information is hardly available in many developing countries, yet such measurements are needed in order to meet the energy needs from this environmentally friendly and clean source of energy. Therefore, it is vitally important to propose an efficient alternative for a solar radiation estimator based on more readily available meteorological parameters. Unfortunately, very few stations in India measure solar irradiance; for such stations where no measured data of solar irradiance are available, the common practice is to calculate solar irradiance from easily measurable metrological parameters like sunshine duration, temperature, etc. Accurate modeling depends upon the quality and quantity of the measured data used and is a better tool for estimating the solar irradiance of locations where measurements have not been done.

A number of models for estimating solar energy are available in the literature, including mathematical, regression, and intelligent models based on fuzzy logic, neural networks, wavelet transformation, etc. A few of these models are discussed in this chapter.

3.4.1 Mathematical Models of Solar Irradiance

The accurate estimation of solar energy is essential in many solar energy applications—for example, in photovoltaic, in solar heating of buildings, etc. Beam and diffuse energy on a horizontal surface should be known to accurately estimate solar energy. However, solar energy on the top of the atmosphere is known with enough accuracy. If scattering and absorption of solar energy by different constituents of the atmosphere can be understood precisely, one can predict terrestrial beam and diffuse energy. These are the different phenomena of attenuation of solar energy, which are showing improvement in their understanding continuously and are resulting in new estimation models. Also there is continuous improvement in solar energy instruments and the ability to measure atmospheric constituents like ozone, aerosols and water vapor, etc., which scatter and absorb solar energy. All this enhances the estimation accuracy. Of course, measured data are the best data, but they have temporal and spatial limitations. In advanced countries even, there are many stations where data are not being measured, and hence such stations' solar energy atlas and typical metrological year (TMY) are based on modeled data.

3.4.1.1 CPCR2 (Code for Physical Computation of Radiation, 2 Bands) Model

This model includes the parameterization of extinction processes, and it requires commonly available input parameters. In this two-band, clear-sky solar energy modeling technique, the solar spectrum is divided into two bands: an ultraviolet/visible band, B_1, (0.29 to 0.7 μm) and an infrared band, B_2, (0.7 to 2.7 μm). The model uses the extraterrestrial solar spectrum, as proposed by the World Radiometric Center (WRC), with a solar constant value of 1367 Wm^{-2}. The UV/visible and infrared bands considered in the CPCR2 model account for 46.04% and 50.57% of the solar constant, respectively. Therefore, coefficients

$f_1 = 0.4604$ and $f_2 = 0.5057$ are to be applied to the total extraterrestrial energy, E_{on1} and E_{on2}, corresponding to bands B_1 and B_2.

3.4.1.1.1 Direct Normal Energy

It is assumed that the direct rays entering the atmosphere encounter extinction processes, which are limited to ozone absorption, molecular scattering, uniformly mixed gases absorption, water vapor absorption, aerosol scattering, and aerosol absorption. Separated extinction layers are considered so that each band atmosphere transmittance for beam energy may be obtained by the simple product of transmittances. Thus, for each of these two bands, the beam energy at normal incidence is given by Equation 3.2:

$$E_{\text{bni}} = T_{\text{Ri}} T_{\text{gi}} T_{\text{oi}} T_{\text{wi}} T_{\text{ai}} E_{0\text{ni}} \quad (3.2)$$

where E_{bn} is the beam solar energy at normal surface (W/m^2), T_R is the Rayleigh transmittance (dimensionless), T_g is the transmittance due to mixed gases (dimensionless), T_o is the transmittance of ozone (dimensionless), T_w is the transmittance of water vapor (dimensionless), T_a is the transmittance of aerosols (dimensionless), $i = 1$ for band B_1, and $i = 2$ for band B_2.

The total beam energy at ground level is given by Equation 3.3:

$$E_{\text{bn}} = E_{\text{bn1}} + E_{\text{bn2}} \quad (3.3)$$

3.4.1.2 Diffuse and Global Energy

The diffuse energy at ground level is modeled as a combination of three individual components corresponding to the two scattering layers (molecules and aerosols) and to a backscattering process between the ground and sky. The diffuse energy can be found by using Equation 3.4:

$$E_{\text{di}} = E_{\text{dRi}} + E_{\text{dAi}} + E_{\text{ddi}} \quad (3.4)$$

where E_d is the diffuse energy (W/m^2), E_{dR} is the scattering due to molecules (W/m^2), E_{dA} is the scattering due to aerosols (W/m^2), and E_{dd} is the backscattering between the ground and sky (W/m^2).

The global energy is given by Equation 3.5:

$$E_{\text{g}} = E_{\text{d}} + E_{\text{b}} \quad (3.5)$$

3.4.2 REST2 (Reference Evaluation of Solar Transmittance, 2 Bands) Model

Previous results from in-depth performance assessment studies have shown that the CPCR2 two-band model was a top performer when compared to simpler broadband models. A recent and thorough study demonstrated that CPCR2 performed consistently well to predict direct normal energy (DNI) under both ideal and realistic conditions.

The present contribution describes some new features of the model, includes its latest algorithmic improvements, and proposes a benchmark dataset for the performance assessment of this or any other similar model. The general structure of REST2 is almost identical to that of CPCR2, with a band separation at 0.7 μm. Band 1 covers the UV and visible light, from 0.29 to 0.70 μm. It is characterized by strong absorption of ozone in the UV and strong scattering by molecules and aerosols over the whole band. Band 2 covers the near-infrared, from 0.7 to 4 μm, and is characterized by strong absorption by water vapor, carbon dioxide, and other gases, along with only limited scattering. This approach has been used in a few other models in the literature and has been shown to have two interesting advantages: (i) it improves accuracy compared to regular single-band models and (ii) it simplifies the derivation of energy, whose spectral ranges correspond almost perfectly to band 1. Using the latest extraterrestrial spectral energy distribution and latest solar constant value of 1366.1 W/m^2, the extra-atmospheric energy at the mean sun–earth distance is E_{0n1} = 635.4 W/m^2 (or 46.51%) and E_{0n2} = 709.7 W/m^2 (or 51.95%) in the two bands, respectively. To the difference of CPCR2, the parameterizations for direct and diffuse energy take into account the circumsolar energy subtended in the typical total field of view of tracking devices.

3.4.2.1 Direct Energy

The formalism is essentially the same as in CPCR2, except that an additional provision is made for nitrogen dioxide absorption, as in REST. For each of the two bands, *i*, the band direct normal energy, E_{bni}, is obtained from a product of individual transmittances:

$$E_{bni} = T_{Ri} T_{gi} T_{oi} T_{ni} T_{wi} T_{ai} E_{0ni} \tag{3.6}$$

where T_{Ri}, T_{gi}, T_{oi}, T_{ni}, T_{wi}, and T_{ai} are the band transmittances for Rayleigh scattering, uniformly mixed gases absorption, ozone absorption, nitrogen dioxide absorption, water vapor absorption and aerosol extinction, respectively.

The broadband DNI is simply obtained as the sum of the two-band components:

$$E_{bn} = E_{bn1} + E_{bn2} \tag{3.7}$$

3.4.2.2 Diffuse and Global Energy

As with CPCR2, the formalism here is also based on a two-layer scattering scheme. The top layer is assumed to be the source for all Rayleigh scattering, as well as for all ozone and mixed gas absorption. Similarly, the bottom layer is assumed to be the source for all aerosol scattering, as well as for aerosol, water vapor, and nitrogen dioxide absorption. After scattering has occurred in the top layer, the down-welling diffuse energy is assumed to behave as direct energy at an effective air mass, m = 1.66. This is the air mass value that is used in calculating the transmittances dealing with absorption in the bottom layer.

The incident diffuse energy on a perfectly absorbing ground (i.e., with zero albedo) is defined as:

$$E_{dpi} = T_{oi} T_{gi} T_{0ni} T_{0wi} \left[B_{Ri} \left(1 - T_{Ri}\right) T_{ai}^{0.25} + B_a F_i T_{Ri} \left(1 - T_{ai}^{0.25}\right) \right] E_o \tag{3.8}$$

where

$$E_{0i} = E_{0ni} \cos Z \tag{3.9}$$

where Z is the zenith angle. Function F_i is a correction factor introduced to compensate for multiple scattering effects and other shortcomings in the simple transmittance approach used here. B_{R1} and B_{R2} are the forward-scattering fractions for the Rayleigh extinction. In the absence of multiple scattering, they would be exactly 0.5 because molecules scatter equally in the forward and backward directions. Multiple scattering is negligible in Band 2 (so that B_{R2} = 0.5), but not in Band 1. Using a simple spectral model to describe this effect, B_{R1} is obtained after spectral integration and parameterization as:

$$B_{R1} = 0.5\left(0.89013 - 0.0049558 m_R + 0.000045721 m_R^2\right) \tag{3.10}$$

The aerosol forward scatterance factor, B_a, is the same as in CPCR2:

$$B_a = 1 - \exp\left(-0.6931 - 1.8326 \cos Z\right) \tag{3.11}$$

The beam, diffuse, and global broadband energy incident on a horizontal and ideally black surface are finally given as:

$E_b = E_{bn} \cos Z$,
$E_{dp} = E_{dp1} + E_{dp2}$, and
$E_{gp} = E_b + E_{dp}$,

Under normal conditions, a backscattered contribution must be added because of the interaction between the reflecting earth surface and the scattering layers of the atmosphere. This contribution is usually small (e.g., <10% of E_{gp}) but may become far more significant over snowy regions. The ground albedo to consider here, ρ_{gi}, refers to an average over a 5- to 50-km radius around the site under scrutiny. For each band, the sky albedo, ρ_{si}, is obtained as a function of α_i and β_i, (angstrom turbidity coefficient [dimensionless]), and the backscattered diffuse component, E_{ddi}, is derived by considering multiple reflections between the ground and the atmosphere:

$$E_{ddi} = \rho_{gi}\rho_{si}\left(E_{bi} + E_{dpi}\right)/\left(1 - \rho_{gi}\rho_{si}\right) \tag{3.12}$$

where $E_{bi} = E_{bni} \cos Z$. Finally, the total diffuse energy in each band is $E_{di} = E_{dpi} + E_{ddi}$, so that the broadband diffuse energy is obtained as $E_d = E_{d1} + E_{d2}$ and the broadband global energy as $E_g = E_b + E_d$.

3.4.2.2.1 REST (Reference Evaluation of Solar Transmittance) Model

Gueymard has proposed this model. In this model, the total NO_2 absorption is not taken into account. The proposed form of the model is as follows:

$$E_{bn} = E_{on} T_R T_g T_o T_w T_a \tag{3.13}$$

where E_{bn} is direct normal irradiance, E_{on} is the extraterrestrial irradiance, T_R is the Rayleigh transmittance, T_o is the ozone transmittance, T_g is the uniformly mixed gas transmittance, T_w is the water vapor transmittance, and T_a is the aerosol transmittance.

3.4.2.3 Estimation of Global Irradiance

The global solar irradiance can be predicted as follows:

$$E_g = E_{bn} \cos Z + I_d \tag{3.14}$$

where E_{bn} is the computed beam irradiance at normal incidence, E_g is the predicted global solar irradiance at the horizontal surface, and E_d is the predicted diffuse solar irradiance at the horizontal surface.

3.4.2.4 Estimation of Diffuse Irradiance

The diffuse irradiance is estimated on the basis of the parametric model, which was proposed by Iqbal. The horizontal diffuse irradiance E_d at the ground level is a combination of three individual components corresponding to the Rayleigh scattering after the first pass through the atmosphere, D_r; the aerosols scattering after the first pass through the atmosphere, D_a; and multiple reflection processes between the ground and sky, D_m:

$$E_d = D_r + D_a + D_m \tag{3.15}$$

3.4.2.5 Regression Models

A number of models are available in the literature based on linear, multiple quadratic, etc., regression approaches. Few of the regression models for the Indian scenario are presented in this section. Using the same approach, one can develop the model for any scenario.

In this section, the monthly average solar irradiance, sunshine duration hours, temperature, and other data on meteorological parameters are obtained from various sources, including the National Institute of Solar Energy and the Indian Meteorological Department (IMD), New Delhi, for seven Indian stations, namely New Delhi, Nagpur, Ahmedabad, Jodhpur, Kolkata, Vishakhapatnam, and Shillong. These were considered based on their geographical and climatic zones. The geographical features of the Indian stations considered in this work are shown in Table 3.1.

TABLE 3.1

Geographical Features of the Stations in India Considered in This Work

Station	Latitude (⁰N)	Longitude (⁰E)	Height above Mean Sea Level (m)	Climate Zone
New Delhi	28.58	77.20	216	Composite
Nagpur	21.10	79.05	31	Composite
Ahmedabad	23.07	72.63	55	Hot and dry
Jodhpur	26.30	73.02	224	Hot and dry
Kolkata	22.65	88.45	6	Warm and humid
Vishakhapatnam	17.72	83.23	3	Warm and humid
Shillong	25.57	91.88	1600	Cold and cloudy

Monthly average data of S/S_0 and H/H_0 for Jodhpur and New Delhi are shown in Table 3.2 and Table 3.3, respectively. The estimations of diffuse irradiance data of S/S_0 and H_d/H_g for Jodhpur and New Delhi are shown in Table 3.4 and Table 3.5, respectively.

Figures 3.1 and 3.2 indicate the trend of global and diffuse solar irradiance at Jodhpur and New Delhi, respectively, for different months of the year.

Because measurements of solar irradiance are often not available, many authors have attempted to establish relationships linking the values of solar irradiance (global or diffuse) with meteorological parameters like number of sunshine hours, cloud cover, and precipitation. The various meteorological parameters are related to solar irradiance to varying degrees. From the few meteorological stations where both irradiance and sunshine duration are recorded, it is possible to derive a relationship between these two variables in the

TABLE 3.2

Monthly Average Data of S/S_0 and H/H_0 for Jodhpur, India

Month	S/S_0	H (MJ/m²)	H_0 (MJ/m²)	H/H_0
January	0.89	15.89	23.98	0.787
February	0.89	18.09	28.18	0.642
March	0.80	21.16	32.96	0.642
April	0.89	22.57	37.07	0.608
May	0.89	23.56	32.29	0.599
June	0.83	22.73	39.94	0.570
July	0.72	19.21	39.49	0.486
August	0.73	17.91	37.78	0.474
September	0.81	19.84	34.34	0.577
October	0.87	19.35	29.5	0.656
November	0.90	16.68	24.92	0.670
December	0.90	15.04	22.72	0.662

TABLE 3.3

Monthly Average Data of S/S_0 and H/H_0 for New Delhi, India

Month	S/S_0	H (MJ/m²)	H_0 (MJ/m²)	H/H_0
January	0.836	13.32	20.18	0.660
February	0.857	16.42	23.76	0.691
March	0.857	20.64	31.18	0.662
April	0.868	24.07	40.52	0.594
May	0.878	24.43	45.66	0.535
June	0.847	22.54	45.81	0.492
July	0.697	19.07	39.64	0.481
August	0.697	17.79	39.01	0.456
September	0.847	18.90	37.64	0.502
October	0.900	16.80	23.17	0.725
November	0.853	14.13	17.86	0.791
December	0.804	11.93	16.63	0.717

TABLE 3.4
Monthly Average Data of S/S_0 and H_d/H_g for Jodhpur, India

Month	S/S_0	H_d (MJ/m²)	H_g (MJ/m²)	H_d/H_g
January	0.89	4.23	15.89	0.266
February	0.89	5.33	18.09	0.295
March	0.8	6.67	21.16	0.315
April	0.89	7.97	22.57	0.353
May	0.89	9.56	23.56	0.406
June	0.83	10.77	22.73	0.474
July	0.72	11.90	19.21	0.620
August	0.73	11.00	17.91	0.614
September	0.81	7.43	19.84	0.374
October	0.87	19.35	29.5	0.660
November	0.90	16.68	24.92	0.670
December	0.90	15.04	22.72	0.662

TABLE 3.5
Monthly Average Data of S/S_0 and H_d/H_g for New Delhi, India

Month	S/S_0	H_d (MJ/m²)	H_g (MJ/m²)	H_d/H_g
January	0.836	5.21	13.32	0.391
February	0.857	6.22	16.42	0.379
March	0.857	7.56	20.64	0.366
April	0.868	8.83	24.07	0.369
May	0.878	10.68	24.43	0.437
June	0.847	11.66	22.54	0.571
July	0.697	11.83	19.07	0.620
August	0.697	10.27	17.79	0.577
September	0.847	8.27	18.90	0.437
October	0.900	6.37	16.80	0.380
November	0.853	4.92	14.13	0.348
December	0.804	4.87	11.93	0.410

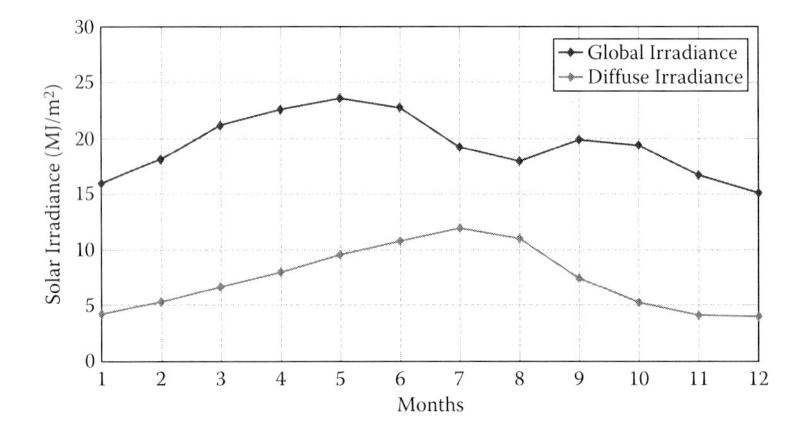

FIGURE 3.1
Global and diffuse solar irradiance at Jodhpur.

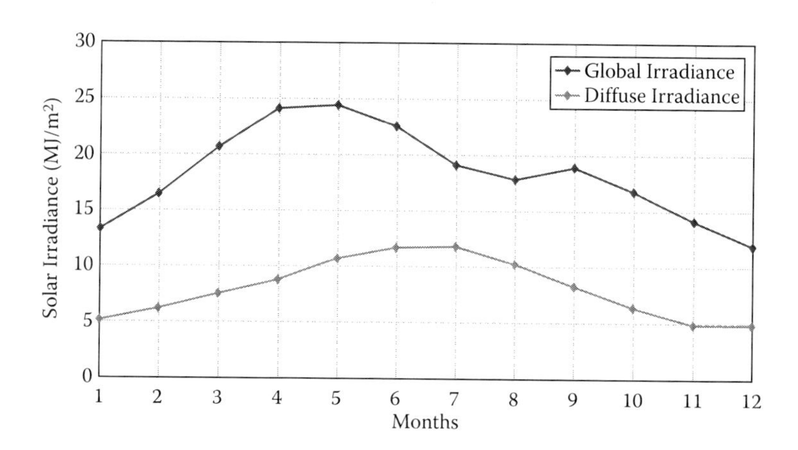

FIGURE 3.2
Global and diffuse solar irradiance at New Delhi.

form of the model first proposed by Angstrom, which has been used as a reference for this modeling. This relation is of the form:

$$\frac{\overline{H}}{\overline{H}_0} = a + b\frac{\overline{S}}{\overline{S}_0} \tag{3.16}$$

where H is the monthly mean of the daily global irradiance on the horizontal surface, H_0 is the extraterrestrial solar irradiance on the 15th of the month, S is the monthly mean of the daily hours of bright sunshine, S_0 is the maximum daily hours of sunshine (or day length), and a and b are the regression coefficients.

The proposed models for the estimation of global and diffuse solar irradiance for Jodhpur, India, are given in Equations 3.17 and 3.18, respectively.

$$\frac{\overline{H}}{\overline{H}_0} = 0.3 + 0.4556\frac{\overline{S}}{\overline{S}_0} \tag{3.17}$$

$$\frac{\overline{H}_d}{\overline{H}_g} = 1.45 - 1.62\frac{\overline{S}}{\overline{S}_0} \tag{3.18}$$

The values of regression coefficients for New Delhi, Nagpur, Ahmedabad, Kolkata, Vishakhapatnam, and Shillong are presented in Table 3.5.

The ratio of S/S_0 is the fraction of maximum possible numbers of bright sunshine hours, and H/H_0 is the atmospheric transmission coefficient, commonly known as the clearness index. The value of S_0 can be computed from Cooper's formula:

$$\overline{S}_0 = \left(\frac{2}{15}\right)\omega \tag{3.19}$$

The extraterrestrial solar irradiance on a horizontal surface H_0 is obtained from the following expression:

$$H_0 = \frac{24 \times 3.6 \times 10^{-3} \times I_{SC}}{\pi}\left(1 + 0.033\cos\left(360\frac{\overline{D}}{365}\right)\right)\cos\varphi\cos\delta\sin\omega + \omega\sin\varphi\sin\delta \tag{3.20}$$

$$\delta = 23.45 \sin\left(360 \frac{248 + \bar{D}}{365}\right) \tag{3.21}$$

$$\omega = \cos^{-1}\left(-\tan\varphi \tan\delta\right) \tag{3.22}$$

where D is the day number, I_{sc} = 1367 Wm^{-2} is the solar constant, φ is the latitude of the location, δ is the declination angle, and ω is the sunset hour angle.

An estimate of extraterrestrial solar irradiance H_0 and relative sunshine duration at different Indian stations can be computed using the MATLAB® program. In calculations, the average day of the month is considered. The average day is the day that has the extraterrestrial solar irradiance closest to the average for the month.

3.4.3 Intelligent Modeling

There are number of models for the assessment of solar irradiance under cloudless skies. Three broadband models—namely REST, REST2, and CPCR2—along with the regression model are explained in previous sections. These models are not suitable for estimating solar irradiance during the monsoon months (June to September) or with cloudy skies, depending upon the geographical location. A regression model to estimate the solar energy considering sunshine duration as an input was presented in the previous sections. The results obtained were satisfactory, but it is only applicable for clear-sky weather condition. However, due to uncertainty in weather conditions, a model based on fuzzy logic, artificial neural networks, and generalized neural networks is presented in the next sections. The uncertainty in the atmosphere may occur due to the existence of dust, moisture, aerosols, clouds, or temperature differences in the lower atmosphere. Of all these factors, clouds can cause the most losses in terms of extraterrestrial solar irradiance reaching the surface at the ground level. The atmosphere causes a reduction of the extraterrestrial solar input by about 30% on a very clear day to nearly 100% on a very cloudy day.

In India around 50 to 100 days in a year are cloudy, so it is very difficult to predict the accurate results of solar irradiance using a mathematical or regression model. Therefore, fuzzy logic and a neural network–based model could be a better option for estimating solar irradiance at a given location using sunshine duration, temperature, and other meteorological parameters, depending upon their availability.

3.4.3.1 Fuzzy Logic–Based Modeling of Solar Irradiance

In this section, seven Indian stations have been considered on the basis of their climate conditions. In India some climatic parameters like solar irradiance, sunshine hours, temperature, rainfall, atmospheric pressure, and humidity are measured by IMD. However, these parameters are not being measured at all stations, particularly solar irradiance. Most of the stations are relatively new and therefore have not accumulated long-term data. The raw data for the proposed study have been obtained through personal communication from the IMD, the National Institute of Solar Energy, and the Ministry of New and Renewable Energy (MNRE). In this section, only two parameters that can be easily measured have been considered as inputs, and the remaining parameters are assumed to be constant. However, each parameter has its own effects on the solar irradiance, and their negligence introduces some error in the estimation. Hence, in this analysis a simple technique is used where all the uncertainties and model complications related to solar irradiance and sunshine duration and temperature are treated linguistically using fuzzy sets. In the fuzzy approach there is neither any assumption nor any model constant. Therefore, this method could minimize errors in the estimation of solar irradiance. In the present analysis seven Indian stations—New Delhi, Kolkata, Mumbai, Jodhpur, Pune, Ahmedabad, and Vishakhapatnam and Shillong—are selected to estimate monthly mean solar irradiance based on sunshine duration and temperature. The MATLAB toolbox (FIS) is used to relate the solar irradiance with sunshine duration and temperature. The inputs for all of the stations are fuzzified into five subsets, such as very low, low, medium, high, and very high values of input data. However, all the data used in this analysis are expressed in the normalized form. Similarly, the solar irradiance values are also subdivided into five groups according to increasing magnitude. The following methodology has been adopted in the development of a fuzzy-based model for solar energy estimation in the MATLAB toolbox. The fuzzy inference system editor is presented in Figure 3.3.

The next step is to define membership functions associated with each of the variables. The membership function editor is the tool used to display and edit all the membership functions for the entire FIS, including input and output variables. In this work, a triangular membership function has been chosen for the input and output variables. The membership functions for sunshine duration, temperature, and global irradiance are shown in Figures 3.4 through 3.6, respectively.

All the data used in the proposed work are expressed in the normalized form. Membership functions and their ranges are presented in Table 3.6.

For the estimation of solar irradiance at a particular station, a set of multiple-antecedent fuzzy rules have been established. The inputs to the rules are the ratio of sunshine duration and ratio of temperature, and the consequent output is the solar irradiance as given in Table 3.7. The consequents of the rules are in the shaded part of the matrix. The input

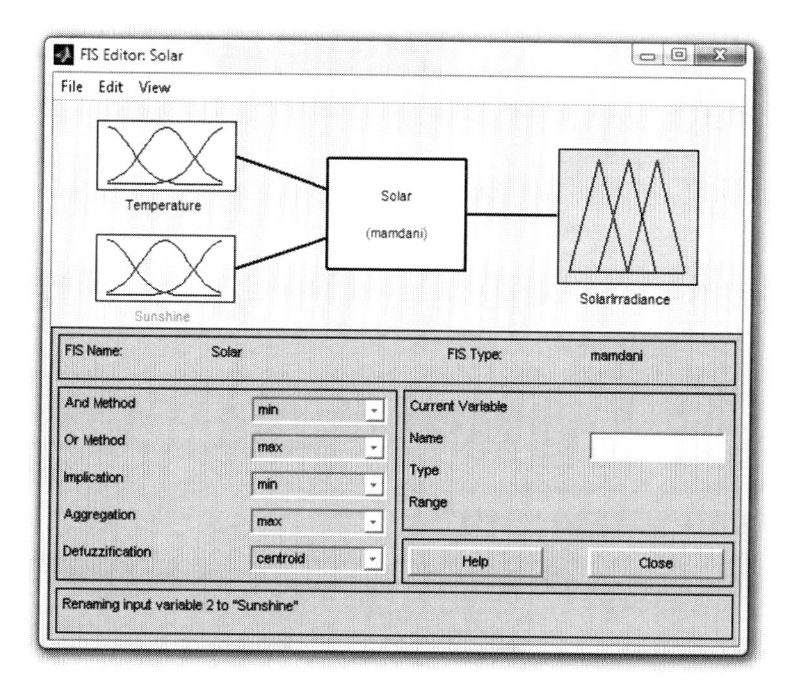

FIGURE 3.3
Fuzzy inference system editor.

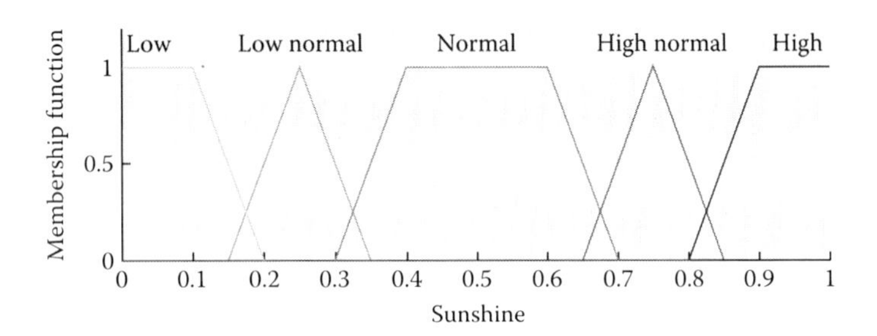

FIGURE 3.4
Fuzzy subset membership functions for sunshine duration.

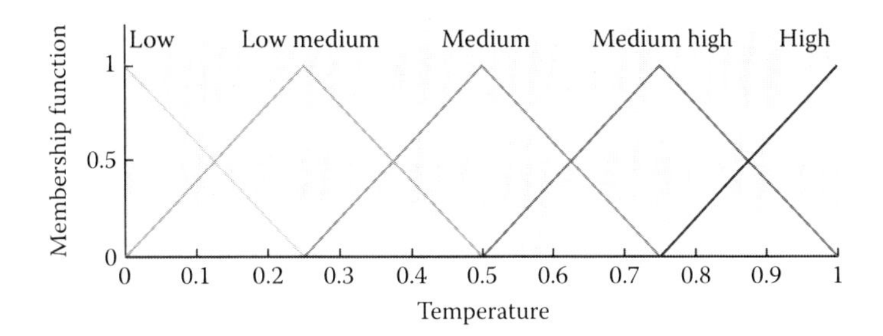

FIGURE 3.5
Fuzzy subset membership functions for temperature.

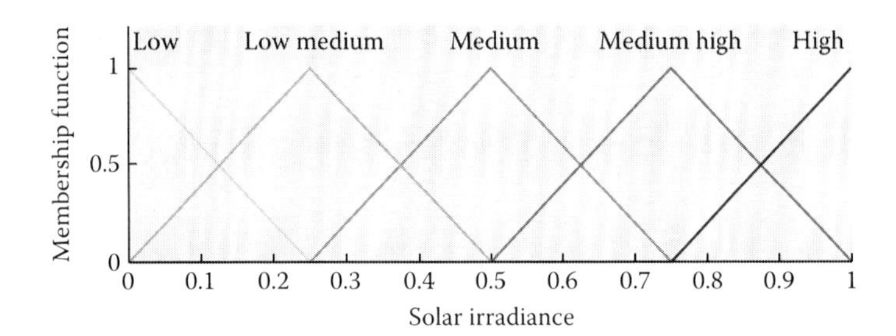

FIGURE 3.6
Fuzzy subset membership functions for solar irradiance.

TABLE 3.6
Membership Functions and Range of Parameters

Inputs				Output	
Membership function	T/T_0	**Membership function**	S/S_0	**Membership function**	H/H_0
Low	[0.00–0.25]	Low	[0.00–0.20]	Low	[0.00–0.20]
Low-Med	[0.00–0.50]	Low-Nor	[0.15–0.35]	Low-Nor	[0.15–0.35]
Medium	[0.25–0.75]	Normal	[0.30–0.70]	Normal	[0.30–0.70]
Med-High	[0.50–1.00]	High-Nor	[0.65–0.85]	High-Nor	[0.65–0.85]
High	[0.75–1.00]	High	[0.80–1.00]	High	[0.80–1.00]

TABLE 3.7

Decision Matrix for Determining Global Solar Irradiance

	AND	S_0/S Low	Low-Nor	Nor	High-Nor	High
Temperature	Low	Low-Med	Low-Med	Low	Low-Med	Med
	Low-Med	Medium	Low-Med	Low-Med	Low-Med	Med
	Medium	High-Med	Med	Low-Med	Low-Med	High-Med
	Med-High	High-Med	High-Med	Med	High-Med	High-Med
	High	High	High-Med	Med	High-Med	High

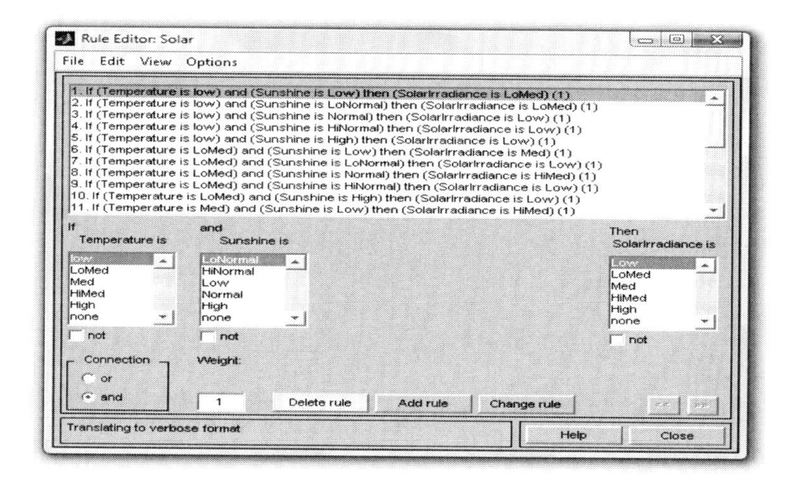

FIGURE 3.7
Fuzzy rules.

variables—temperature ratio and sunshine duration ratio—are described by the fuzzy linguistic variables such as high, medium high, medium, low medium, and low, and the output variable (solar irradiance) is described as low, low normal, normal, high normal, and high. To estimate the solar irradiance at any station in India, fuzzy rules have been defined, as shown in Figure 3.7.

The rule viewer shown in Figure 3.8 presents a sort of micro view of the fuzzy inference system by showing one calculation at a time in detail.

The surface viewer presents the entire output surface of the system, based on the entire span of the input set. Three-dimensional surfaces of output of the fuzzy model are described in Figure 3.9.

3.4.3.2 Datasets

The monthly mean solar irradiance at the seven Indian stations is estimated on the basis of sunshine duration and temperature. The estimated values are compared with the measured values on the basis of absolute relative error (ARE). Table 3.8 gives the monthly average values of the clearness index, which is the ratio of the average daily solar irradiance to the daily maximum irradiance, sunshine ratio (i.e., the ratio of the average daily sunshine hours and the theoretical sunshine duration), and temperature ratio for the New Delhi and Jodhpur stations.

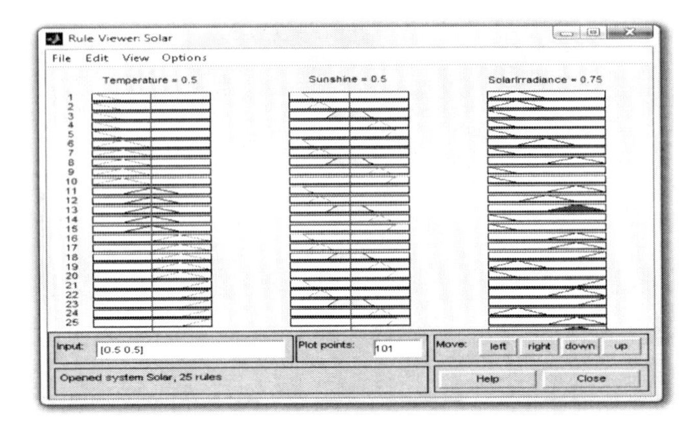

FIGURE 3.8
Fuzzy rules viewer.

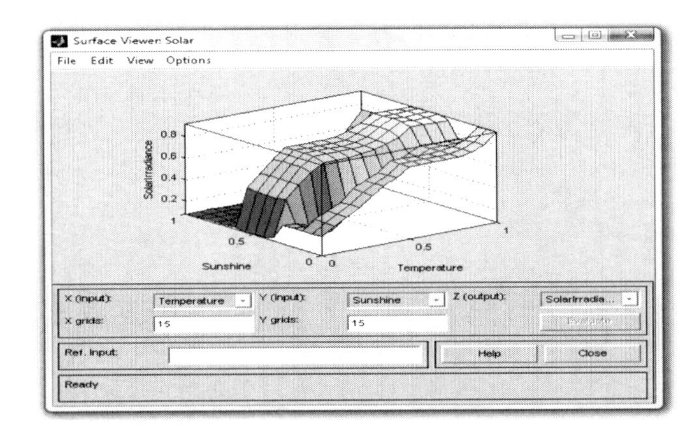

FIGURE 3.9
Three-dimensional surfaces of output of the fuzzy model.

TABLE 3.8
Monthly Average Values of H/H_0, S/S_0, and T/T_0 for New Delhi and Jodhpur

	New Delhi			Jodhpur		
Month	H/H_0	S/S_0	T/T_0	H/H_0	S/S_0	T/T_0
January	0.189	0.836	0.334	0.180	0.89	0.50
February	0.387	0.857	0.509	0.386	0.89	0.60
March	0.657	0.857	0.639	0.675	0.80	0.76
April	0.877	0.868	0.809	0.807	0.89	0.91
May	0.900	0.878	0.896	0.900	0.89	1.00
June	0.779	0.847	0.900	0.822	0.83	0.99
July	0.557	0.697	0.841	0.492	0.72	0.90
August	0.475	0.697	0.813	0.370	0.73	0.86
September	0.546	0.847	0.673	0.551	0.81	0.84
October	0.412	0.900	0.727	0.505	0.87	0.81
November	0.241	0.853	0.596	0.254	0.90	0.65
December	0.100	0.804	0.464	0.100	0.90	0.54

The ARE obtained in the estimation of global and diffuse solar irradiance for New Delhi and Jodhpur stations are presented in Tables 3.9 and 3.10, respectively. It is clearly seen from these tables that ARE varies from 3.6% to 7.4% in the estimation of global solar irradiance, whereas it varies from 6.3% to 8.35% for diffuse solar irradiance. Obtained ARE in the estimation of solar irradiance is found in desired limits because the fuzzy logic technique is an approximate technique.

Results obtained using the fuzzy logic model are also presented graphically. Figures 3.10 through 3.16 show the measured and estimated values of global solar irradiance for New Delhi, Jodhpur, Nagpur, Vishakhapatnam, Kolkata, Ahmedabad, and Shillong, respectively.

TABLE 3.9

ARE Percentages in the Estimation of Monthly Mean Solar Irradiance in Comparison with Measured Data for New Delhi

	H (Measured) (MJ/m^2)		*H* (Estimated) (MJ/m^2)		ARE (%)	
Month	Global	Diffuse	Global	Diffuse	Global	Diffuse
January	13.32	5.21	**12.90**	**4.89**	3.2	**6.2**
February	16.42	6.22	**15.80**	**5.82**	3.8	**6.5**
March	20.64	7.56	**19.71**	**7.96**	4.5	**5.3**
April	24.07	8.83	**23.18**	**8.39**	3.7	**5.0**
May	24.43	10.68	**23.67**	**10.16**	3.1	**4.9**
June	22.54	11.66	**23.20**	**12.16**	2.9	**4.3**
July	19.07	11.83	**18.20**	**11.01**	4.6	**6.9**
August	17.79	10.27	**16.86**	**9.53**	5.2	**7.2**
September	18.90	8.27	**19.95**	**7.64**	5.6	**7.6**
October	16.80	6.37	**16.07**	**5.92**	4.3	**7.0**
November	14.13	4.92	**13.73**	**4.61**	2.8	**6.2**
December	11.93	4.87	**11.56**	**5.18**	3.1	**6.4**

TABLE 3.10

ARE Percentages in the Estimation of Monthly Mean Solar Irradiance in Comparison with Measured Data for Jodhpur

	H (Measured) (MJ/m^2)		*H* (Estimated) (MJ/m^2)		ARE (%)	
Month	Global	Diffuse	Global	Diffuse	Global	Diffuse
January	15.89	4.23	**15.41**	**3.97**	3.0	**6.2**
February	**18.09**	5.33	**17.51**	**4.97**	3.2	**6.7**
March	**21.16**	6.67	**20.42**	**6.28**	3.5	**5.9**
April	**22.57**	7.97	**21.71**	**8.36**	3.8	**5.0**
May	**23.56**	9.56	**23.32**	**9.13**	3.1	**4.5**
June	**22.73**	10.77	**022.1**	**10.25**	2.8	**4.8**
July	**19.21**	11.90	**20.30**	**12.76**	5.7	**7.2**
August	**17.91**	11.00	**16.99**	**10.21**	5.1	**7.2**
September	**19.84**	7.43	**20.81**	**6.85**	4.9	**7.8**
October	**19.35**	5.25	**18.62**	**4.90**	3.8	**6.7**
November	**16.68**	4.13	**16.18**	**4.38**	3.0	**6.0**
December	**15.04**	3.98	**14.60**	**3.72**	2.9	**6.6**

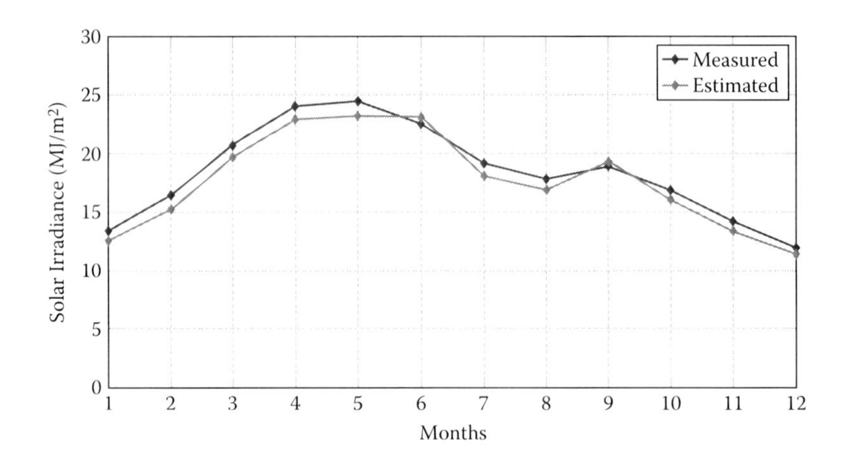

FIGURE 3.10
Measured and estimated global solar irradiance for New Delhi.

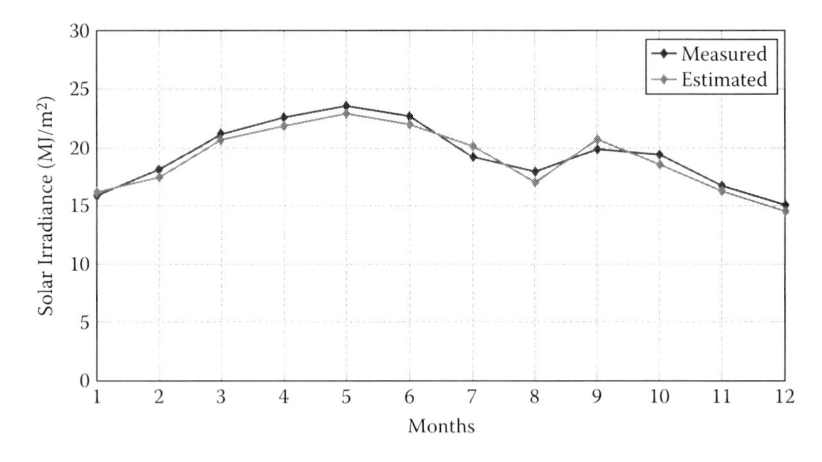

FIGURE 3.11
Measured and estimated global solar irradiance for Jodhpur.

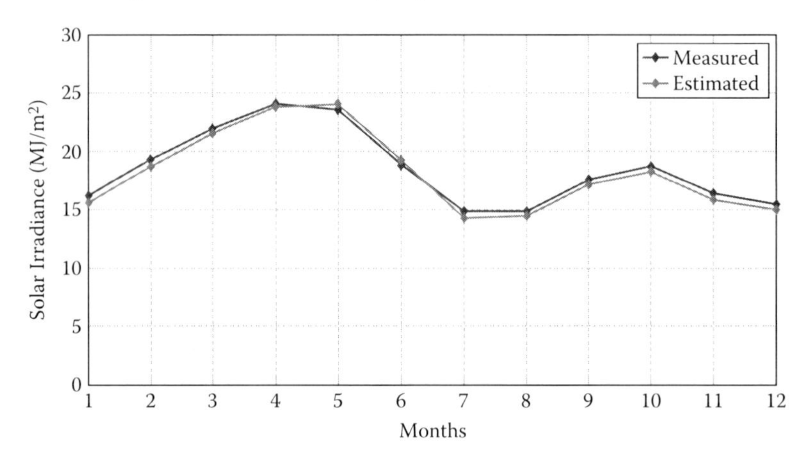

FIGURE 3.12
Measured and estimated global solar irradiance for Nagpur.

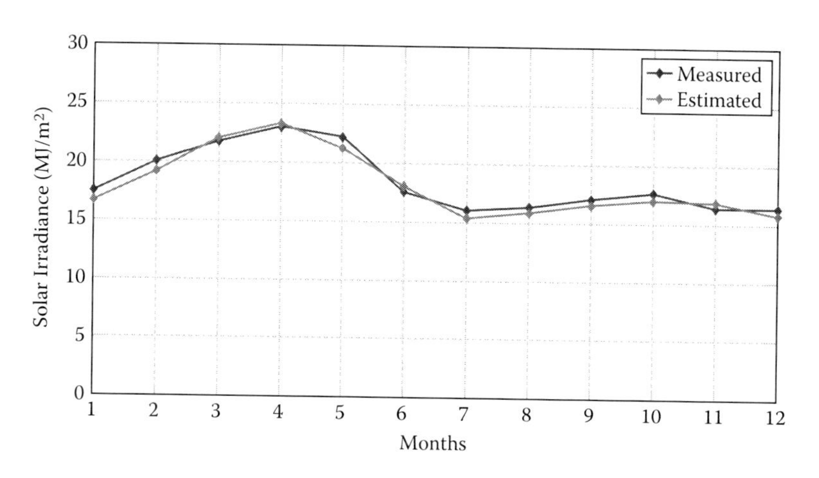

FIGURE 3.13
Measured and estimated global solar irradiance for Vishakhapatnam.

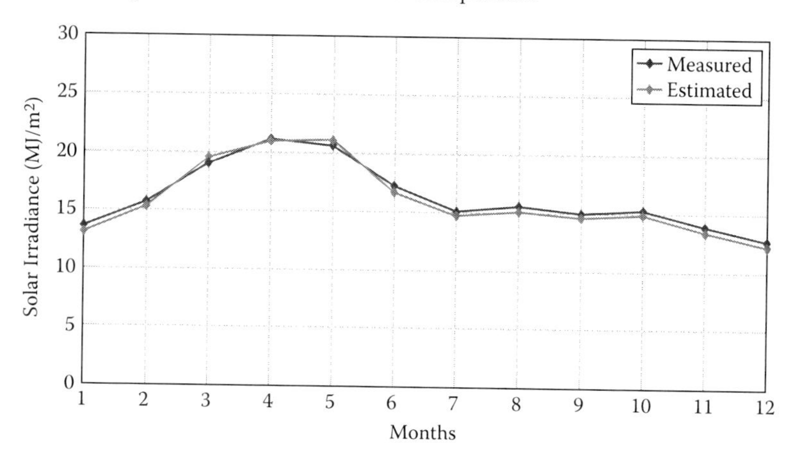

FIGURE 3.14
Measured and estimated global solar irradiance for Kolkata.

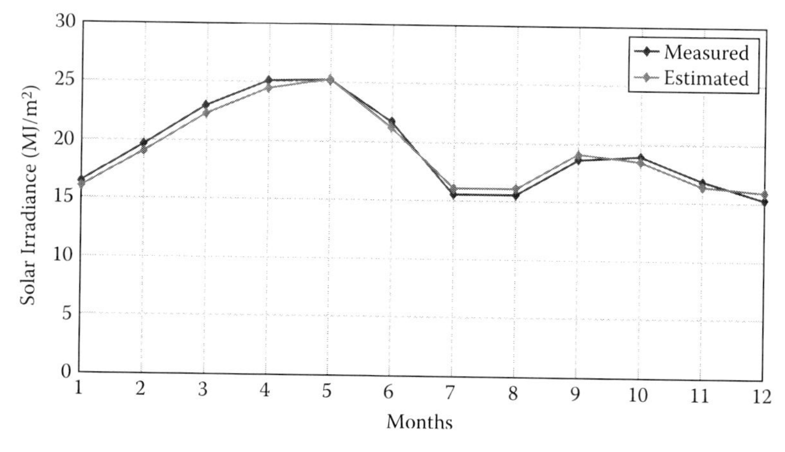

FIGURE 3.15
Measured and estimated global solar irradiance for Ahmedabad.

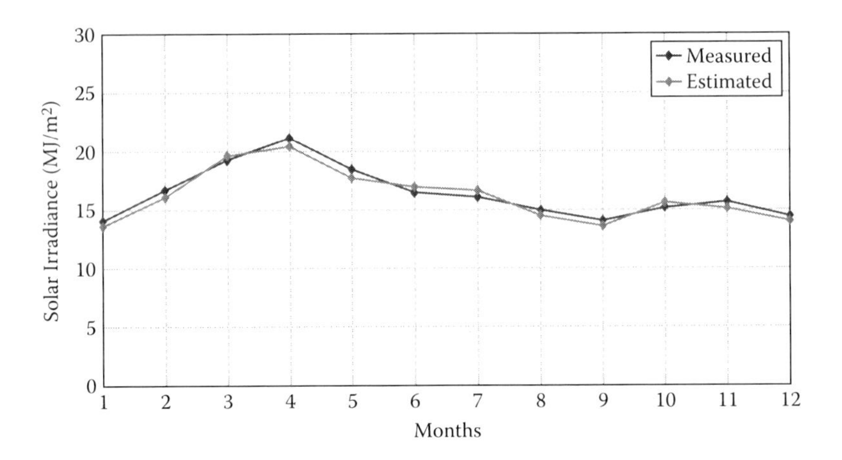

FIGURE 3.16
Measured and estimated global solar irradiance for Shillong.

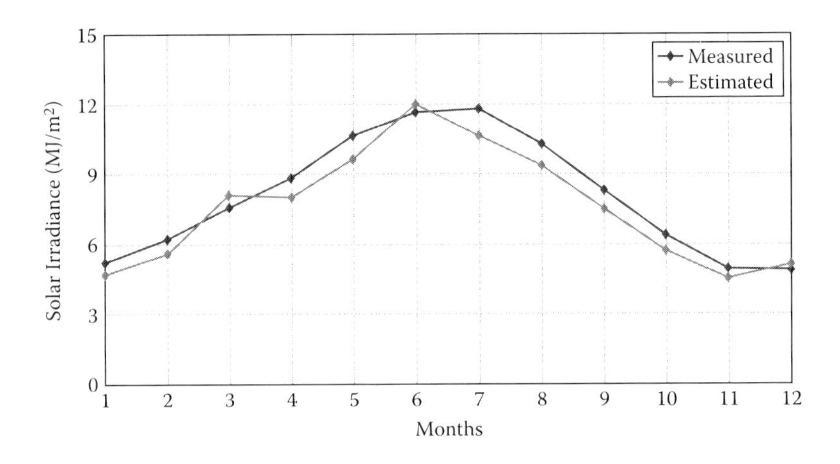

FIGURE 3.17
Measured and estimated diffuse solar irradiance for New Delhi.

However, estimated values of diffuse solar irradiance using fuzzy model in comparison with the measured values are also presented graphically in Figures 3.17 through 3.23 for New Delhi, Jodhpur, Nagpur, Vishakhapatnam, Kolkata, Ahmedabad and Shillong, respectively.

3.4.3.3 Artificial Neural Network for Solar Energy Estimation

Neurons are the basic elements of the human brain and they provide us with the ability to apply previous experiences to our actions. Artificial neural networks (ANNs) are computing algorithms that mimic the four basic functions of biological neurons: receive inputs from other neurons or sources, combine them, perform operations on the result, and output the final result. What makes ANNs exciting is the fact that once a network has been set up, it can learn in a self-organizing way that emulates brain functions such as pattern recognition, classification, and optimization. ANNs have the ability to handle large and

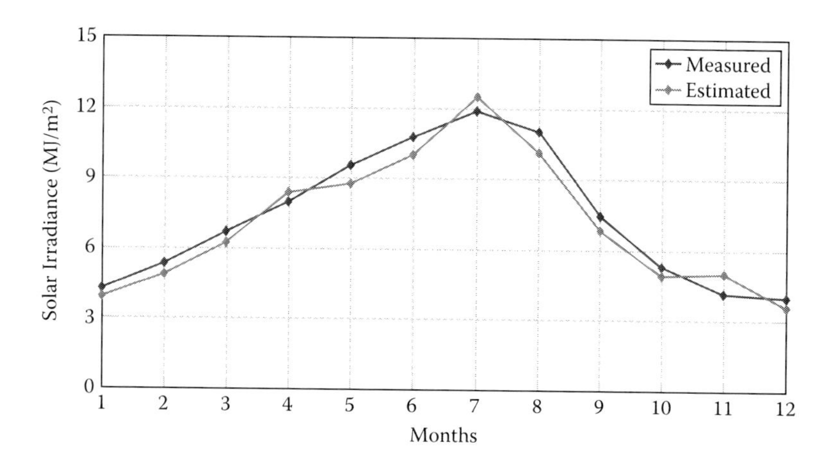

FIGURE 3.18
Measured and estimated diffuse solar irradiance for Jodhpur.

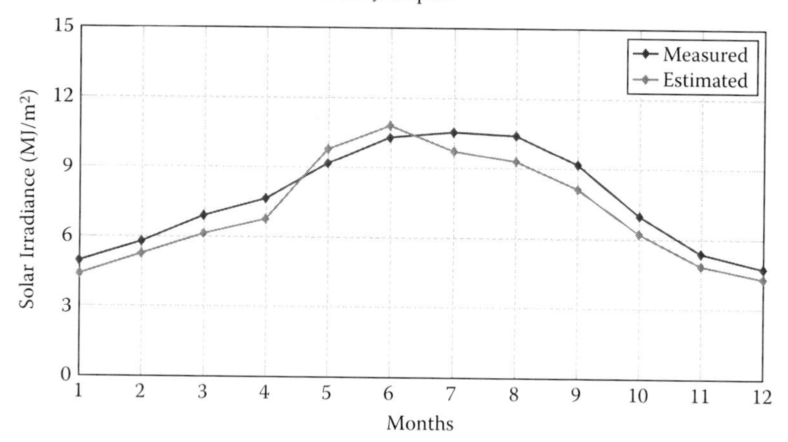

FIGURE 3.19
Measured and estimated diffuse solar irradiance for Nagpur.

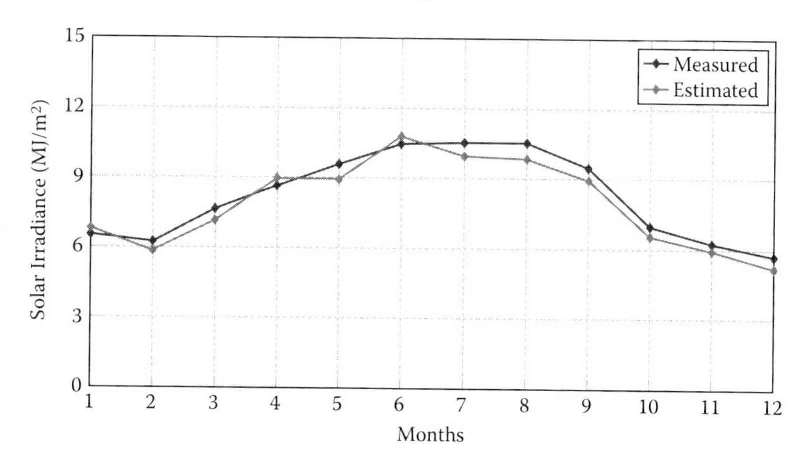

FIGURE 3.20
Measured and estimated diffuse solar irradiance for Vishakhapatnam.

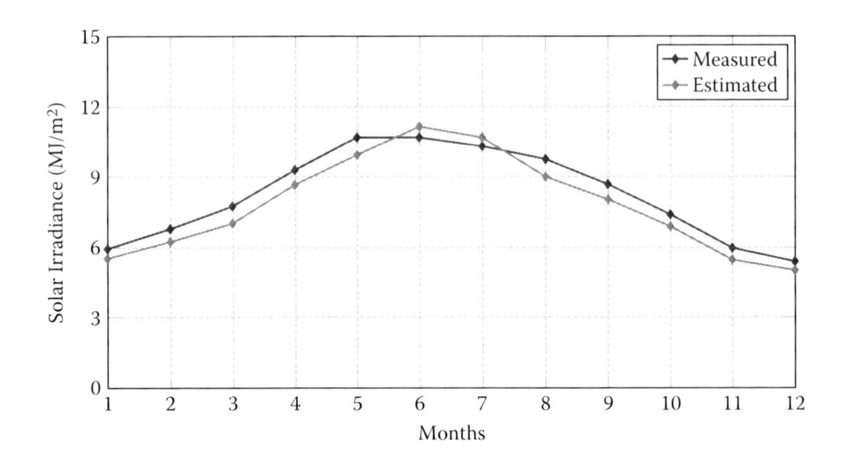

FIGURE 3.21
Measured and estimated diffuse solar irradiance for Kolkata.

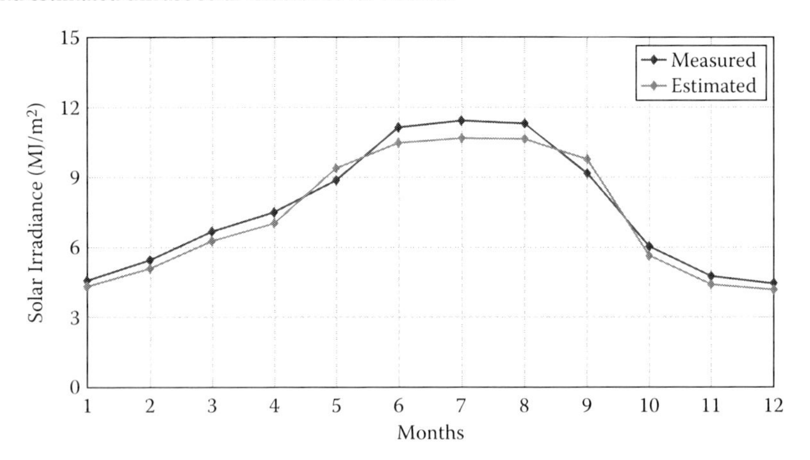

FIGURE 3.22
Measured and estimated diffuse solar irradiance for Ahmedabad.

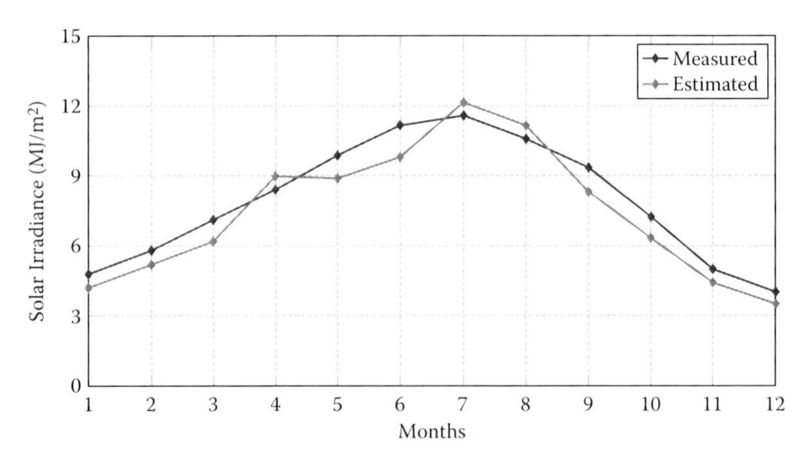

FIGURE 3.23
Measured and estimated diffuse solar irradiance for Shillong.

complex systems with many interrelated parameters. They ignore excess data that are less significant and concentrate on the more important inputs.

The most popular and powerful learning algorithm in neural networks is the back-propagation and its variants. This algorithm is based on the error correction learning rule. Basically, the error-correcting back-propagation process consists of two passes through the different layers of a network: a forward pass and a backward pass. In the forward pass, an activity pattern is applied to the sensory nodes of the network, and its effect propagates through the network layer by layer. Finally a set of outputs is produced as the actual response of the network. During the backward pass, on the other hand, the synaptic weights of the network are all adjusted in accordance with the error correction rule. Specifically, the actual response of the network is subtracted from a desired response to produce an error signal. The error signal is then propagated backward through the network.

An ANN is characterized by its architecture, training or learning algorithm, and activation function. The architecture describes the connections between the neurons. It consists of an input layer, an output layer, and generally, one or more hidden layers in between as depicted in Figure 3.24a and b.

3.4.3.3.1 *Artificial Neuron Model*

The model of an artificial neuron is shown in Figure 3.24. In this model the processing element (neuron) computes the weighted sum of its inputs and outputs according to whether this weighted input sum is above or below a certain threshold, θ_k. The externally applied bias has the effect of lowering the net input of the activation function:

$$y_k = f(u_k \theta_k)$$

$$u_k = \Sigma w_{kj} x_j$$

Here $x_1, x_2 \ldots x_p$ are input signals and $w_{k1}, w_{k2} \ldots w_{kp}$ are interconnection weights of the neuron k. u_k is the linearly combined output.

The interconnection of several layers forms a multilayer feed-forward network as shown in Figure 3.24c. The layer between the input and output layers is called a hidden layer, and its function is to intervene between the external input and output. The hidden layer had no direct contact with the external environment. The radial basis function network is a special class of multilayer feed-forward networks. The hidden layer in this employs a radial basis function, such as a Gaussian kernel, as the activation function. Feedback networks that have closed loops are called recurrent networks. In a single-layer recurrent network, the processing element output is feedback to itself or to another processing element or to both. In the multilayer recurrent network, the neuron output can be directed back to the nodes in the preceding layers.

The architecture used in this model shown in Figure 3.25 has an input layer of six inputs and one hidden layer with a tan-sigmoid activation function, o_j, defined as:

$$o_j = 1/(1+e^{-n}) \tag{3.23}$$

where n is the corresponding input.

The MATLAB neural network toolbox is used for the implementation of the feed-forward back-propagation network. The flowchart for the same algorithm is shown in Figure 3.26.

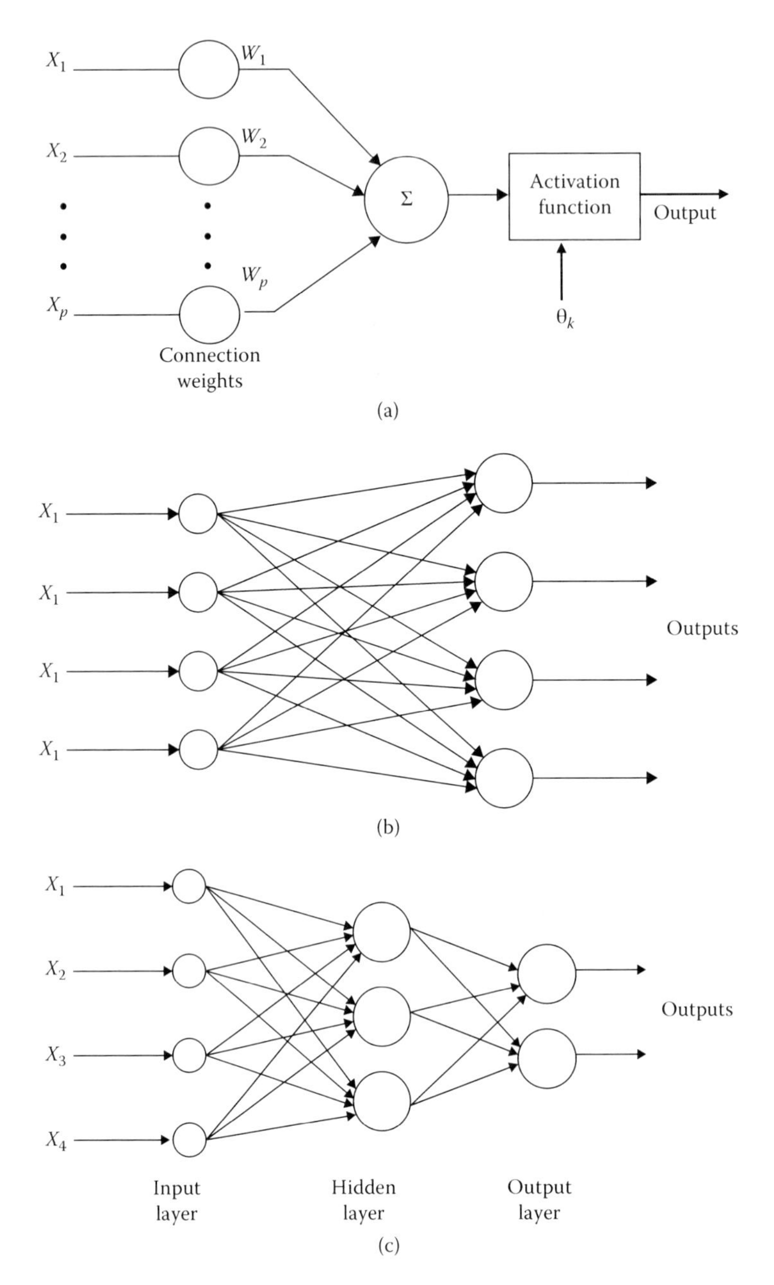

FIGURE 3.24
(a) Artificial neuron model. (b) Single-layer feed-forward network. (c) Multilayer feed-forward network.

3.4.3.3.2 Normalization of Meteorological Data

The input and output data for the neural network may have different ranges if actual monthly data are used. This may cause convergence problems during the learning process. To avoid such problems, the input and output data are scaled such that they remain within the range of 0.1 to 0.9. The lower limit is 0.1, so that during testing it could not go far beyond the lower extreme limit, which is 0. Similarly, the upper limit 0.9, so that the data

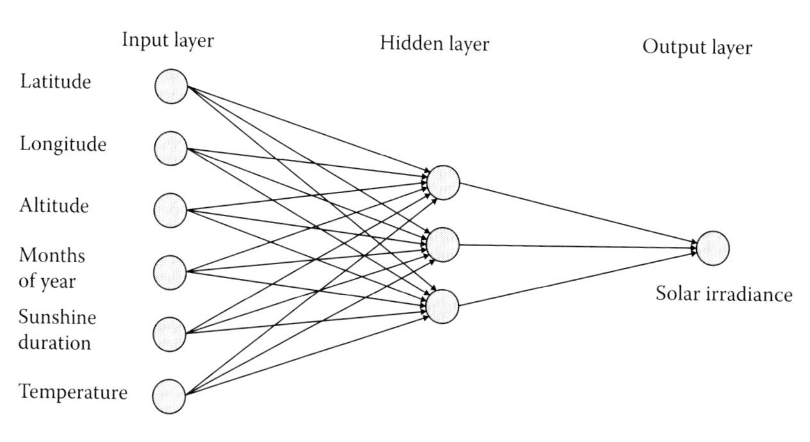

FIGURE 3.25
Proposed ANN architecture.

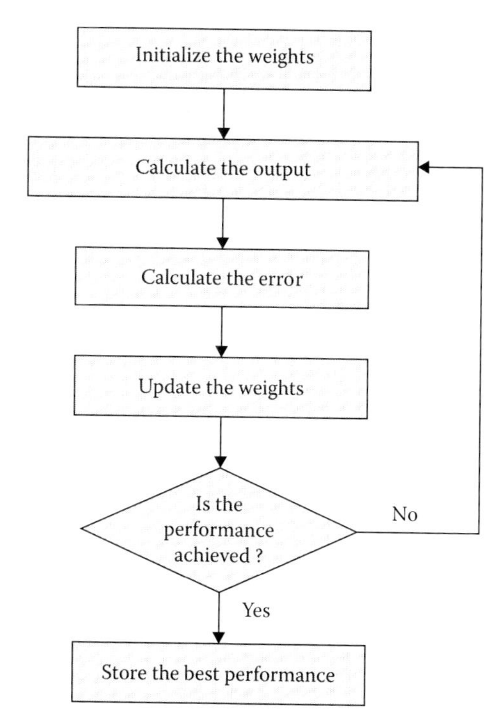

FIGURE 3.26
Flowchart for the proposed ANN model.

could go up to the upper extreme limit in testing, which is 1.0. These margins of 0.1 on both sides are called safe margins. The actual data are scaled using the following expression for the solar energy estimation problem:

$$L_S = \frac{(Y_{max} - Y_{min})}{(L_{max} - L_{min})}(L\text{-}L_{min}) + Y_{min} \tag{3.24}$$

where

L = The actual data
L_s = The scaled data, which are used as input to the network
L_{max} = Maximum value of particular dataset
L_{min} = Minimum value of particular dataset
Y_{max} = Upper limit (0.9) of normalization range
Y_{min} = Lower limit (0.1) of normalization range

The data used in the proposed work are the global irradiance, diffuse irradiance, air temperature, and sunshine duration. These data are normalized by dividing them by according to extraterrestrial value. For the purposes of analysis, seven Indian stations—New Delhi, Kolkata, Mumbai, Jodhpur, Pune, Ahmedabad, Vishakhapatnam, and Shillong—are selected to assess the monthly mean global and diffuse solar irradiance based on the sunshine duration and temperature.

3.4.3.3.3 Implementation and Results

Data from the seven stations were used to train and validate the neural networks. The input parameters were latitude, longitude, altitude, sunshine ratio (S/S_0), temperature ratio (T/T_0), and month. The output parameter is the clearness index (H/H_0). The estimated solar energy has been obtained by multiplying the estimated clearness index by H_0.

Using the normalized data, the ANN model has trained for New Delhi, Kolkata, Ahmedabad, and Vishakhapatnam. It is clearly seen from Figures 3.27 to 3.30 that most of the measured data overlaps with the trained data; this means model is trained successfully. Using the developed model, data are tested, as shown in Figures 3.31 through 3.40 for New Delhi, Jodhpur, Nagpur and Shillong, respectively. It is found that the maximum ARE is in the range of 3.1% to 4.7%.

Similarly, the normalized data of input and output parameters for New Delhi and Kolkata stations are presented in Table 3.11.

The ANN model has been trained for New Delhi, Kolkata, Ahmedabad, and Vishakhapatnam using the normalized input and output data (Table 3.12). It is clearly seen from Figures 3.35 to 3.38 that most of the measured data are very close to the

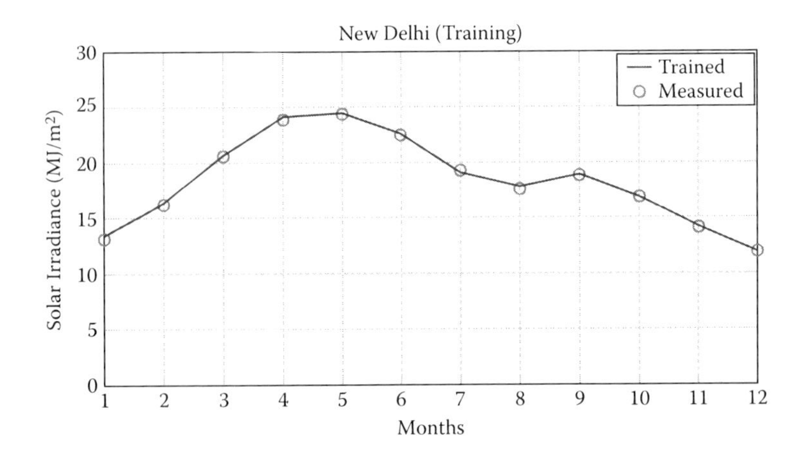

FIGURE 3.27
Training results for the assessment of global solar energy for New Delhi.

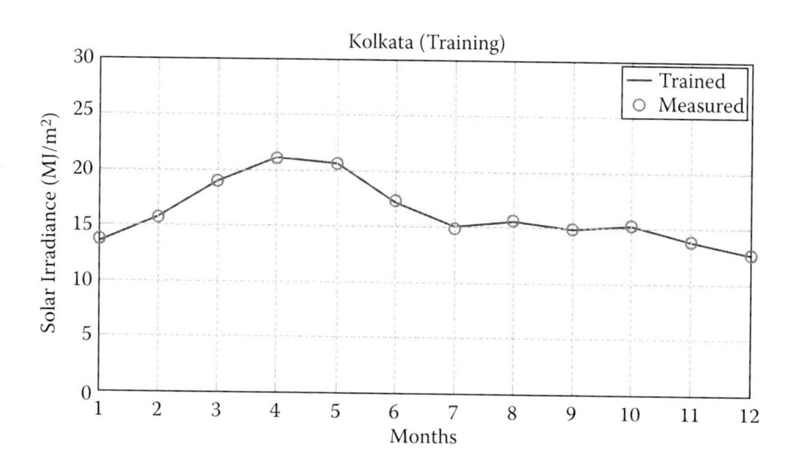

FIGURE 3.28
Training results for the assessment of global solar energy for Kolkata.

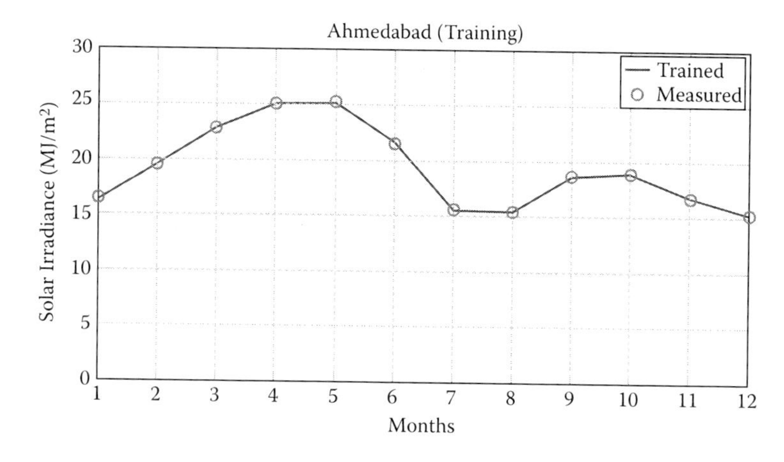

FIGURE 3.29
Training results for the assessment of global solar energy for Ahmedabad.

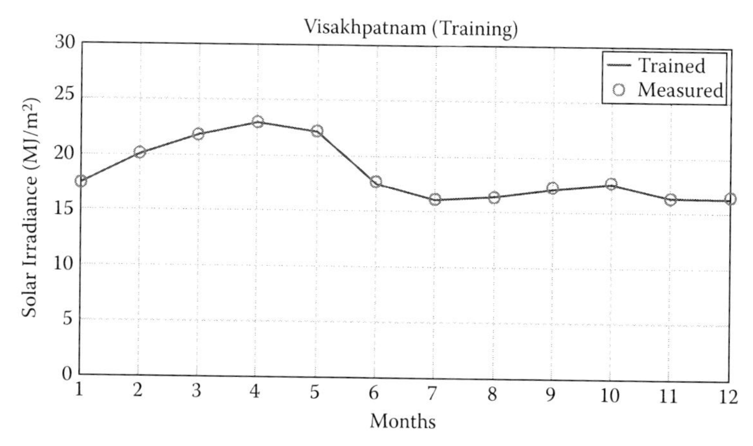

FIGURE 3.30
Training results for the assessment of global solar energy for Vishakhapatnam.

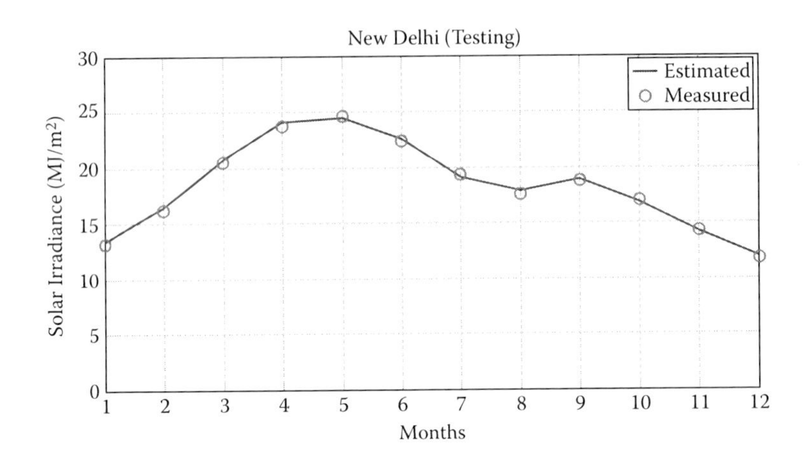

FIGURE 3.31
Testing results for the assessment of global solar energy for New Delhi.

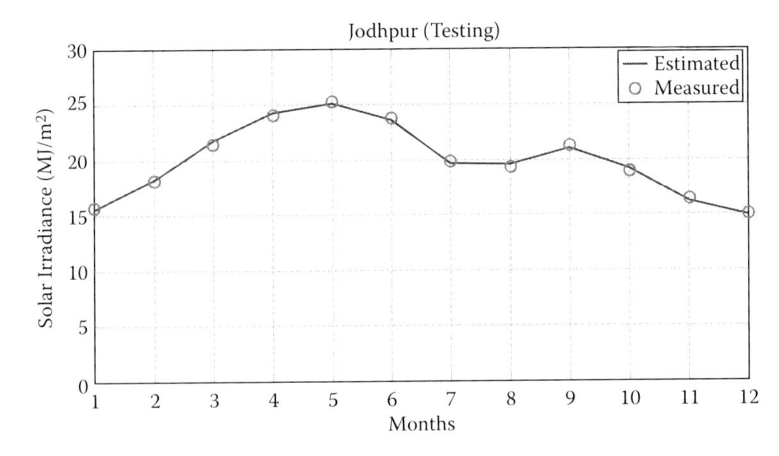

FIGURE 3.32
Testing results for the assessment of global solar energy for Jodhpur.

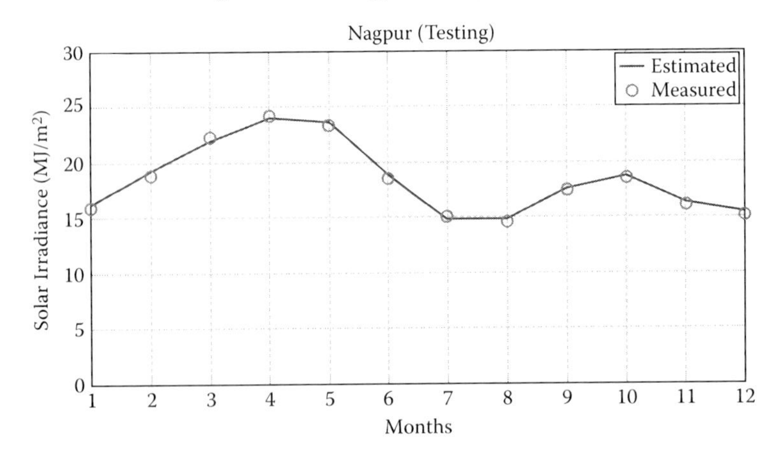

FIGURE 3.33
Testing results for the assessment of global solar energy for Nagpur.

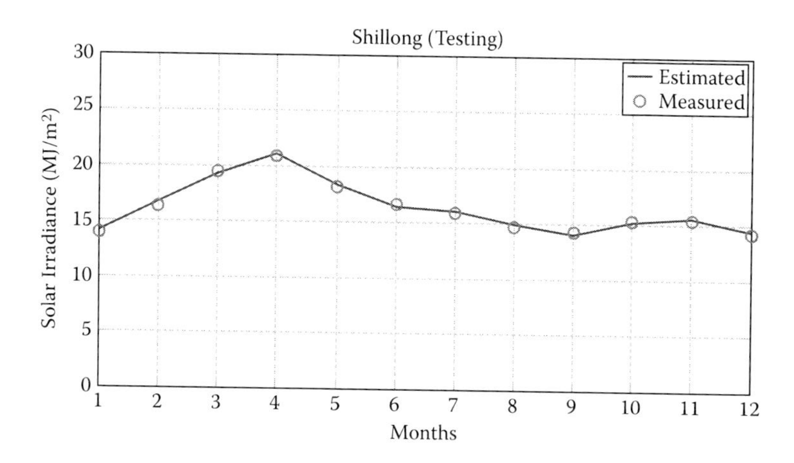

FIGURE 3.34
Testing results for the assessment of global solar energy for Shillong.

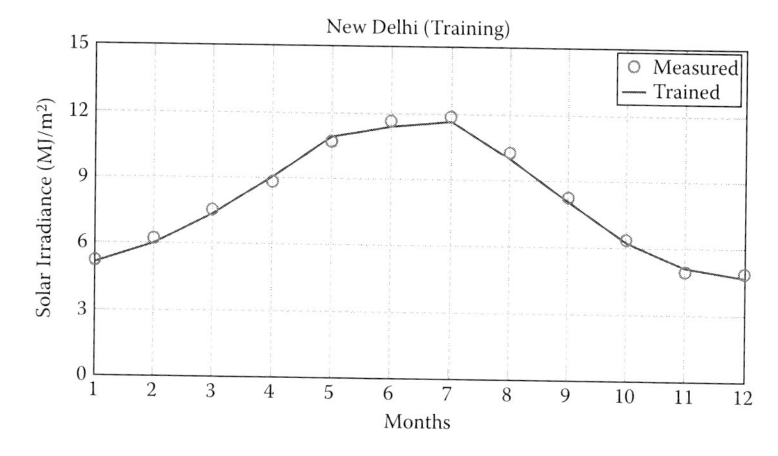

FIGURE 3.35
Training results for the assessment of global solar energy for New Delhi.

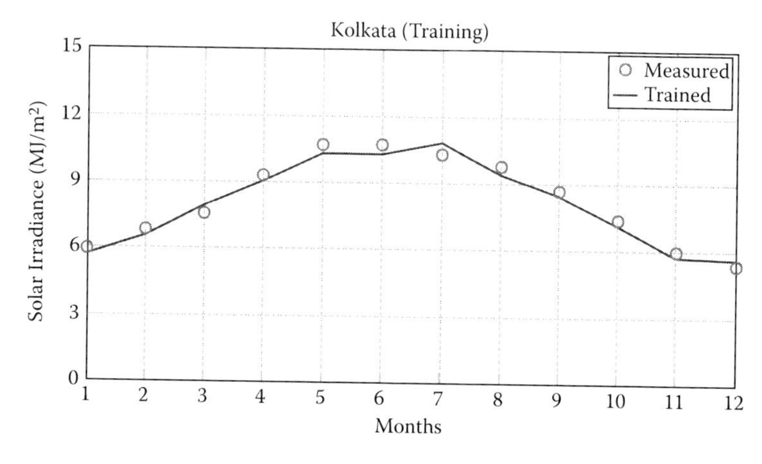

FIGURE 3.36
Training results for the assessment of global solar energy for Kolkata.

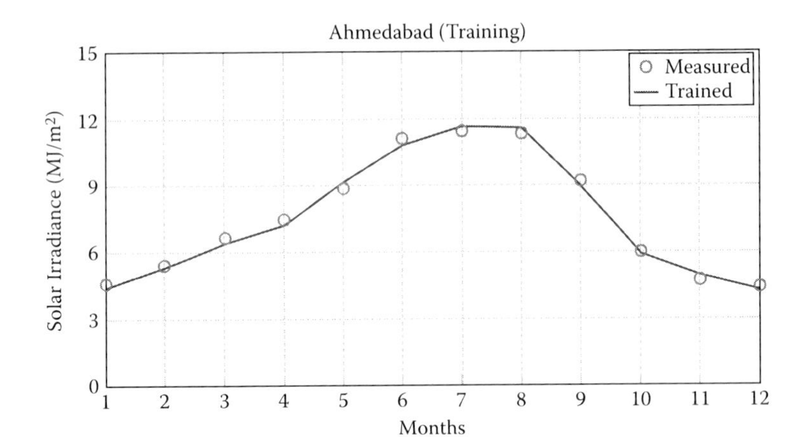

FIGURE 3.37
Training results for the assessment of global solar energy for Ahmedabad.

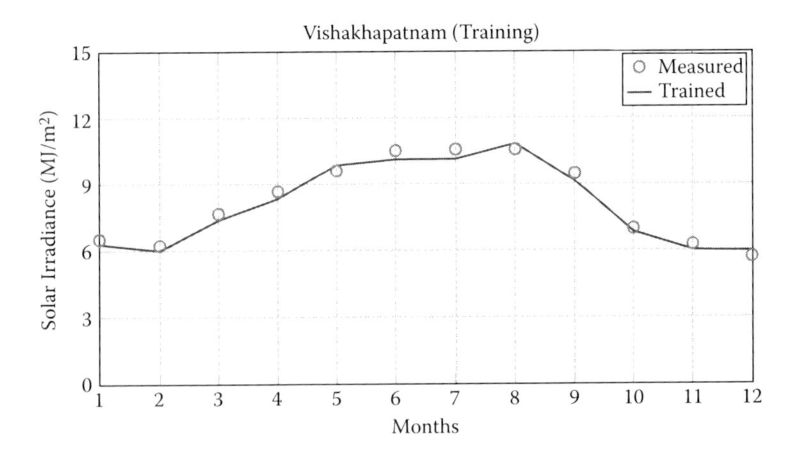

FIGURE 3.38
Training results for the assessment of global solar energy for Vishakhapatnam.

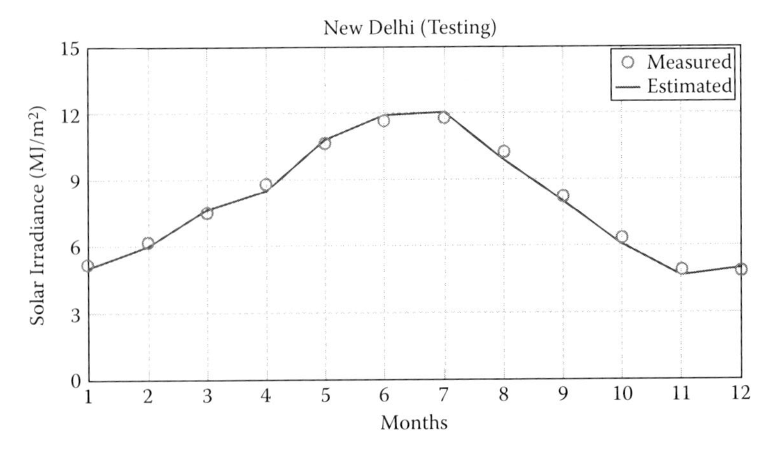

FIGURE 3.39
Testing results for the assessment of global solar energy for New Delhi.

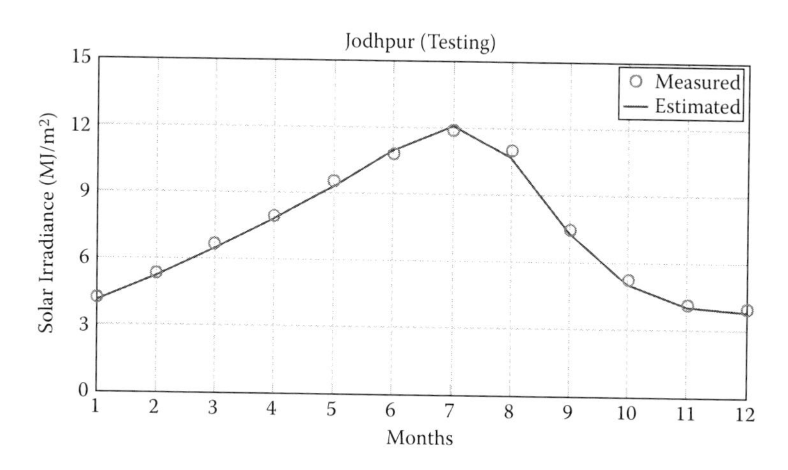

FIGURE 3.40
Testing results for the assessment of global solar energy for Jodhpur.

TABLE 3.11
Monthly Average Values of H_d/H_g, S/S_0, and T/T_0 for New Delhi and Kolkata for the Assessment of Diffuse Solar Energy

	New Delhi			Kolkata		
Month	H_d/H_g	S/S_0	T/T_0	H_d/H_g	S/S_0	T/T_0
January	0.391	0.836	0.334	0.437	0.900	0.646
February	0.378	0.857	0.509	0.431	0.900	0.734
March	0.366	0.857	0.639	0.406	0.900	0.841
April	0.367	0.868	0.809	0.440	0.889	0.900
May	0.437	0.878	0.896	0.517	0.878	0.899
June	0.517	0.847	0.900	0.621	0.776	0.867
July	0.620	0.697	0.841	0.681	0.641	0.835
August	0.577	0.697	0.813	0.627	0.652	0.832
September	0.438	0.847	0.673	0.582	0.754	0.830
October	0.379	0.900	0.727	0.483	0.878	0.822
November	0.348	0.853	0.596	0.429	0.900	0.764
December	0.408	0.804	0.464	0.424	0.889	0.680

trained data, which means that the model is trained successfully. Using the developed model, data for stations in Jodhpur, Nagpur, Shillong, and New Delhi are tested, as shown in Figures 3.39 through 3.42. The maximum ARE is found to be in the range of 4.3% to 6.2%.

The size of the ARE was used to determine the results. The ARE evaluates the performance of the model developed, which is determined by using the following formula:

$$\text{ARE} = \frac{|P_{\text{desired}} - P_{\text{forecasted}}|}{P_{\text{desired}}} \times 100 \tag{3.25}$$

The computed AREs are shown in Table 3.13 and Table 3.14 for New Delhi and Jodhpur, respectively.

TABLE 3.12

Monthly Average Values of H/H_0, S/S_0, and T/T_0 for New Delhi and Kolkata for the Assessment of Global Solar Energy

	New Delhi			Kolkata		
Month	H/H_0	S/S_0	T/T_0	H/H_0	S/S_0	T/T_0
January	0.189	0.836	0.334	0.181	0.900	0.646
February	0.387	0.857	0.509	0.386	0.900	0.734
March	0.657	0.857	0.639	0.702	0.900	0.841
April	0.877	0.868	0.809	0.900	0.889	0.900
May	0.900	0.878	0.896	0.860	0.878	0.899
June	0.779	0.847	0.900	0.529	0.776	0.867
July	0.557	0.697	0.841	0.330	0.641	0.835
August	0.475	0.697	0.813	0.376	0.652	0.832
September	0.546	0.847	0.673	0.312	0.754	0.830
October	0.412	0.900	0.727	0.347	0.878	0.822
November	0.241	0.853	0.596	0.212	0.900	0.764
December	0.100	0.804	0.464	0.100	0.889	0.680

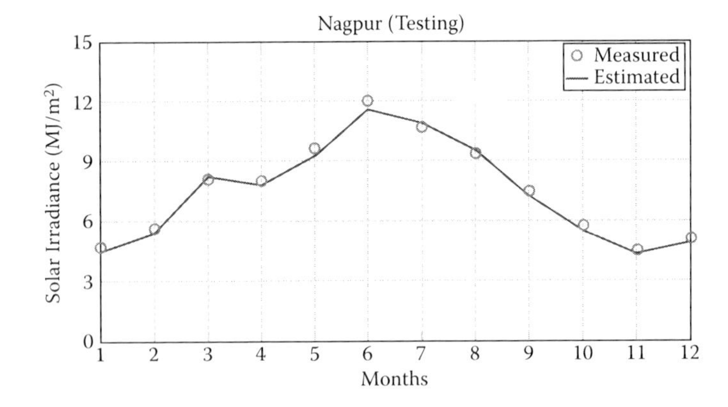

FIGURE 3.41
Testing results for the assessment of global solar energy for Nagpur.

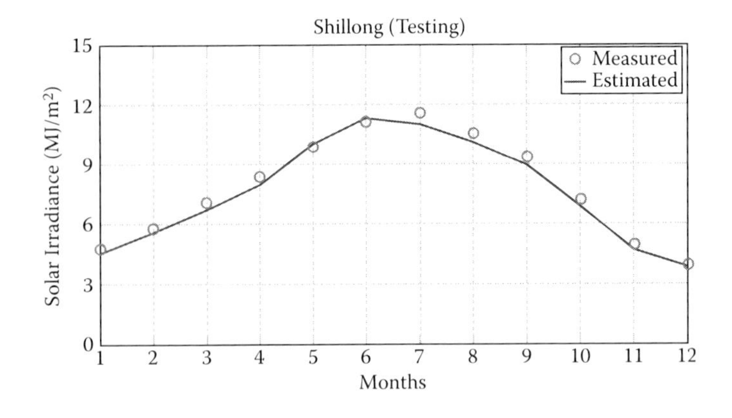

FIGURE 3.42
Testing results for the assessment of global solar energy for Shillong.

TABLE 3.13

ARE Percentages in the Assessment of Monthly Mean Solar Irradiance Compared with Measured Data for New Delhi

	H (Measured) (MJ/m²)		*H* (Estimated) (MJ/m²)		ARE (%)	
Month	**Global**	**Diffuse**	**Global**	**Diffuse**	**Global**	**Diffuse**
January	13.32	5.21	**12.93**	**5.43**	2.9	**4.2**
February	16.42	6.22	**15.91**	**6.44**	3.1	**3.5**
March	20.64	7.56	**19.90**	**7.33**	3.6	**3.0**
April	24.07	8.83	**23.18**	**9.10**	3.7	**3.1**
May	24.43	10.68	**25.24**	**10.30**	3.3	**3.5**
June	22.54	11.66	**21.98**	**11.18**	2.5	**4.1**
July	19.07	11.83	**19.83**	**11.27**	4.0	**4.7**
August	17.79	10.27	**17.02**	**10.74**	4.3	**4.6**
September	18.90	8.27	**19.69**	**8.66**	4.2	**4.8**
October	16.80	6.37	**17.52**	**6.66**	4.3	**4.6**
November	14.13	4.92	**14.44**	**5.10**	2.2	**3.3**
December	11.93	4.87	**11.60**	**4.70**	2.7	**3.5**

TABLE 3.14

ARE Percentages in the Assessment of Monthly Mean Solar Irradiance Compared with Measured Data for Jodhpur

	H (Measured) (MJ/m²)		*H* (Estimated) (MJ/m²)		ARE (%)	
Month	**Global**	**Diffuse**	**Global**	**Diffuse**	**Global**	**Diffuse**
January	15.89	4.23	**15.57**	**4.06**	2.0	**3.9**
February	18.09	5.33	**17.71**	**5.53**	2.1	**3.7**
March	21.16	6.67	**20.71**	**6.91**	2.1	**3.6**
April	22.57	7.97	**22.02**	**8.20**	2.4	**2.9**
May	23.56	9.56	**24.15**	**9.83**	2.5	**2.8**
June	22.73	10.77	**23.30**	**10.42**	2.5	**3.2**
July	19.21	11.90	**18.61**	**11.44**	3.1	**3.9**
August	17.91	11.00	**17.34**	**11.52**	3.2	**4.7**
September	19.84	7.43	**20.73**	**7.75**	3.7	**4.5**
October	19.35	5.25	**18.58**	**5.48**	4.0	**4.4**
November	16.68	4.13	**16.23**	**4.25**	2.7	**2.9**
December	15.04	3.98	**14.62**	**4.10**	2.8	**3.1**

3.4.3.3.4 Drawbacks of Conventional ANN

The conventional neural network model suffers from the following serious drawbacks:

1. The training time for the conventional neural network is too long, which results in a slower response of the system.
2. The number of hidden layers and hidden neurons can't be predicted accurately, and they are large in terms of complex function approximation.
3. The existing neuron model performs only the operation of summation of its weighted inputs; it does not perform the operation of product on its weighted inputs.

4. The threshold (activation) function has an effect on training time; also the accuracy of the test result depends on the threshold function.
5. Back-propagation learning also has some shortcomings; for example, learning can be slow and the problem of local minima may occur in the system.
6. The normalization range has an effect on training data. Hence selection of suitable range (i.e., maximum and minimum values) is of great importance as it affects the results of the neural network training.
7. The training time of the neural network depends on the mapping of the input-output pattern (I/O mapping) presented to the network.
8. The training time of the network also depends on the sequence in which data are presented.

To overcome these drawbacks, a number of variants have been developed in the past decades. Most of the variants are either place a burden on the learning algorithms or/and increase the computational labor. In this thesis, a new generalized neuron model is used that overcomes the drawbacks of the conventional neural network by performing various possible variations and modifications on the previous model to solve the problem of solar energy estimation. The model should incorporate nonlinearities present in the system. The model should also incorporate the following features:

1. The generalized neural network should consist of characteristics of both simple neurons and high-order neurons.
2. There is no need to select the number of hidden layers and the number of neurons; as a result, the complexity of the network should reduce.
3. The input-output mapping should not affect the response of the network.
4. The normalizing effect should not be there.

3.4.3.4 Generalized Neural Model

The general structure of the common neuron is an aggregation function and its transformation through a filter. It is shown in the literature that ANNs can be universal function approximators for given input-output data. The common neuron structure shown in Figure 3.43 has summation as the aggregation function with sigmoidal, radial basis, tangent hyperbolic, or linear limiters as the thresholding function. The aggregation operators used in the neurons are generally crisp. However, they overlook the fact that most of

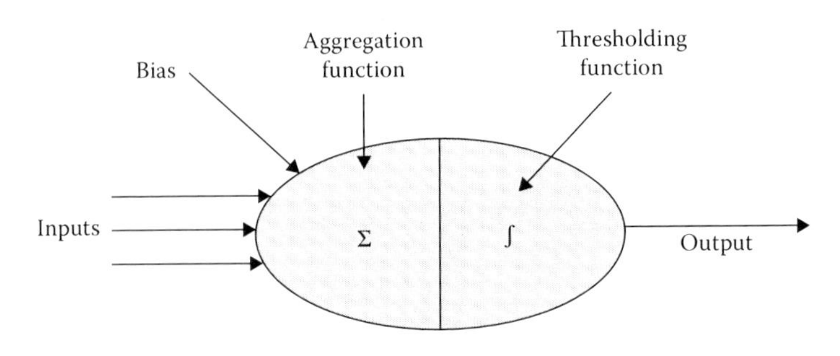

FIGURE 3.43
Simple neuron model.

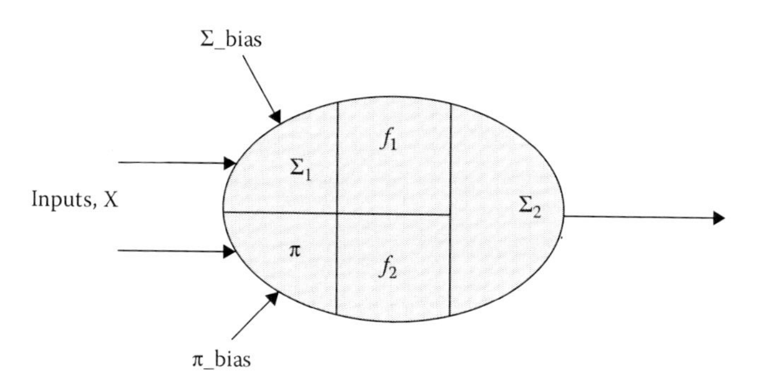

FIGURE 3.44
Generalized neuron model.

the processing in the neural networks is done with incomplete information at hand. Thus, a GN model approach has been adopted that uses the fuzzy compensatory operators that are partly sum and partly product to take into account the vagueness involved.

Use of the sigmoidal thresholding function and ordinary summation or product as an aggregation function in the existing model fails to cope with the nonlinearities involved in real-life problems. To deal with these, the proposed model has both sigmoidal and Gaussian functions with weight sharing. The GNN model has flexibility at both the aggregation and threshold function levels to cope with the nonlinearity involved in the type of applications dealt with, as shown in Figure 3.44.

The neuron model described earlier is known as the summation type compensatory neuron model because the outputs of the sigmoidal and Gaussian functions are summed up. Similarly, the product type compensatory neuron models may also be developed. It is found that in most of the applications, the summation type compensatory neuron model works well.

3.4.3.4.1 *GNN Model for Solar Energy Estimation*

Existing models of neurons in the structure of an ANN use the sigmoidal activation function and ordinary summation as aggregation functions. These models face problems in training when nonlinearity of real-life problems is involved. To deal with this, the proposed model has both summation (Σ) and product (π) as aggregation functions. The generalized neuron model has flexibility at both the aggregation and activation function levels to cope with the nonlinearity present in the types of applications usually dealt with. The product and power nonlinearity in problems added complexity in terms of training, but with the help of the product aggregation function, it is quite easy to train. In this chapter we have tried product neuron layers along with summation neuron layers in the ANN and found that the training time is drastically reduced for mapping the nonlinear system if one layer is the product later and the other is the summation layer.

The generalized neural network is developed on the basis of Boolean algebra. It is well known that with the help of sum of product and product of sum, one can implement any given function. Similarly in the generalized neuron structure, summation and product have been incorporated as aggregation functions, and the aggregated outputs pass through a nonlinear squashing/thresholding function as shown in Figure 3.45. The Σ-part has the summation of weighted input with sigmoidal activation function f_1, whereas the part has the product of weighted input with Gaussian activation function f_2. The final output of the neuron is a function of the weighted outputs O_Σ and O_π.

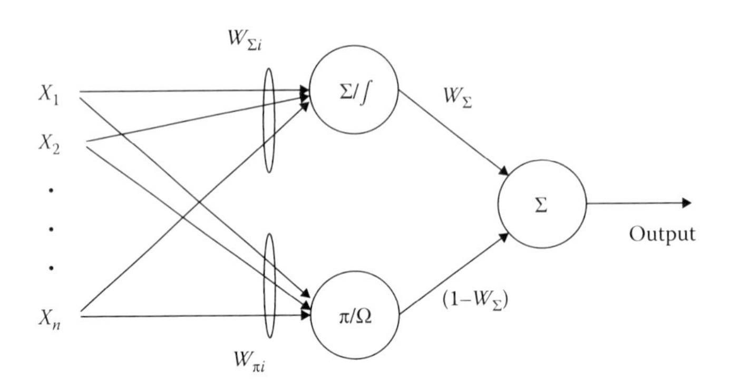

FIGURE 3.45
Proposed generalized neuron model.

The output of the summation (Σ) part of the generalized neuron is:

$$O_{\Sigma} = f_1(\Sigma W_{\Sigma i} X_i + X_{o\Sigma}) \tag{3.26}$$

The output of the product (π) part of the generalized neuron is:

$$O_{\pi} = f_2(\pi W_{\pi i} X_i + X_{o\pi}) \tag{3.27}$$

Finally, the outputs are summed to get the neuron output. The output of the neuron can be mathematically written as:

$$O_i = O_{\Sigma} * W_{\Sigma} + O_{\pi}(1 - W_{\Sigma}) \tag{3.28}$$

The existing model of the neuron in the structure of the ANN uses the sigmoidal activation function and ordinary summation as aggregation functions. These models face problems in terms of training when the nonlinearity in real-life problems is present.

The data for New Delhi, Kolkata, Ahmedabad, and Vishakhapatnam are trained using GNN model. It is clearly seen from Figures 3.46 to 3.49 that during training both the values (measured and trained) are overlapping each other or are very close to each other. This shows that the network is trained successfully.

Now the network is tested for the New Delhi, Jodhpur, Nagpur, and Shillong stations, and the results are within the desired limits. The testing performance of the generalized neural network model is plotted as shown in Figures 3.50 through 3.53 for New Delhi, Jodhpur, Nagpur, and Shillong stations.

Using GNN, the maximum ARE in the estimation of global solar energy is found to be in the range of 2.1% to 3.4% for the various Indian stations considered in the proposed work. However, it is 2.1% and 2.3% for Jodhpur and New Delhi, respectively. Similarly, the GNN model has been trained for New Delhi, Kolkata, Ahmedabad, and Vishakhapatnam using the normalized input and output data for the estimation of diffuse energy. It is clearly seen from Figures 3.54 to 3.57 that most of the measured are is very close to the trained data; this means that the model is trained successfully. Using the developed model, data for stations in Jodhpur, Nagpur, Shillong, and New Delhi are tested as shown in Figures 3.58 through 3.61. It is found that the maximum ARE is

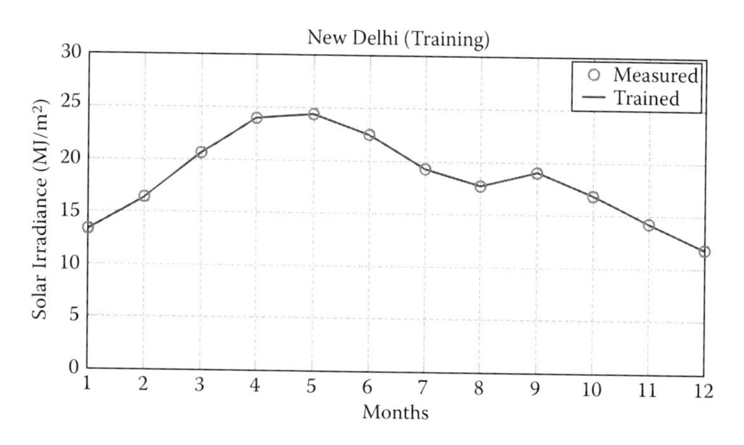

FIGURE 3.46
Training results for the estimation of global solar energy for New Delhi.

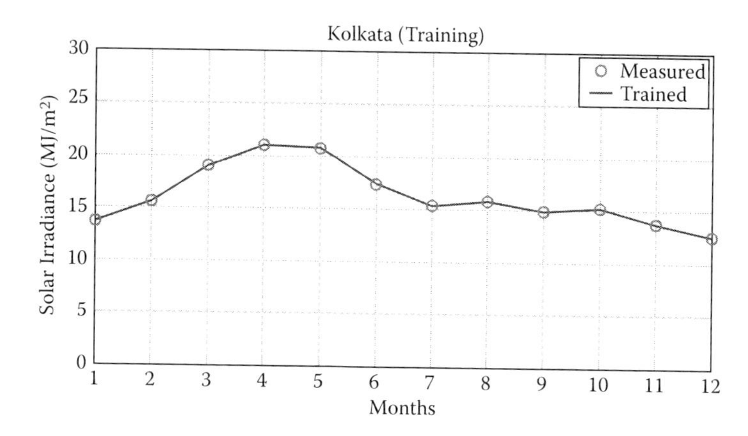

FIGURE 3.47
Training results for the estimation of global solar energy for Kolkata.

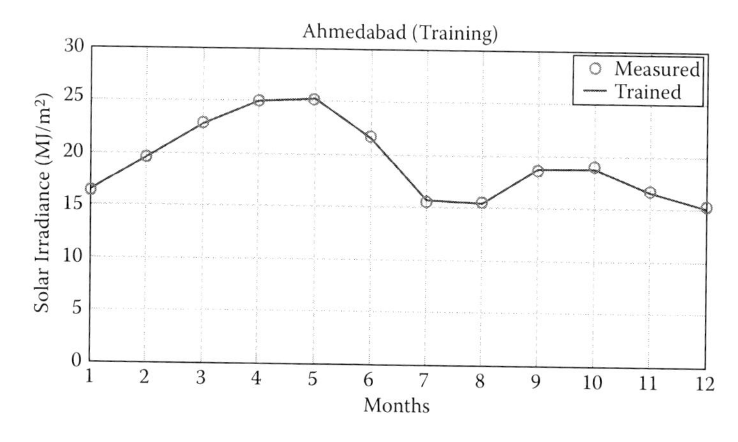

FIGURE 3.48
Training results for the estimation of global solar energy for Ahmedabad.

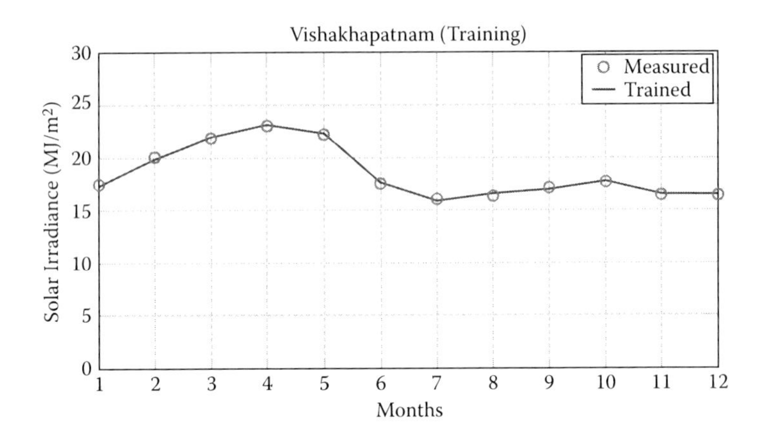

FIGURE 3.49
Training results for the estimation of global solar energy for Vishakhapatnam.

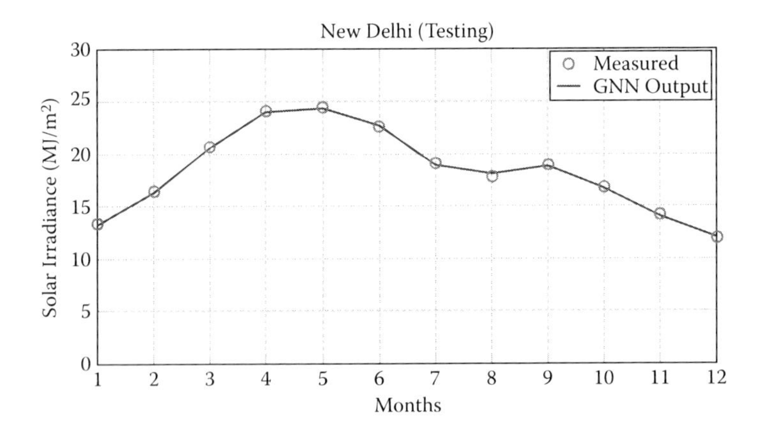

FIGURE 3.50
Testing results for the estimation of global solar energy for New Delhi.

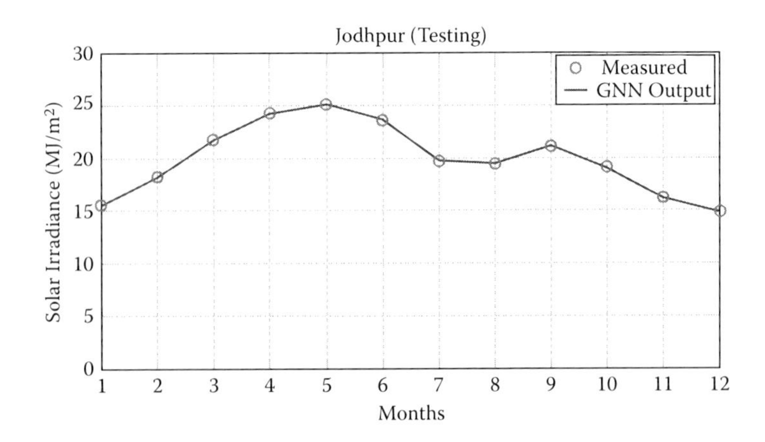

FIGURE 3.51
Testing results for the estimation of global solar energy for Jodhpur.

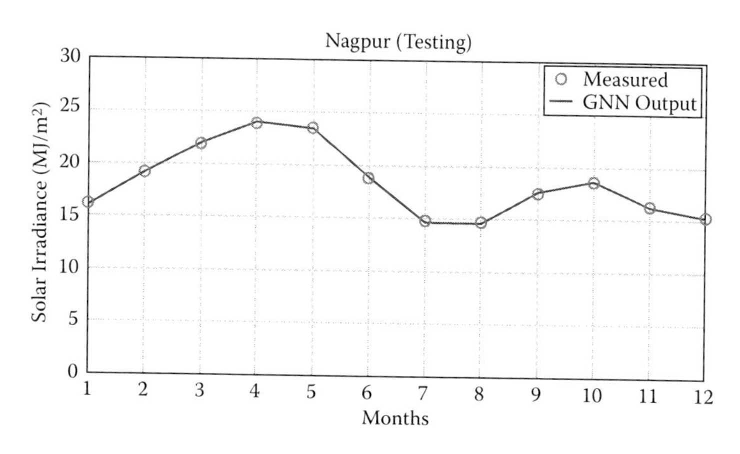

FIGURE 3.52
Testing results for the estimation of global solar energy for Nagpur.

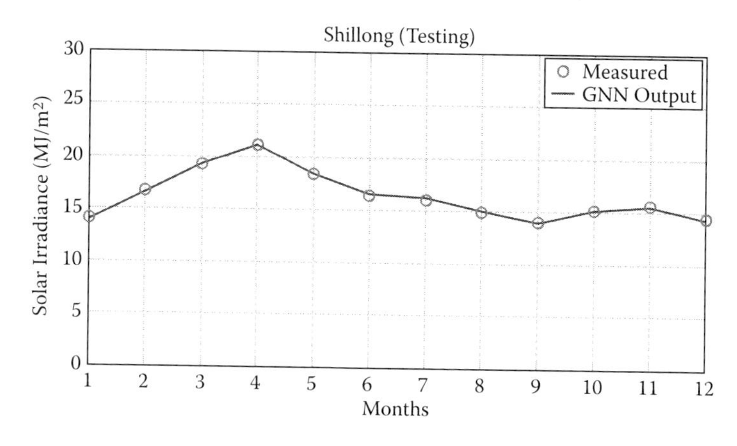

FIGURE 3.53
Testing results for the estimation of global solar energy for Shillong.

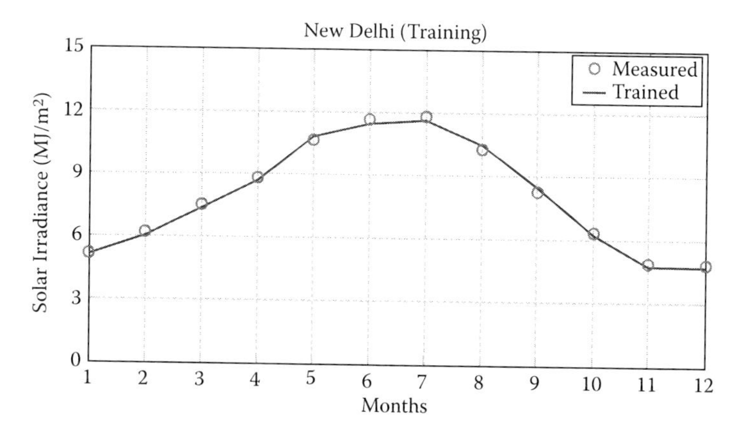

FIGURE 3.54
Training results for the estimation of diffuse solar energy for New Delhi.

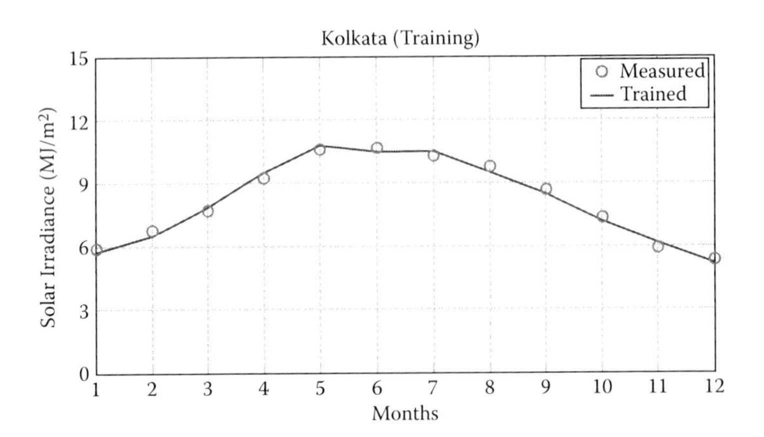

FIGURE 3.55
Training results for the estimation of diffuse solar energy for Kolkata.

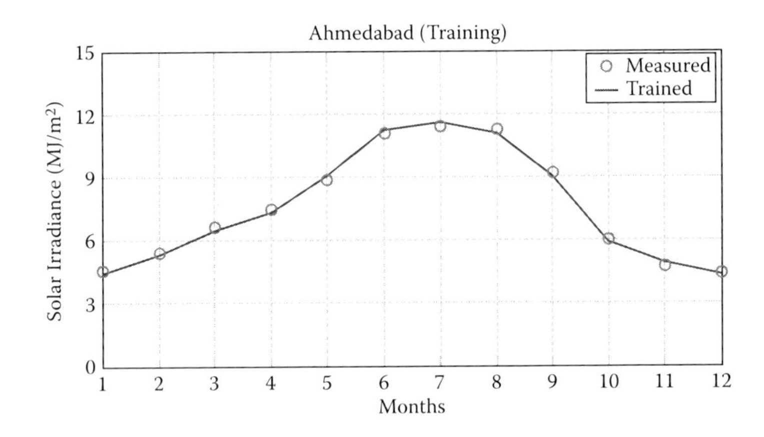

FIGURE 3.56
Training results for the estimation of diffuse solar energy for Ahmedabad.

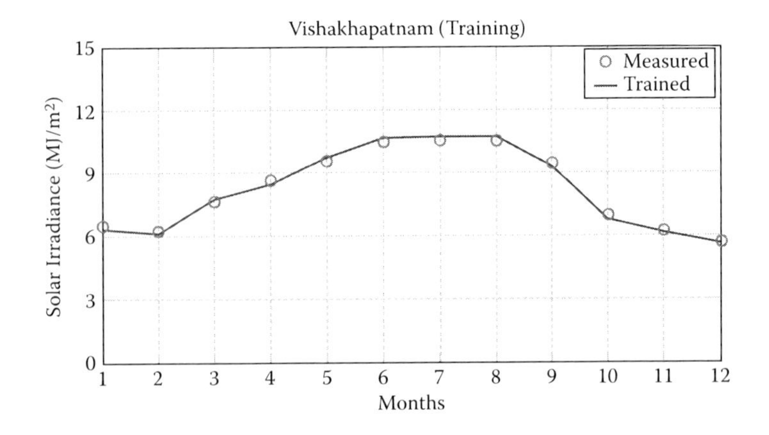

FIGURE 3.57
Training results for the estimation of diffuse solar energy for Vishakhapatnam.

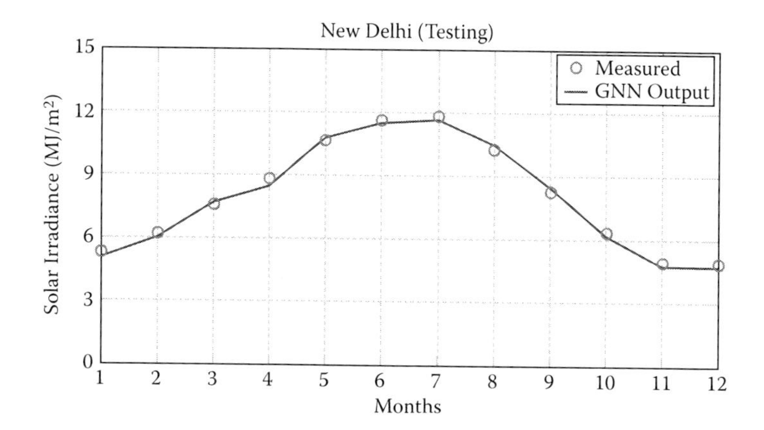

FIGURE 3.58
Testing results for the estimation of diffuse solar energy for New Delhi.

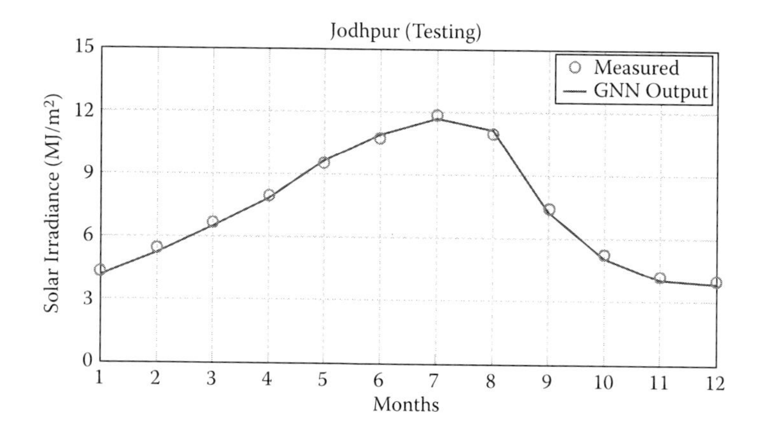

FIGURE 3.59
Testing results for the estimation of diffuse solar energy for Jodhpur.

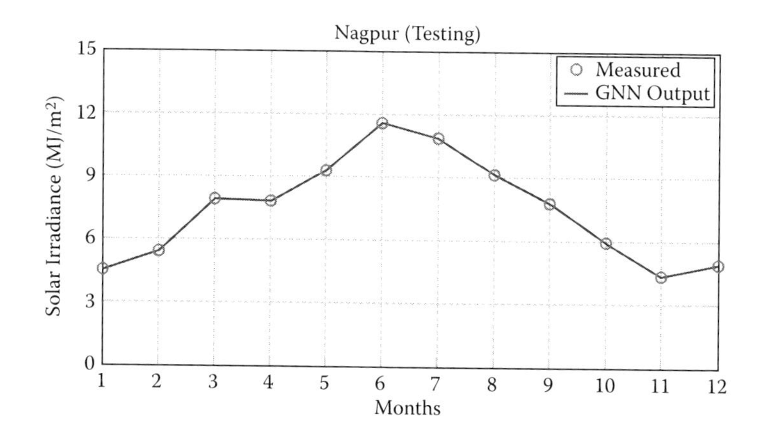

FIGURE 3.60
Testing results for the estimation of diffuse solar energy for Nagpur.

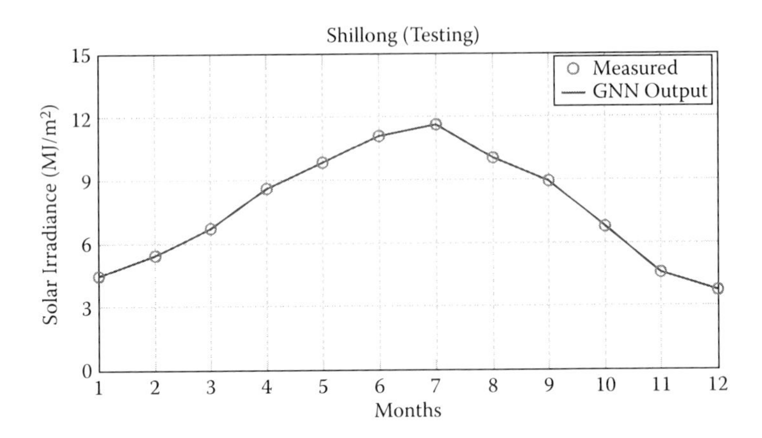

FIGURE 3.61
Testing results for the estimation of diffuse solar energy for Shillong.

in the range of 3.2% to 5.6%. However, the ARE percentages for New Delhi and Jodhpur are 4.7% and 3.7%, respectively.

The results obtained for solar energy estimation from the generalized neural network model have been compared with the actual data, and it is found that the generalized neural network model is most efficient in terms of training and most accurate in terms of estimation.

The GNN model has been trained for New Delhi, Kolkata, Ahmedabad, and Vishakhapatnam. The results for these stations are presented in Figures 3.54, 3.55, 3.56, and 3.57, respectively.

Using the GNN model, data for Jodhpur, Nagpur, Shillong, and New Delhi stations are tested as shown in Figures 3.58 through 3.61.

3.5 Summary

In this chapter different types of methods, including mathematical, regression, and intelligent models based on fuzzy logic, artificial neural networks, and generalized neural networks, have been explained with case studies. The case studies for the Indian scenario have been discussed and presented. The same can be implemented for other locations as well. From this chapter, it can be concluded that the mathematical models and regression models are only applicable for clear skies; however, the intelligent models can be used for clear and cloudy weather conditions. Further, the results obtained from generalized neural networks are better compared to other intelligent models and mathematical models.

Questions:

1. What is the necessity of solar energy estimation for solar power generation?
2. Discuss the importance of intelligent techniques for the estimation of solar irradiance.

3. Present a comparative study of conventional and intelligent methods for solar energy estimation.
4. Discuss various factors governing the availability of solar energy on the earth. Write the applications for the following solar radiation–measuring instruments:

   Pyranometer

   Pyrheliometer

   Sunshine recorder

4

Solar Photovoltaic Cells, Modules, and Arrays

4.1 Introduction

The technology used for to convert sunlight into electricity is called photovoltaics (PV). The term comes from *photo* meaning "light" and *voltaic* meaning "electricity." PV cells are also called "solar cells." These are basically semiconductor devices that can generate electricity based on the application of sunlight. Solar cells were discovered by a French scientist named Edmund Becquerel in 1839, but the concept was not comprehended completely until the early to middle 1900s when quantum physics and solid-state physics were developed in theory of light. From the time when solar cells were commercially used for the first time in the United States to power space satellites in the 1950s, there has been remarkable progress in PV cells as a total photovoltaic module. Most of the PV cells used today are silicon based, and very soon there will be PV cells made up of other semiconductor materials that will surpass silicon cells in terms of their cost and performance and so will become viable competitors in the PV cell market.

4.2 The Solar Cell

The main part of a PV system is its PV cells. This is generally a semiconductor device with a p-n junction that is exposed to sunlight. PV cells can be made by using different types of semiconductor materials, depending upon the manufacturing process. The Si type PV cells have a thick film of bulky silicon and a thinner layer of silicon, both connected to electric terminals. To get a p-n junction to form, one side of the silicon layer must be doped. Also a metallic grid should be placed on the sun-facing surface of the semiconductor of a PV cell. Figure 4.1 shows the basic structure of a PV cell. When light is incident on the PV cell, charge carriers are generated and so an electric current will start flowing in the circuit. This charge generation depends on the photon energy of the incident light. It should be large enough to break the covalent bonds of the electrons of the semiconductor, and this depends on two things: the incident light wavelength and the type of semiconductor material used. So, the operation of a PV cell can be described as the phenomenon of solar radiation absorption by the semiconductor materials, generation and transportation of free charge carriers at the p-n junction, and then collection of these carriers at the terminals of the PV cells. The strength of the electrical charge generated by this process depends upon the absorption capacity of the semiconductor material used and the lux of the incident light.

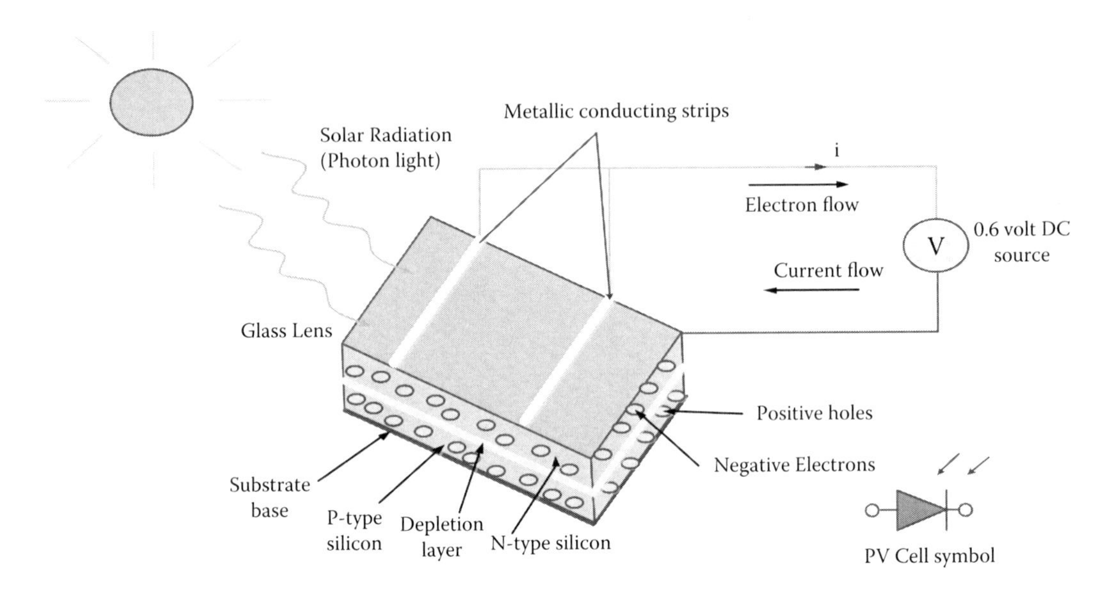

FIGURE 4.1
Physical structure of a PV cell.

The semiconductor absorption capacity depends on the following factors:

1. Energy band gap of the semiconductor materials
2. The shape of the surface and its reflectance
3. The semiconductor intrinsic concentration of the charge carriers
4. The electronic mobility
5. The rate of recombination
6. Temperature

Solar radiation has different photon energy levels. Lower-energy photons cannot generate energy if the strength is less than the band gap of the PV cell. The photons with energy that is greater than the band gap energy of the PV cell can generate electricity, but only an amount corresponding to the band gap energy is used and rest will be dissipated in the body of the PV cell as heat. Semiconductors that have lower-energy band gaps have an advantage in that they have a larger radiation spectrum but a disadvantage in that the generated voltage is typically lower. One PV cell can generate around 0.5 to 0.8 volts—the exact amount depends on the type of semiconductor used and the development technique. Such a low voltage cannot be used for any practical application. One PV cell generates up to 2 watts of power, which cannot even operate wristwatches or pocket calculators. So, to get a higher power output, these cells are connected to form modules, and the modules are assembled to make a PV array.

Thirty-six to 72 PV cells are connected to create a module for electric power applications. The modules should be connected in either a series or parallel configuration. When modules are connected in series, then the current will be the same and the voltage will be the sum of the voltages of all modules; if connected in parallel the voltage will be the same, but the current will be the summation of all the currents of all the modules.

Usually when forming solar modules, the modules are connected in series to form a string. To form a PV array, the strings are connected in parallel. The number of modules in each string is specified according to the required voltage level of the array, and the number of strings is specified according to the required current rating of the array. Most PV arrays have a power diode, called a bypass diode, connected in parallel with each individual module or with a number of modules. The function of this diode is to conduct the current when one or more of these modules are damaged or shaded.

4.3 Material for the Solar Cell

Semiconductor materials are primarily used in PV cells. The major materials are thin film and crystalline types, and they vary with respect to manufacturing methods, light absorption capacity, energy conversion efficiency, and production cost. Table 4.1 shows the three major types of PV cells based on the technology used: monocrystalline, polycrystalline, and thin-film. Microelectronics are used in both monocrystalline and polycrystalline technologies. The efficiency of monocrystalline technology is about 10% to 15%, whereas that of polycrystalline is 9% to 12%, and in thin films for Si-based 10%, $CuInSe_2$ 12%, and CdTe 9%. So, PV cell of monocrystalline type is having the highest efficiency. Silicon is not the best material for a PV cell, so many others are used as well. Si is used because its fabrication process is economical and it can be used to produce PV cells on a large scale, whereas other materials may have better conversion efficiency, but commercially higher and often unfeasible costs.

The models recorded in the literature vary in terms of accuracy and complexity, and thus, appropriateness for different studies. The single-diode model shown in Figure 4.1 is one of the most popular models used to represent the electric characteristics of a single PV cell.

Monocrystalline or single crystalline: This type of PV cell is the most efficient and most widely used compared to other types. This type of module can produce a lot of power per square foot. Here a single crystal is used to make each cell of the PV cell. The wafers are cut into rectangular-shaped cells so that the solar panel can have the maximum number of cells.

TABLE 4.1

A Comparison of Three Types of Solar Cells

Specifications	Monocrystalline panels	Polycrystalline panels	Thin-film panels
Type			
Life span	25–30 years	20–25 years	15–20 years
Tolerance of temperature	0 to +5%	−5 to +5%	−3 to +3%
Efficiency	10% to 15%	9% to 12%	9% to 12%

Polycrystalline cells: With this type, instead of making cells from a single crystal, similar types of silicon materials are melted and poured into a mold to form a PV cell. A square block is created and then cut into square wafers so that as compared to single-crystal wafers it has less wastage of space and material.

Thin-film type PV panels: One of the newest technologies in PV cells is thin-film technology. In this type, various materials can be used such as gallium arsenide, cadmium telluride, or copper indium diselenide. These materials are directly deposited on stainless steel, glass, or any other substrate. This type of PV cell may perform better in low-light conditions than crystalline modules. These are as thin as a few micrometers or even less.

Amorphous silicon: This is another of the newest technologies used in thin-film technology. In this type amorphous silicon vapor is deposited on a couple of stainless steel rolls that have micro-meter thickness. This technology requires only 1% of the material that would be used to create crystalline silicon cells (Table 4.1).

Gallium arsenide (GaAs): This semiconductor material is made up of two elements: gallium (Ga) and arsenic (As), and these materials have a crystal structure similar to silicon. This material has an advantage over Si in that is has higher light absorption capacity. So, it requires only few micrometer-thick layers as compared to crystalline silicon, which requires a wafer with a thickness of 200 to 300 micrometers to absorb the same amount of light. These also have the added advantage of higher energy conversion efficiency—about 20% to 30% in comparison to Si crystal.

Gas has high heat resistance, so it is a good choice for concentrators where cell temperatures are high. It is also popularly used in space applications where higher resistance to radiation damage is needed and greater efficiency is required. The drawback is that is very costly, which is why it is used in applications where small areas of GaAs cells are needed.

Cadmium telluride (CdTe): Cadmium and telluride are used to make polycrystalline semiconductor modules. CdTe has a high absorption capacity—almost 90% of the solar spectrum can be absorbed using a layer only a few micrometers thick. Also its manufacturing process is very cheap and easy like spraying or screen printing, and it has a high rate of evaporation. Its conversion efficiency is similar to Si—about 7%. It has major disadvantages, however, in terms of stability and performance; in addition, cadmium is a toxic material, so extra precautions are needed during the manufacturing process. Because of that, it is used in small amounts.

CIS (i.e., a semiconductor compound material made up of copper, indium, and selenium) is a major thin-film research area as it has the highest conversion capacity of any other existing thin-film materials—about 17.7% in 1996, but now it is closer to 18% as compared to polycrystalline Si PV cells (a prototype demonstrated a capacity of 10%). As it has such high energy conversion capacity with no problem with outdoor degradation, it seems apparent that thin-film PV cells will be the future choice of the solar industry. It also has the highest light absorbent capacity—for example, 90% of solar spectrum can be absorbed by only 0.5 micrometer of CIS. It does have drawbacks, however, as manufacturing CIS is very difficult and there are safety issues due to the highly toxic gas hydrogen selenite in it.

4.4 Solar PV Modules

Solar modules and arrays: Solar modules are made by connecting solar cells in series. The number of solar cells used in module formation depends on its voltage rating. Solar cells, modules, panels, and arrays are shown in Figure 4.2. For example, to design a 12-volt PV module, 33 to 36 cells need to be connected in series. The cells are mounted on a rigid,

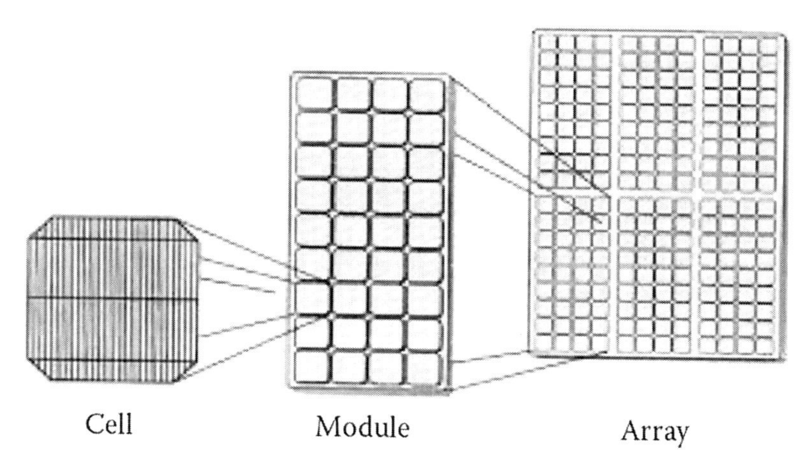

FIGURE 4.2
Solar PV cell, module, and array formation.

transparent, airtight cover. An array in a PV cell can be made by connecting a number of modules in series or in parallel as per the voltage rating and current desired in our application. To make a complete PV system, solar modules are required, but also some other parts like balance of system (BOS), wiring, conversion devices, storage devices, and support structures.

4.5 Bypass Diodes

If, for any reason, one of the modules fails to operate or some object creates a shadow over it, then the current of that module will reduce and will limit the currents of other modules that pass through it, so the string current will be reduced. For such cases there is a provision so that the current can pass through the module of the string by bypassing it, with the diode attached with the module. For example, if in Figure 4.3 C2 fails to operate, then S2 will pass the current that is generated by C1 and C3.

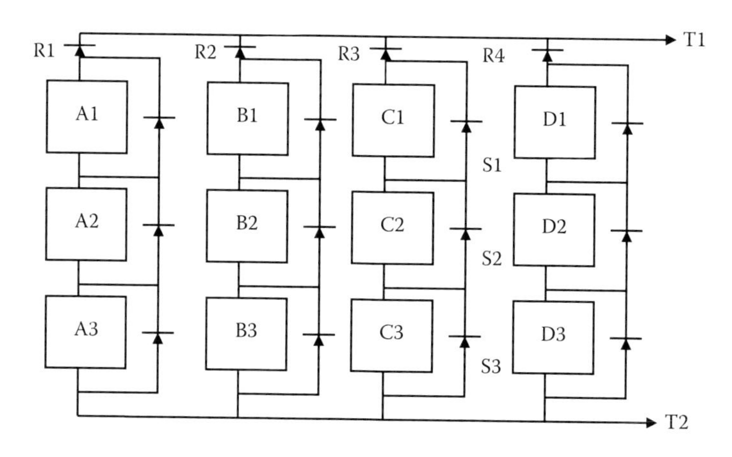

FIGURE 4.3
Series–parallel connection of PV module.

4.6 Hot Spot Formation

A solar module has a large number of cells connected in series. As shown in Figure 4.4, if one of the cells is shaded and another is not shaded and connected to a heavy load, then current from unshaded cells causes I^2R loss and heat generation in the shaded cell. Thus, shaded cells have higher resistance.

4.7 Fill Factor

Fill factor is the ratio of peak power to the product of V_{oc} and I_{sc}.

So, $FF = \dfrac{V_m * I_m}{V_{oc} * I_{sc}}$

$$P_m = I_{sc} * V_{oc} * FF$$

$$P_m = \text{Area of solar cell} \times \text{Incident solar radiation}$$

$$\text{Efficiency} = \frac{V_{oc} * I_{sc} * FF}{\text{incident solar radiation } x \text{ area of cell}}$$

The fill factor value varies between 0.7 and 0.8.

4.8 Solar Cell Efficiency and Losses

Solar cell efficiency is the ratio of power output of the cell or array to sunlight power over the total cell exposed area. A solar cell can have a maximum efficiency of 47%. The efficiency of a solar array is less than that of individual cells, as there is a larger area between individual cells. Packing factor here is fraction of the total cell area. The solar cell has a lower efficiency because of the following losses:

- Some portion of energy is absorbed by the non-PV surface of the cell.
- A portion of the solar energy is reflected back to the environment.

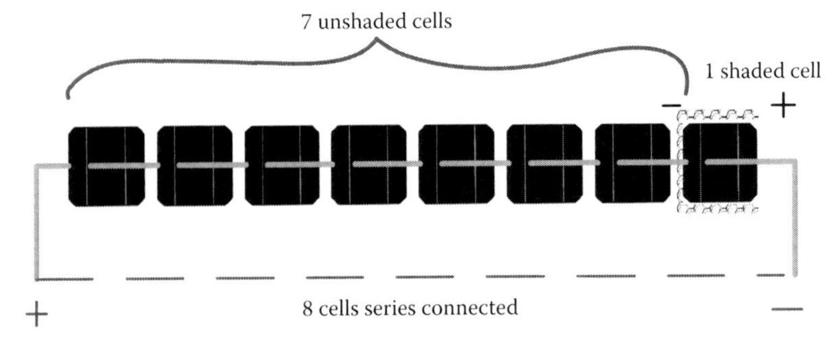

FIGURE 4.4
A string of shaded and unshaded cells.

- Some portion of energy is converted into heat dissipation.
- Various electrical losses.
- Mass production of modules.
- If there is mismatch between different cells in a module or modules in an array, then losses can take place.
- If used at higher temperatures, electron and hole pairs are recombined. In the laboratory they are rated at 100 watt/m^2 and used at 28°C. Practically, they operate at 50°C to 60°C, so efficiency is reduced by 1% to 2%. Finally, the efficiency we get after various losses from an array is 8% only. Research is ongoing to improve the efficiency of solar arrays.

4.9 Methods to Increase Cell Efficiency

Solar cell efficiency can be improved by using following methods:

1. Using better semiconductor materials
2. Using concentrated sunlight
3. Using cascaded systems

4.10 Equivalent Circuit of a PV Cell

PV cells (shown in Figure 4.5) consist of the following:

- A current source, which represents the current generated by the separation of electrons and holes by photons of incident light, so light generated current I_{ph}.
- A diode shunted with a current source, which represents the p-n junction of the PV cell.
- A shunt resistance R_{sh} and a series resistance R_{se}. Shunt resistance represents the leakage of current due to the presence of impurities in the p-n junction (ideally this should be maintained as high as possible), and series resistance represents the distributed resistance of semiconductor and metallic contacts (ideally it should be zero).

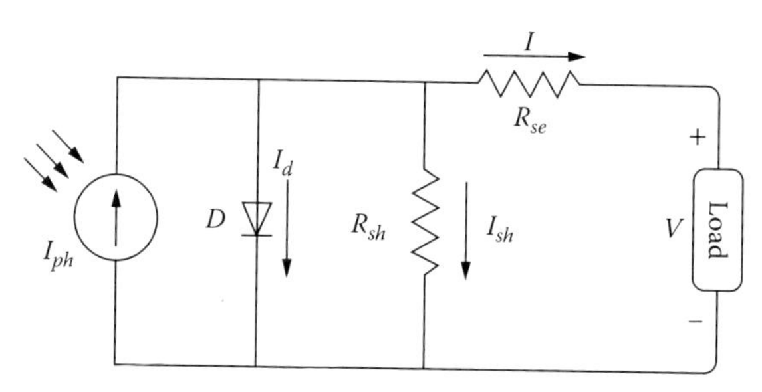

FIGURE 4.5
Equivalent PV cell with series and parallel resistance.

We can improve the accuracy of this single-diode model by using two diodes in parallel—one representing the diffusion current of the quasi-neutral region junction with ideality factor 1, and the other one representing a junction space-charge region generation recombination with ideality factor 2. But the drawback is that it makes the model more complex in terms of output voltage and current, as it will have two diode equations. The panel undergoes series losses when current flows through it, and that is represented by series resistance R_{se}. As the series resistance becomes larger, so do the losses and the efficiency of output from the solar panel decreases. Practically, the series resistance has a low value of 10^{-3} to 10^{-10} ohms. Also during the manufacturing process there may be tiny cracks and scratches in the edges of the cell and electrode, which also create leakage of current, so the photocurrent flowing through the load may diminish.

This problem is solved by connecting a shunt resistance R_{sh} which is a much larger resistance than the series one—more than 1 KΩ. So the current relationship from the diagram we get is:

$$I = I_{ph} - I_{\mathrm{d}} - I_{sh}$$

where photocurrent I_{ph} is,

$$I_{ph} = [I_{sc} + K_i\left(T_{op} - T_{ref}\right)]I_{rr}$$

I_d = Diode current

I_{sh} = Current through shunt resistor R_{sh}

I = Output current

From the theory of semiconductors, the basic equation for I-V characteristics of an ideal PV cell is as follows:

$$I = I_{ph} - I_d\{\exp\left(\left(qV + IR_{se}\right) / AkT\right) - 1\} - \left(V + IR_{se}\right) / R_{sh}$$

In this equation I_{ph} is the current generated by a PV cell due to incident light, I_{d} is current leakage through the diode, k is the Boltzmann constant (1.3806503 × 10–23 J/K), q is the electron charge (1.60217646 × 10–19 C), T is the p-n junction temperature in Kelvin, and A is the diode ideality constant.

4.11 Electric Characteristics of a Solar Cell

The V-I characteristics of a solar cell are shown in Figure 4.6 for different irradiances of solar radiation, where:

V_{oc} is the open circuit voltage

I_{sc} is the short circuit current

The PV module equation for the relationship between output voltage and current is:

$$I = I_L - I_0\left[\exp\left(\frac{V + IR_{se}}{\propto}\right) - 1\right]$$

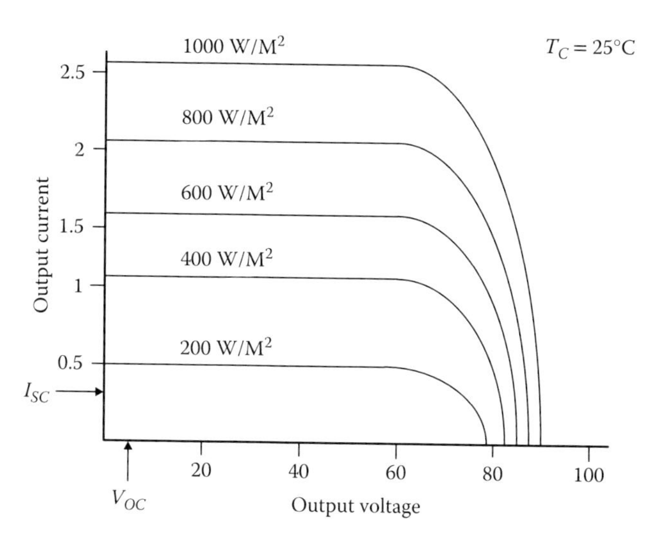

FIGURE 4.6
V-I characteristics of solar cell.

where:

I_L = Light current of PV cell
I_0 = Saturation current
I = Current across load
V = Output voltage of PV cell
A = Timing completion of thermal voltage factor of PV cell
R_{se} = PV cell series resistance

In Figure 4.7 V-I characteristics of the PV cell are shown at 25°C with different irradiance characteristics. It is seen that the value of irradiance increases, the short circuit current and open circuit voltage of PV cell increases, so output power will be also large. The parameters I_L, I_0, R_{se}, and α all are functions of temperature, so PV cell performance has a temperature effect.

In a DC circuit, power P is calculated by the product of voltage V and current I.

Mathematically,

$$P \text{ (watts)} = I \text{ (amperes)} \times V \text{ (volts)}$$

There is a maximum power output point (MPP) in the V-I characteristics. The maximum voltage and maximum current at these points are expressed as V_{mp} and I_{mp}.

So, $P_{max} = (V_{mp} \times I_{mp})$ watts.

4.12 Standard Test Conditions (STC) of the PV Cell

The environmental conditions for a standard test are as follows:

- Air Mass (AM) = 1.5
- $S = 1000\ W/m^2$

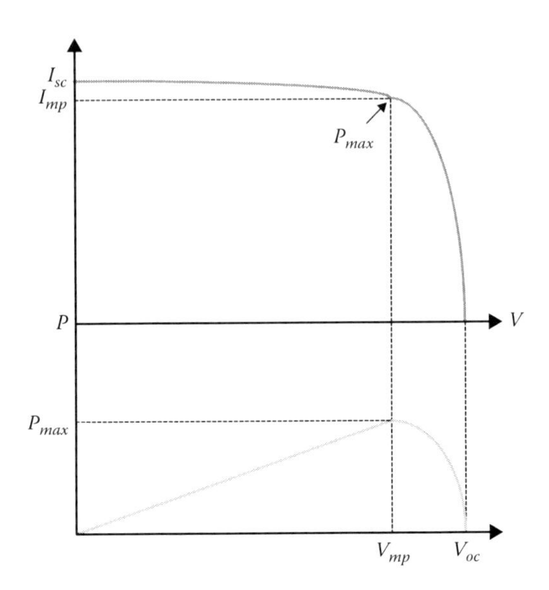

FIGURE 4.7
PV characteristics of solar cell.

- Wind Speed = 1 m/s
- $T = 273 + 25$ K

where S represents solar insolation; Kelvin temperature is T; and air mass is the length of the path that the solar beam needs to travel through the atmosphere, which is represented as ratio of the length of the path traveled by the beam in the atmosphere to the solar beam to reach the ground in a vertical path through atmosphere. Air mass is unity when the sun is at the highest or zenith position at sea level and $\alpha = 90$ degrees. These are the ideal conditions for testing a PV cell.

4.13 Factors Affecting PV Output

4.13.1 Tilt Angles

A PV module has incident power from the sunlight, which depends on the power contained in sunlight and on the angle between the module and sun. When both the sun and the absorbing surface are perpendicular to each other, than the power density on the surface is equal to incident sunlight, or we can say it will be maximum at this condition. But as the angle between the sun and the fixed surface changes, then power density will be less than the incident sunlight.

In Figure 4.8 the PV module is titled at angle β. The surface azimuth angle is represented by γ, which is the angle in the horizontal plane between the inclined plane surface and the line of south and horizontal plane projection, and it is positive as measured from south to west.

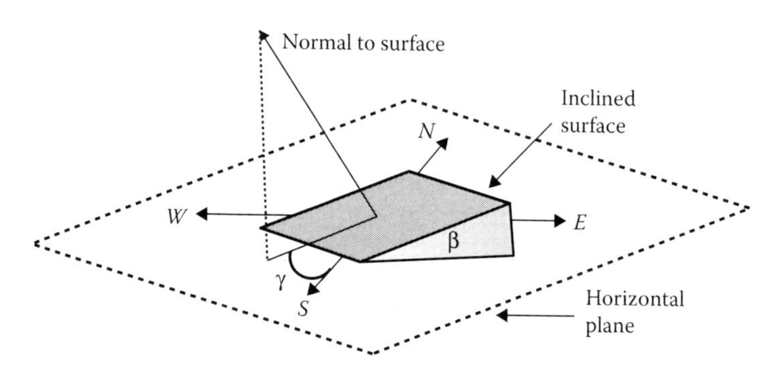

FIGURE 4.8
Module tilted at an angle.

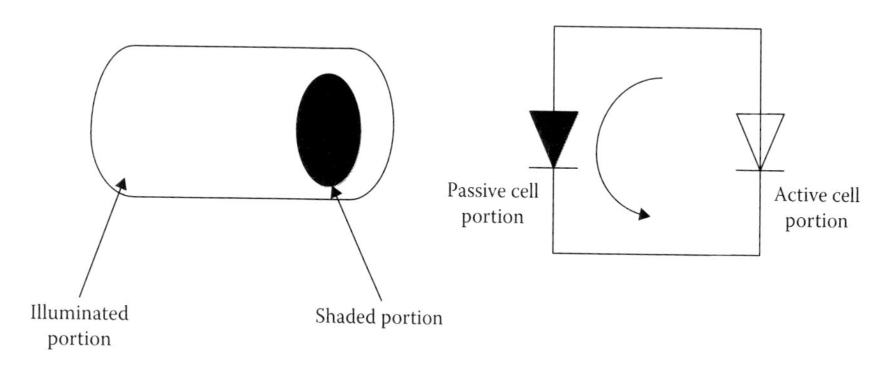

FIGURE 4.9
Partial shading of cell.

4.13.2 Partial Shading

A PV system with partial shading is a major issue because fill factor, short circuit current, open circuit voltage, and some other parameters are changed as a result. The shaded cells become reverse biased by other cells and thus exhibit higher resistance and nonuniform temperatures in the entire area of the solar cell. High dissipation of power in a shaded pole in a small area can cause breakdown in localized regions of the PV cell's p-n junction and can damage an entire module by creating hot spots. So the partial shading effect, like temperature variations and irradiance, needs to be analyzed to understand the total effect on PV cell parameters.

As shown in Figure 4.9 the effect of partial shading and the formation of hot spots does not produce power, whereas the remaining producing power and the generated voltage will forward-bias a parallel rectifier connected with the shaded portion.

When the shaded portion is very small, then a large circulating current will flow through it, which results in excessive heating and formation of hot spots; if used for a long period in this condition, it can damage the PV module.

4.13.3 Effect of Light Intensity

Changing the incident light intensity on a solar cell, changes all solar cell parameters, including the short circuit current, the open circuit voltage, the fill factor (FF), the efficiency, and the impact of series and shunt resistances. The light intensity on a solar cell is referred to based on the number of suns, where 1 sun corresponds to standard illumination at 1 kW/m^2. For example, a system with 10 kW/m^2 incident on the solar cell would be operating at 10 suns. A PV module designed to operate under 1 sun condition is called a "flat-plate" module, whereas those using concentrated sunlight are called "concentrators."

4.13.4 Concentrators

A concentrator is a solar cell designed to operate under illumination conditions greater than 1 sun. The incident sunlight is focused or guided by optical elements such that a high-intensity light beam shines on a small solar cell. Concentrators have several potential advantages, including a higher efficiency potential than a 1-sun solar cell and a possible lower cost.

4.14 PV Module Testing and Standards

High-quality PV modules are subject to a number of requirements. First, they have to deliver the guaranteed rated power reliably while withstanding an extremely wide range of environmental conditions. They must also be safe and durable, ensuring the system's high yield over the long term. And they should also be able to generate the total amount of energy that was used to manufacture them in the shortest possible time. To top things off, they need to be commercially viable. The PV modules must conform to the latest edition of any of the following standards set by the International Electrotechnical Commission (IEC) or their equivalents established by Bureau of Indian Standards (BIS) for PV module design qualification and type approval:

- Crystalline Silicon Terrestrial PV Modules IEC 61215/IS14286
- Thin Film Terrestrial PV Modules IEC 61646/Equivalent IS
- Concentrator PV Modules and Assemblies IEC 62108
- In addition, the modules must conform to IEC 61730 Part 1 (requirements for construction) and Part 2 (requirements for testing) for safety qualification or equivalent IS.
- PV modules to be used in a highly corrosive atmosphere (coastal areas, etc.) must qualify for Salt Mist Corrosion Testing as per IEC 61701/IS 61701.

4.14.1 Identification and Traceability

Each PV module must use a radio-frequency identification tag (RFID), which must contain the following information:

- Name of the manufacturer of the PV module
- Name of the manufacturer of the solar cells
- Month and year of manufacture (separately for solar cells and modules)
- Country of origin (separately for solar cells and modules)
- I-V curve for the module
- Peak wattage, I_m, V_m, and *FF* for the module
- Unique serial number and model number of the module
- Date and year the IEC PV module qualification certificate was obtained

- Name of the test lab issuing the IEC certificate
- Other relevant information on the traceability of solar cells and the module as per the ISO 9000 series

4.14.2 Authorized Testing Laboratories/Centers

PV modules must qualify (enclose test reports/certificate from IEC/NABL-accredited laboratory) as per the relevant IEC standard. Additionally the performance of the PV modules at Standard Test Conditions (STC) must be tested and approved by one of the testing laboratories accredited by the IEC and the National Accreditation Board for Testing and Calibration Laboratories (NABL), including the Solar Energy Centre in Huryana. For small-capacity PV modules up to 50 Wp capacity, STC performance will be sufficient. However, a qualification certificate from an IEC/NABL-accredited laboratory as per the relevant standard for any of the higher-wattage regular modules should be accompanied by the STC report/certificate.

- When applying for testing, the manufacturer has to give the following details:
- A copy of the registration of the company, particularly for the relevant product/component/PV system to be tested
- Adequate proof from the manufacturer, actually showing that they are manufacturing the product by way of production, testing, and other facilities
- Certification as per Jawaharlal Nehru National Solar Mission (JNNSM) standards for other bought-out items used in the system

Without such proof, testing centers are advised not to accept the samples.

4.14.3 Warranty

PV modules used in solar power plants/systems must be warranted for their output peak watt capacity, which should not be less than 90% at the end of 10 years and 80% at the end of 25 years

Test Standards:

- Quality Testing
- Crystalline PV (IEC61215)
- Thin-Film PV (IEC61646)
- Concentrator PV (IEC61208)
- Safety Testing
- Thin-Film and Cr Si ANSI/UL 1703
- Thin-Film and Cr Si IEC61730
- Concentrator PV UL8703
- Concentrator PV TS3.80815
- Inverter Standard
- UL1741
- Power and Control Electronics
- Component Testing
- Solar Thermal Testing
- Tracking and Racking Systems

- Engineering Evaluation Testing
- Long-Term Outdoor Exposure
- Module Characterization
- Development Testing
- CEC Performance Testing
- Current Service Offerings
- Product Testing and Certification
- Safety Testing and Certification
- All IEC PV Standards
- UL Standard Certification and Listing
- Performance Testing
- PV, CPV, and CSP Plant Services
- Planning, Commissioning, Monitoring
- Solar Thermal Testing and Certification
- Rebate and FIT verification and listing
- Power Management
- Systems Power Analysis

4.14.4 Quality Certification, Standards, and Testing for Grid-Connected Rooftop Solar PV Systems/Power Plants

Quality certification and standards for grid-connected rooftop solar PV systems are essential for the successful mass-scale implementation of this technology in order to achieve 40-GW target of rooftop solar power dictated by the National Solar Mission program. It is also imperative to put efficient and rigorous monitoring mechanisms in place to provide adherence to these standards. In addition, a few standards that are still under development or in the draft stage need to be introduced in the ongoing rooftop solar PV programs at the earliest opportunity. The relevant standards and certifications for a grid-connected rooftop solar PV system/plant (component-wise, up to the LV side) are given here. Currently, all applicable standards (international and Indian) are listed, and bifurcation between mandatory and advisory is shown.

4.14.5 Guidelines and Best Practices

- **Solar PV Roof Mounting Structure:** Aluminum frames should be avoided for installations in coastal areas.
- **Solar Panels:** Plants installed in high-dust regions like Rajasthan and Gujarat must have the solar panels tested against relevant dust standards. (An applicable standard would be IEC 60068-2-68.)
- **Fuse:** The fuse should have DIN rail-mountable fuse holders and should be housed in thermoplastic IP 65 enclosures with transparent covers.

4.14.6 Cables

- For the DC cabling, XLPE or XLPO insulated and sheathed, UV-stabilized, single-core, flexible copper cables should be used; multicore cables should not be used.

- For the AC cabling, PVC or XLPE insulated and PVC sheathed, single- or multicore flexible copper cables should be used; outdoor AC cables should have a UV-stabilized outer sheath.
- The total voltage drop on the cable segments from the solar PV modules to the solar grid inverter should not exceed 2.0%.
- The total voltage drop on the cable segments from the solar grid inverter to the building distribution board should not exceed 2.0%.
- The DC cables from the SPV module array should run through a UV-stabilized PVC conduit pipe of adequate diameter with a minimum wall thickness of 1.5 mm.
- Cables and wires used for the interconnection of solar PV modules should be provided with solar PV connectors (MC4) and couplers.
- All cables and conduit pipes shall be clamped to the rooftop, walls, and ceilings with thermoplastic clamps at intervals not exceeding 50 cm; the minimum DC cable size shall be 4.0 mm^2 copper; the minimum AC cable size shall be 4.0 mm^2 copper. In three-phase systems, the size of the neutral wire shall be equal to the size of the phase wires.

4.15 Commercially Available Modules

IEC 61215/IS 14286	Design Qualification and Type Approval for Crystalline Silicon Terrestrial Photovoltaic (PV) Modules
IEC 61646/IS 16077	Design Qualification and Type Approval for Thin-Film Terrestrial Photovoltaic (PV) Modules
IEC 62108	Design Qualification and Type Approval for Concentrator Photovoltaic (CPV) Modules and Assemblies
IEC 61701—As applicable	Salt Mist Corrosion Testing of Photovoltaic (PV) Modules
IEC 61853- Part 1/IS 16170: Part 1	Photovoltaic (PV) module performance testing and energy rating—Irradiance and temperature performance measurements, and power rating
IEC 62716	Photovoltaic (PV) Modules—Ammonia (NH3) Corrosion Testing **(Advisory—As per the site condition like dairies, toilets)**
IEC 61730-1, 2	Photovoltaic (PV) Module Safety Qualification—Part 1: Requirements for Construction, Part 2: Requirements for Testing
IEC 62804 (Draft Specifications)	Photovoltaic (PV) modules—Test methods for the detection of potential-induced degradation (PID). IEC TS 62804-1: Part 1: Crystalline silicon **(Mandatory when the system voltage is more than 600 VDC and advised when the system voltage is less than 600 VDC)**
IEC 62759-1	Photovoltaic (PV) modules—Transportation testing, Part 1: Transportation and shipping of module package units
IEC 62109-1, IEC 62109-2	Safety of power converters for use in photovoltaic power systems Safety compliance (Protection degree IP 65 for outdoor mounting, IP 54 for indoor mounting)
EC/IS 61683 **(For standalone system)**	Photovoltaic Systems—Power conditioners: Procedure for Measuring Efficiency (10%, 25%, 50%, 75%, and 90% to 100% loading conditions)
BS EN 50530 (Will become IEC 62891) **(For grid interactive systems)**	Overall efficiency of grid-connected photovoltaic inverters This European Standard provides a procedure for the measurement of the accuracy of the maximum power point tracking (MPPT) of inverters, which are used in grid-connected photovoltaic systems. In that case the inverter energizes a low-voltage grid of stable AC voltage and constant frequency. Both the static and dynamic MPPT efficiency are considered.

(Continued)

IEC 62116/UL 1741/ IEEE 1547	Utility-Interconnected Photovoltaic Inverters—Test Procedure of Islanding Prevention Measures
IEC 60255-27	Measuring relays and protection equipment—Part 27: Product safety requirements
IEC 60068-2 (1, 2, 14, 27, 30, and 64)	Environmental Testing of PV System—Power Conditioners and Inverters
IEC 61000-2, 3, and 5	Electromagnetic Interference (EMI) and Electromagnetic Compatibility (EMC) Testing of PV Inverters (as applicable)
	Fuses
IS/IEC 60947 (Part 1, 2, and 3), EN 50521	General safety requirements for connectors, switches, circuit breakers (AC/DC)
IEC 60269-6	Low-voltage fuses—Part 6: Supplementary requirements for fuse-links for the protection of
	Surge Arrestors
IEC 61643-11:2011/IS 15086-5 (SPD)	Low-voltage surge protective devices—Part 11: Surge protective devices connected to low-voltage power systems—Requirements and test methods
	Cables
IEC 60227/IS 694, IEC 60502/IS 1554 (Parts 1 and 2)	General test and measuring method for PVC insulated cables (for working voltages up to and including 1100 V and UV resistant for outdoor installation)
BS EN 50618	Electric cables for photovoltaic systems (BT(DE/NOT)258), mainly for DC cables
	Earthing/Lightning
IEC 62561 Series (Parts 1 and 2) (Chemical earthing)	IEC 62561-1 Lightning protection system components (LPSC)—Part 1: Requirements for connection components IEC 62561-2 Lightning protection system components (LPSC)—Part 2: Requirements for conductors and earth electrodes IEC 62561-7 Lightning protection system components (LPSC)—Part 7: Requirements for earthing enhancing compounds
	Junction Boxes
IEC 60529	Junction boxes and solar panel terminal boxes shall be of the thermo-plastic type with IP 65 protection for outdoor use and IP 54 protection for indoor use
	Energy Meter
IS 16444 or as specified by the DISCOMs	AC: Static direct connected watt-hour Smart Meter Class 1 and 2—Specification (with import and export/net energy measurements)
	Solar PV Roof Mounting Structure
IS 2062/IS 4759	Material for the structure mounting

4.16 PV Module Reliability

Table 4.2 displays reliability test results for various PV module manufacturers. Overall, most participating manufacturers performed well, with relatively few incidents of outright failure. The mere participation in the PVEL Product Qualification Program indicates already the importance that the participating manufacturers place on the reliability of their products. Because of this, the average and median results presented here may be better than the average and median results of the industry taken as a whole. Results indicate average values of multiple individual PV modules from each manufacturer (Table 4.2). All PV modules are standard 60- or 72-cell crystalline silicon modules. A different number of manufacturers participated in each test. The vertical axis in each chart indicates the power degradation caused by stress testing in percent relative to prestress output (after

TABLE 4.2

Reliability Score Card 2016

Reliability Test	Top Result	Bottom Result	Median Result	Std. Dev.
Thermal Cycling	−1.07%	−34.59%	−4.68%	7.29%
Damp Heat	−0.57%	−58.77%	−3.59%	14.86%
Dynamic Mechanical Load	−0.13%	−4.10%	−2.30%	1.11%
PID (−1 kV)	0.47%	−58.27%	−2.69%	18.60%
Humidity-Freeze	−0.13%	−4.10%	−2.30%	1.11%

light soaking). Top performers are defined as those to the left of the red vertical line indicated on the results charts.

4.17 PV Module Data Sheet

The specification of a Giga Solar module is presented in Table 4.3.

TABLE 4.3

Giga Solar @ Brand Module Data

Electrical Data @ STC* Cell Format		GS-145 4 × 8	GS-163 6 × 6	GS-181 4 × 10	GS-218 6 × 8	GS-272 6 × 10	GS-327 6 × 12	GS-381 6 × 14	GS-436 6 × 16	GS-490 6 × 18
Peak Power	(W)	145	163	181	218	272	327	381	436	490
Max Power Voltage (Vmpp)	(V)	16.4	18.4	20.5	24.5	30.7	36.8	43.0	49.3	55.2
Max Power Current (Impp)	(A)	8.9	8.9	8.9	8.9	8.9	8.9	8.9	8.9	8.9
Open Circuit Voltage (Voc)	(V)	19.6	22.1	24.5	29.4	36.8	44.1	51.5	58.8	66.2
Short Circuit Current (Isc)	(A)	9.4	9.4	9.4	9.4	9.4	9.4	9.4	9.4	9.4
Module Efficiency	(%)	16.9	17.0	17.0	17.0	17.2	17.3	17.4	17.4	17.4

* *STC:* 1000 W/m², 25° C Cell temperature, AMI 1.5. Peak Power +5/−0% (Wp).

Mechanical Data	(Metric)								
Weight (kg)	3.6	4.0	5.3	6.3	7.8	9.2	10.7	12.1	13.6
Width (mm)	660	976	660	976	976	976	976	976	976
Length (mm)	13.02	986	1618	1302	1618	1934	2250	2566	2882
Thickness (mm)	6.5	6.5	6.9	6.9	6.9	6.9	6.9	6.9	6.9

Mechanical Data	(Imperial)								
Weight (lb)	8.0	8.9	11.7	13.9	17.1	20.3	23.5	26.7	30.0
Width (In)	26.0	38.4	26.0	38.4	38.4	38.4	38.4	38.4	38.4
Length (In)	51.3	38.8	63.7	51.3	63.7	76.1	88.6	101.0	113.5
Thickness (In)	0.26	0.26	0.27	0.27	0.27	0.27	0.27	0.27	0.27

Temperature Coefficients		
Power Wp	%/K	–0.40
Voltage Vmpp	%/K	–0.38
Current Impp	%/K	0.05

4.18 Array Junction Boxes

Also referred to as solar PV generators, array junction boxes (AJBs) basically collect DC power from PV strings. The collected power is then transferred either directly or through a main junction box to the power inverter. The power inverter converts the DC power to AC, which after metering, is used for captive consumption or supplied to the grid.

Array junction boxes are provided with one or more of the accessories mentioned here to improve their functionality:

- Fuse/miniature circuit breaker (MCB) overload protection on each string.
- Fuse/MCB overload protection on the output side.
- Disconnector/isolator on the output side to isolate the AJB from the main junction box (MJB)/inverter.
- Plug-in and plug-out PV string input connectors: these special connectors enable quick yet reliable connection and disconnection of the PV string from the array junction box. This is useful during maintenance and site up-gradation work.
- Blocking diodes are used for reverse current flow protection.
- Surge protection device for protection against electrical surges.

Individual string performance monitoring device and device to transfer data to the control room, which is useful in case of large PV installations. This enables easy identification of PV modules that are performing at below optimal level. After identification, such PV panels/modules can be repaired or replaced. Because array junction boxes are installed in outdoor environments near the PV modules, they are normally required to be made from a material that is rugged, fire resistant, electrically insulated, and UV resistant. Polycarbonate is one such material.

Array junction boxes are installed in several MW-size solar PV power plants and also in telecom tower-based solar PV projects.

4.19 Questions

1. Discuss the reasons for the low efficiency of solar cells. What methods are used to improve the solar cell efficiency? Explain how the variation of insolation and temperature affects the I-V characteristics of a solar cell.
2. A solar cell of area 0.9 cm^2 receives solar radiation with photons of 1.8 eV and an intensity of 0.9 mW/cm^2. Measurements show open circuit voltage of 0.6 V/cm^2, short circuit current of 10 mA/cm^2, and the maximum current is 50% of the short circuit current. The efficiency of the cell is 25%. Calculate the maximum voltage that the cell can give and find the fill factor as well.

5

Maximum Power Point Tracking Techniques and Charge Controllers

5.1 MPPT and Its Importance

Maximum power point tracking (MPPT) is used to track the maximum power from the system. This plays a significant role in renewable energy sources because of its intermittent nature. However, there are certain limitations, like performance and efficiency, which depend on various other physical parameters. Solar photovoltaics (PVs) are considered as a promising source, along with wind energy. The performance of a solar photovoltaic system depends on various parameters such as solar irradiance, cell temperature, dust, spectral density, etc. This problem of variable efficiency can be regulated with the help of various control techniques. The maximum power point (MPP) is the point at which maximum power can be extracted from the solar cell. This point can be easily traced on the characteristic curve of the solar cell. To obtain maximum power from the solar cell, it should always operate at a point on the power-voltage (P-V) curve where the efficiency of the solar photovoltaic system is at its maximum.

5.2 MPPT Techniques

The following are some of the more widely used MPPT techniques applied on various PV applications, such as space satellites, solar vehicles, solar water pumping, etc. The classification could be made on the basis of control strategies. The three types of control strategies are direct, indirect, and probabilistic. In direct control, prior knowledge of PV generator characteristics is not required. The PV panel operating point variations lead directly to the MPP. The techniques of differentiation, perturb and observe, feedback voltage, incremental conductance, fuzzy logic, and neural network are considered direct methods. Direct methods also can be classified as sampling methods and modulation methods. For estimating the MPP, the indirect methods utilize the data of P-V curves of PV systems for different temperatures and insolation, or they make use of mathematical functions that are obtained from empirical data. These indirect methods include the look-up table method, curve fitting, open circuit, and short circuit.

5.2.1 Curve-Fitting Technique

This technique of MPPT is based on predicting the P-V curve. It is already known to us from the mathematics that the maximum point of any curve is its extreme value. For particular solar radiation and temperature values, the extreme value of the P-V curve corresponds to the maximum power point. So at first the characteristic is predicted. To predict the P-V curve of a PV generator, the nonlinear characteristic of the PV generator can be modeled offline based on mathematical equations or numerical approximations. For the fitting of the P-V curve, a polynomial function of the appropriate degree should be assumed. The accuracy of the fitting of the curve depends on the degree of polynomial function.

5.2.2 Fractional Short-Circuit Current (FSCC) Technique

Output power is the multiplication of voltage and current. So the maximum output power at a given environmental condition corresponds to a single operating point called (V_m, I_m). So if one of the operating points is known to us, then the other point can be easily found. Considering that in the FSCC technique, the current corresponding to the maximum power point has an approximately linear relationship with the short-circuit current of the PV array under different environmental conditions, the modeling of the nonlinear characteristics of the PV array could be done. This technique is less complex than the other techniques. So it is used when accuracy is not as much of a concern.

5.2.3 Fractional Open-Circuit Voltage Technique

It is found from the literature that voltage at a maximum power point is approximately proportional to the open-circuit voltage under varying irradiance and temperature levels. Because the proportionality constant depends on the fabrication, solar cell technology, fill factor, and meteorological conditions, it has to be calculated by empirically determining V_{mpp} and V_{oc} at different irradiance and temperature levels.

5.2.4 Look-Up Table Technique

The fractional open-circuit voltage technique cannot seek the MPP accurately under varying environmental conditions, as these techniques do not consider the effect of weather fluctuations. In this technique the MPP corresponding to expected environmental conditions is calculated in advance, and these values are kept in the memory of the MPPT control system. Then the output power is measured with the help of the PV panel voltage and current, and then this output power is compared with the values corresponding to the maximum power, which is stored in the controlling system. However, a great deal of memory is required for data storage when using this technique.

5.3 Direct Method: The "True Seeking"

5.3.1 Perturb and Observe

The simplest and most commonly used technique for tracking the maximum power point is perturb and observe (P&O). This algorithm could be easily implemented by making small changes in the control parameter, which could be an array voltage, array current, and duty ratio of the power converter. This algorithm is based on the fact that after making an incremental change in the array voltage, if the power increases, then the next perturbation

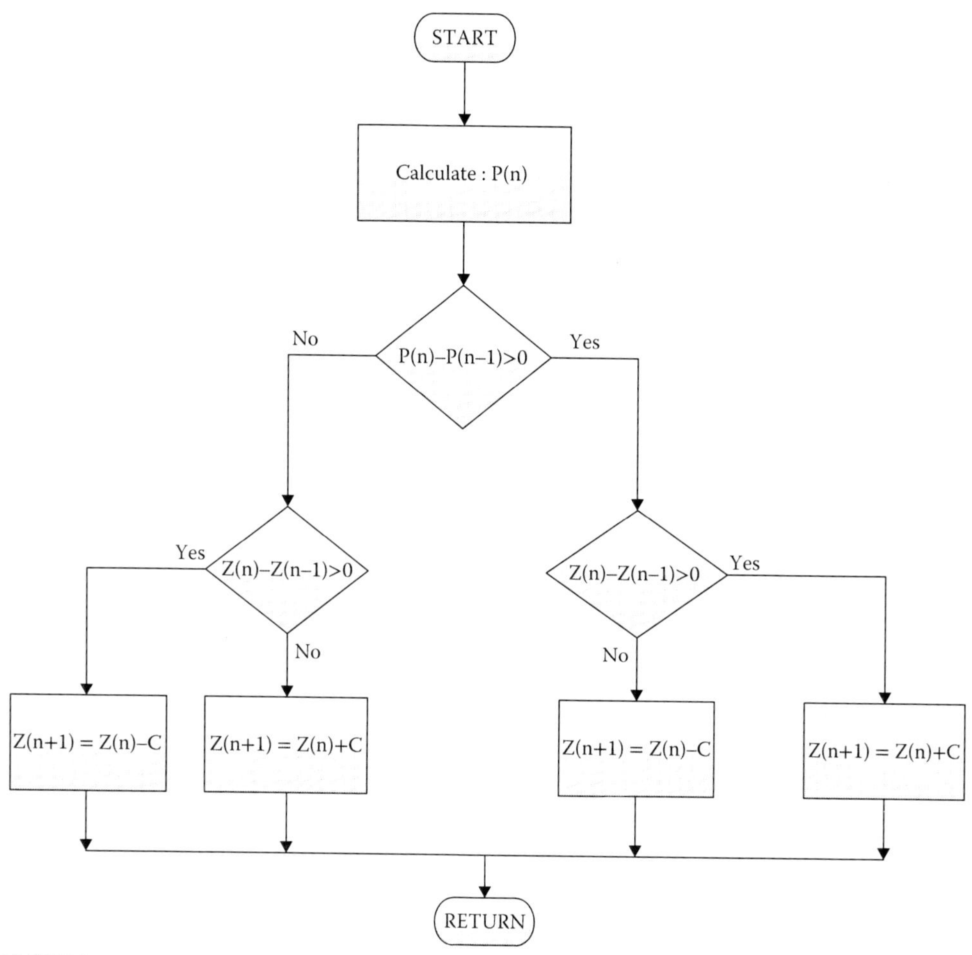

FIGURE 5.1
Flow chart for perturb and observe MPPT technique.

should be in the same direction (i.e., increase in the voltage), but if the power decreases, then the next perturbation should be in the reverse direction. But this algorithm does not give an accurate maximum power point because at the steady state there are oscillations around the maximum power point. In order to avoid these oscillation perturbations, the step size should be reduced, but this will also reduce the speed. For this variable, the perturbation technique has been proposed in order to achieve optimization in terms of speed and accuracy. This technique is again not applicable under variable atmospheric conditions. This method is also known as the "hill climbing method." A flow chart illustrating this method is shown in Figure 5.1.

5.3.2 Incremental Conductance Method

The incremental conductance method is based on the fact that the slope of the PV module curve is zero at the maximum power point. This slope will be positive for values of output power smaller than the MPP and negative for values of output power greater than the MPP.

Maximum output power can be obtained by using the derivative of PV output power with respect to voltage and equating this to zero:

$$\frac{dP}{dV} = I + v\frac{dI}{dV} = 0 \tag{5.1}$$

By using Equation 5.1 the following equation can be obtained:

$$\frac{dI}{dV} \cong \frac{\Delta I}{\Delta V} = -\frac{I_{MPP}}{V_{MPP}} \tag{5.2}$$

$$\frac{dP}{dV} = 0 \qquad \frac{\Delta I}{\Delta V} = -\frac{I}{V} \qquad \text{at MPP} \tag{5.3}$$

$$\frac{dP}{dV} > 0 \qquad \frac{\Delta I}{\Delta V} > -\frac{I}{V} \qquad \text{left side of MPP} \tag{5.4}$$

$$\frac{dP}{dV} < 0 \qquad \frac{\Delta I}{\Delta V} < -\frac{I}{V} \qquad \text{right side of MPP} \tag{5.5}$$

Instantaneous conductance is compared with the incremental conductance in order to track the maximum power point. After achieving the MPP, the operation of the PV module is forced to remain at this point unless a change in current occurs as a result of varying meteorological parameters, which leads to a variation in MPP. The flow chart for the incremental conductance technique is presented in Figure 5.2.

5.4 MATLAB® Simulations

Three MPPT techniques are implemented in the MATLAB®/Simulink environment to study and compare the dynamic performance of a PV system. All simulations are performed under different meteorological conditions. These conditions include constant solar irradiance and temperature, varying irradiance and constant temperature, constant irradiance and varying temperature, and varying irradiance and temperature.

5.4.1 Performance Analysis of a PV System at Constant Solar Irradiance and Temperature

The graphical representation of the perturb and observe (P&O) technique, incremental conductance technique, and constant voltage MPPT technique are shown in Figures 5.3 through 5.8, respectively. Also, the analytical performance of the three MPPT techniques is presented in Table 5.1.

In Figures 5.3 to 5.8, it is observed that the power ripple and voltage fluctuations are highest in the case of P&O and least with the constant voltage, whereas response time and battery charging voltage are almost the same in all three MPPT techniques. The oscillations

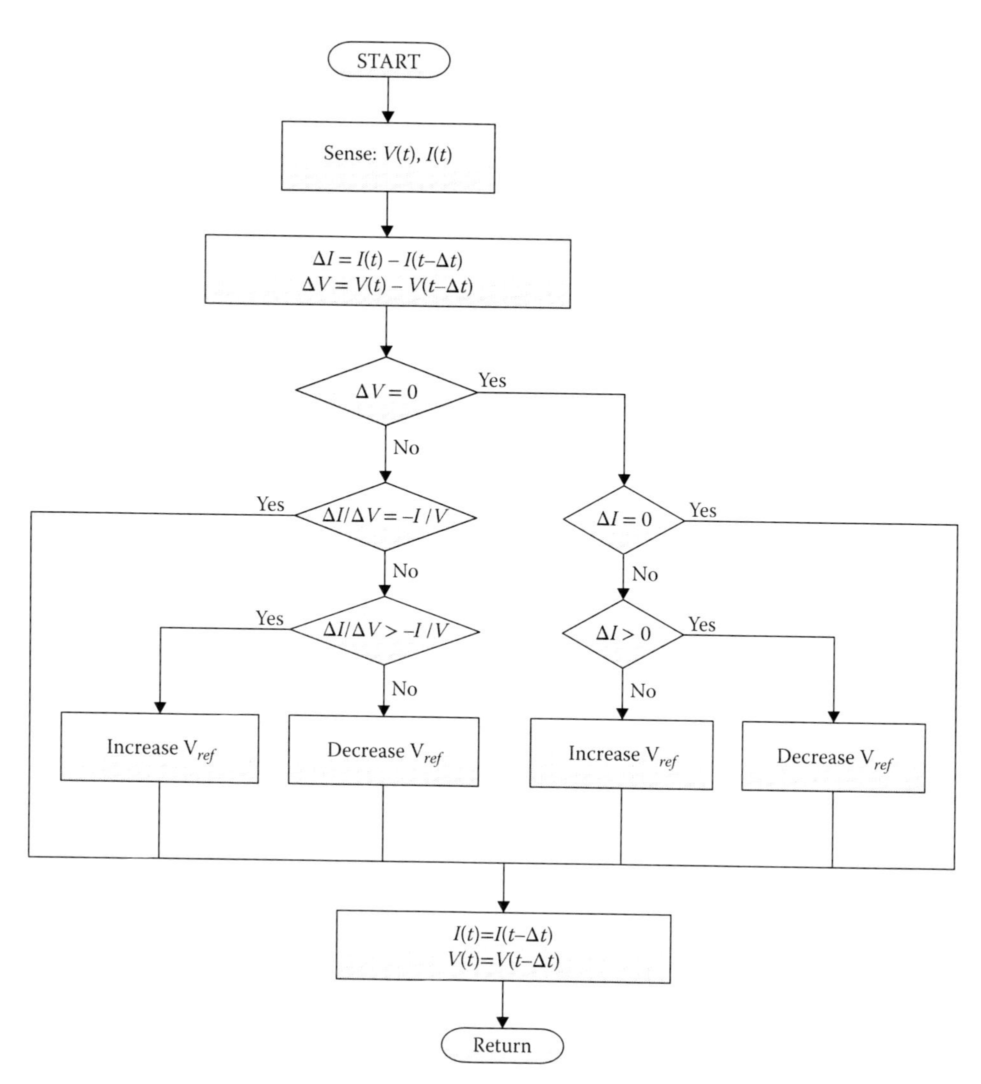

FIGURE 5.2
Flow chart for incremental conductance MPPT technique.

TABLE 5.1
Performance of P&O, Incremental Conductance, and Constant Voltage MPPT Techniques

Parameters	MPPT Techniques		
	P&O	**Incremental Conductance**	**Constant Voltage**
Power ripple (W)	65	0.7	0.3
Voltage fluctuations (V)	4.1	1.47	1.35
Response time (s)	0.0894	0.07	0.081
Battery charging voltage (V)	47.51	47.79	48
Battery charging current (A)	Not constantly oscillating between 0 and 59 A throughout the period	31.6	32.7

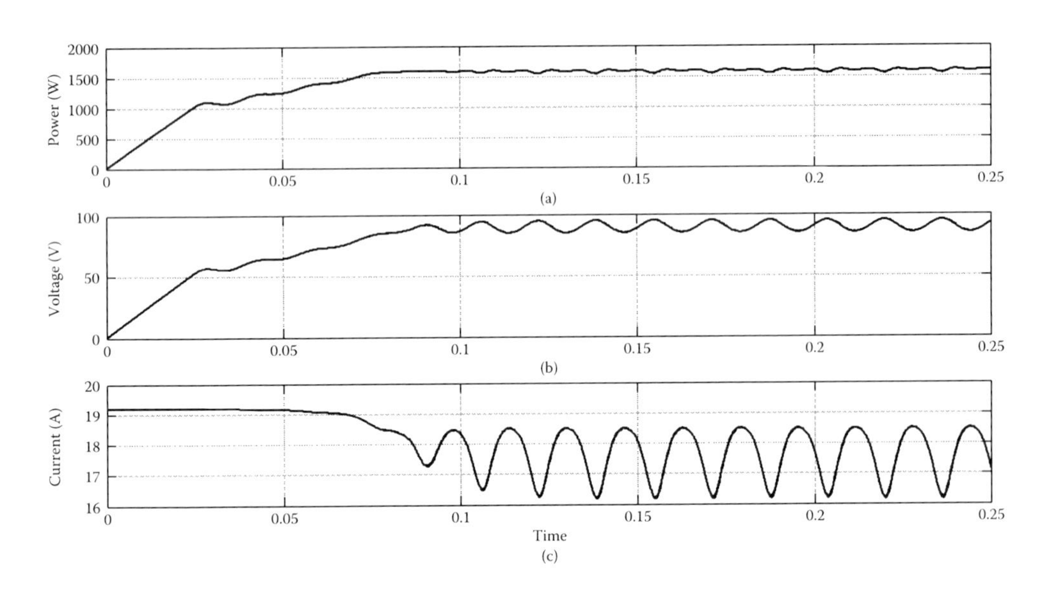

FIGURE 5.3
Performance of a PV array: (a) power, (b) voltage, and (c) current at constant irradiance and temperature using P&O MPPT technique.

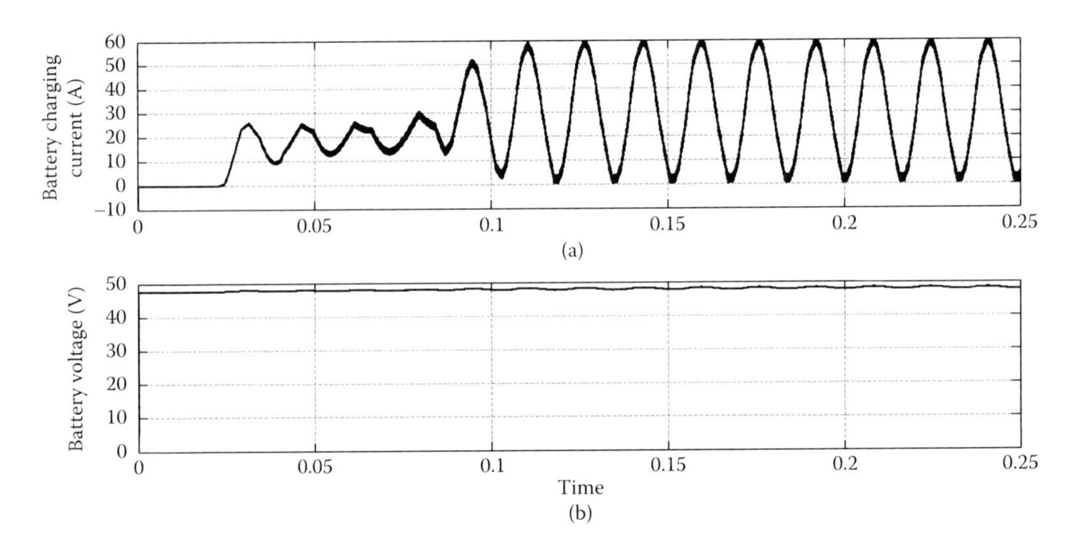

FIGURE 5.4
Performance of PV array: (a) battery charging current and (b) battery charging voltage at constant irradiance and temperature using P&O MPPT technique.

in P&O are not constant and vary between 0 and 59 A. It is concluded that the overall performance of the constant voltage technique is better compared to the other techniques.

It was mentioned earlier that the PV modules are characterized at STC, but meteorological parameters like solar irradiance and temperature are not constant under practical conditions. The amount of solar irradiance reaching the earth's surface varies greatly because of various meterological and atmospheric parameters like water vapor molecules, number

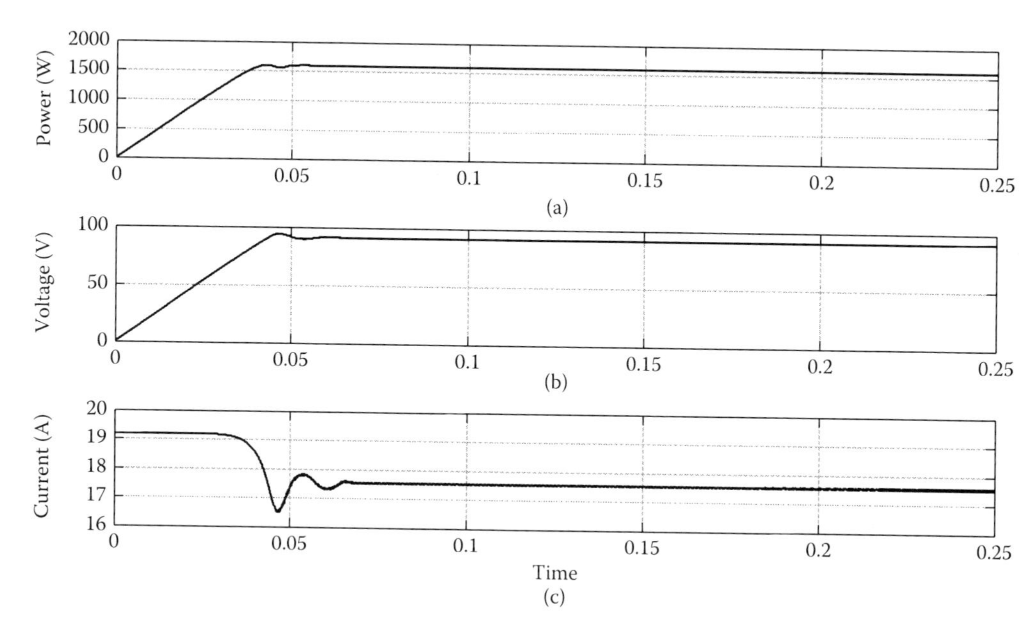

FIGURE 5.5
Performance of PV array: (a) power, (b) voltage, and (c) current at constant irradiance and temperature using incremental conductance MPPT technique.

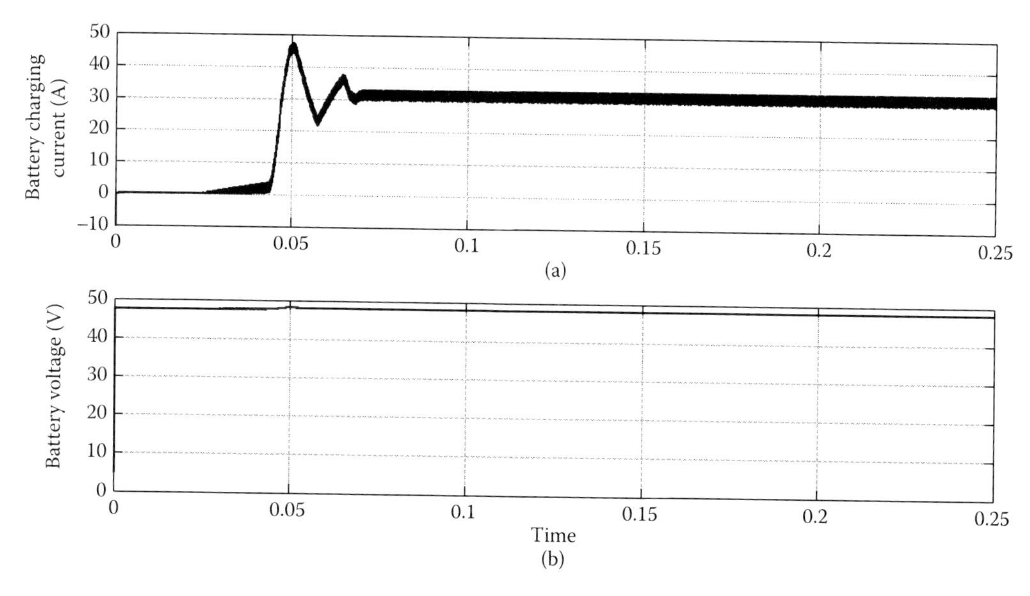

FIGURE 5.6
Performance of PV array: (a) battery charging current and (b) battery charging voltage at constant irradiance and temperature using incremental conductance MPPT technique.

of gaseous molecules, aerosoles, clouds, change in position of sun, etc. Under such practical conditions, the impact of these parameters is of utmost importance on PV power output. To incorporate such practical conditions, performance analysis of PV system at varying irradiance and constant temperature, constant irradiance and varying temperature, and

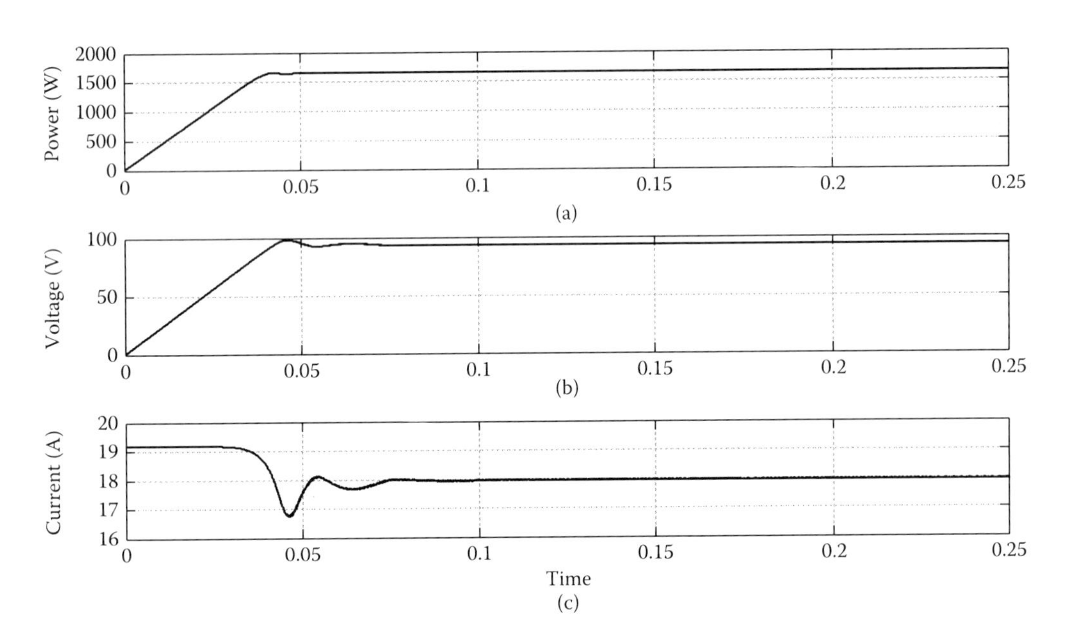

FIGURE 5.7
Performance of PV array (a) power, (b) voltage, and (c) current at constant irradiance and temperature using constant voltage MPPT technique.

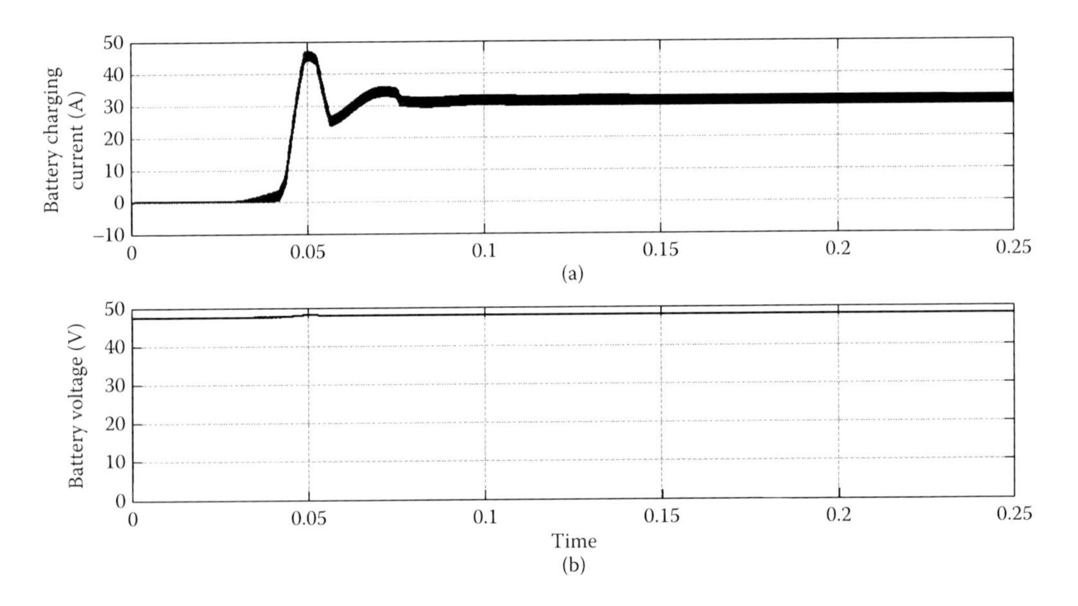

FIGURE 5.8
Performance of PV array (a) battery charging current and (b) battery charging voltage at constant irradiance and temperature using constant voltage MPPT technique.

varying irradiance and temperature is carried out. The effect of solar irradiance and temperature on PV output is presented in Figures 5.9 and 5.10, respectively. It is clearly seen in Figures 5.11 to 5.16 that there is a large impact of these parameters on PV output.

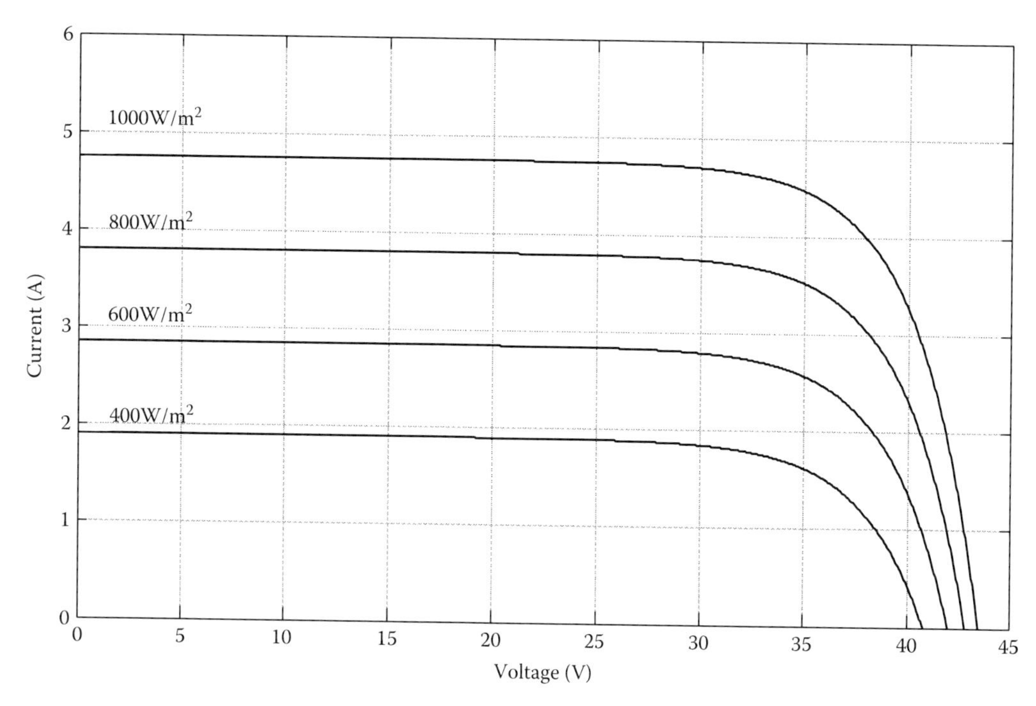

FIGURE 5.9
Performance of PV system with varying irradiance.

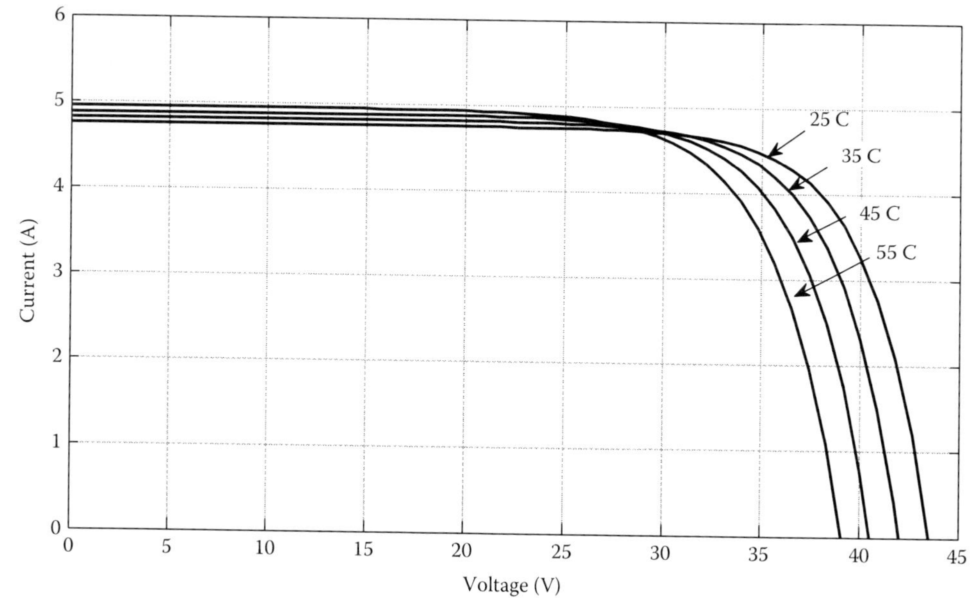

FIGURE 5.10
Performance of PV system with varying temperature.

Further, it is observed that the PV output current varies with irradiance. The operating point changes with the solar irradiance, temperature, and load conditions. With the increase in operating temperature, the output current increases, but the value of the voltage decreases drastically. This results in a reduction of the output power.

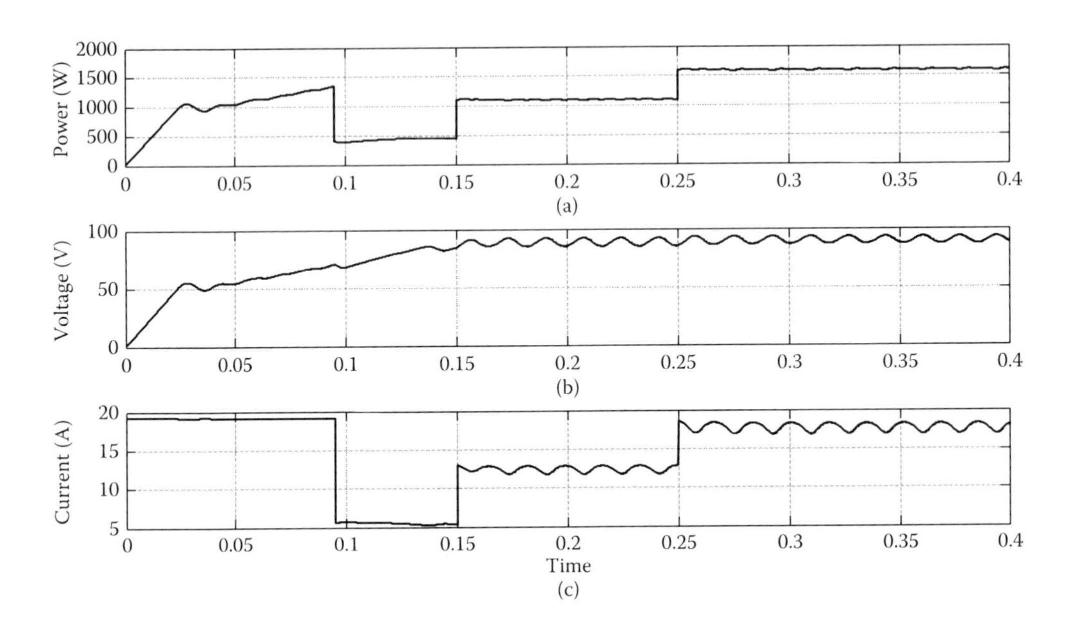

FIGURE 5.11
Performance of PV array: (a) power, (b) voltage, and (c) current at varying irradiance and constant temperature using P&O MPPT technique.

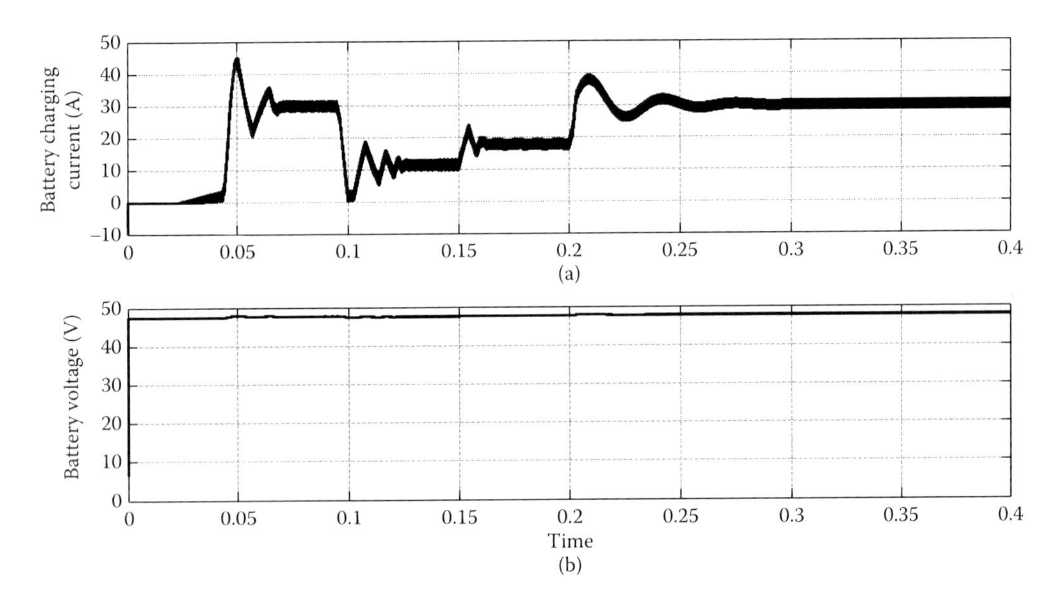

FIGURE 5.12
Performance of PV array: (a) battery charging current and (b) battery charging voltage at varying irradiance and constant temperature using incremental conductance MPPT technique.

5.4.2 Performance Analysis of PV System at Varying Irradiance and Constant Temperature

The performance of PV systems using P&O, incremental conductance, and constant voltage is presented in Figures 5.11, 5.13, and 5.15 for changing irradiance and constant

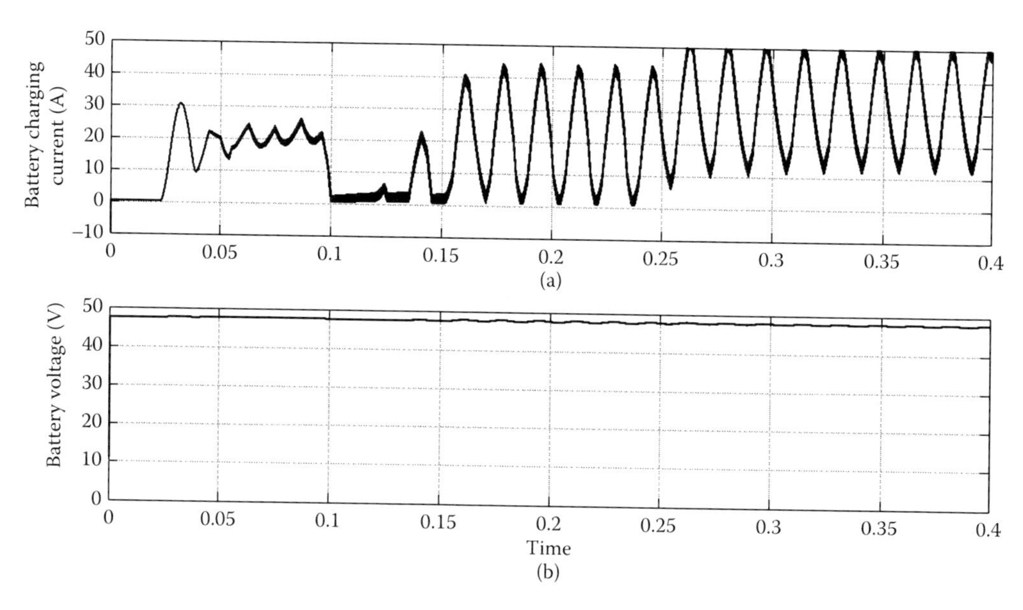

FIGURE 5.13
Performance of PV array: (a) battery charging current and (b) battery charging voltage at varying irradiance and constant temperature using P&O MPPT technique.

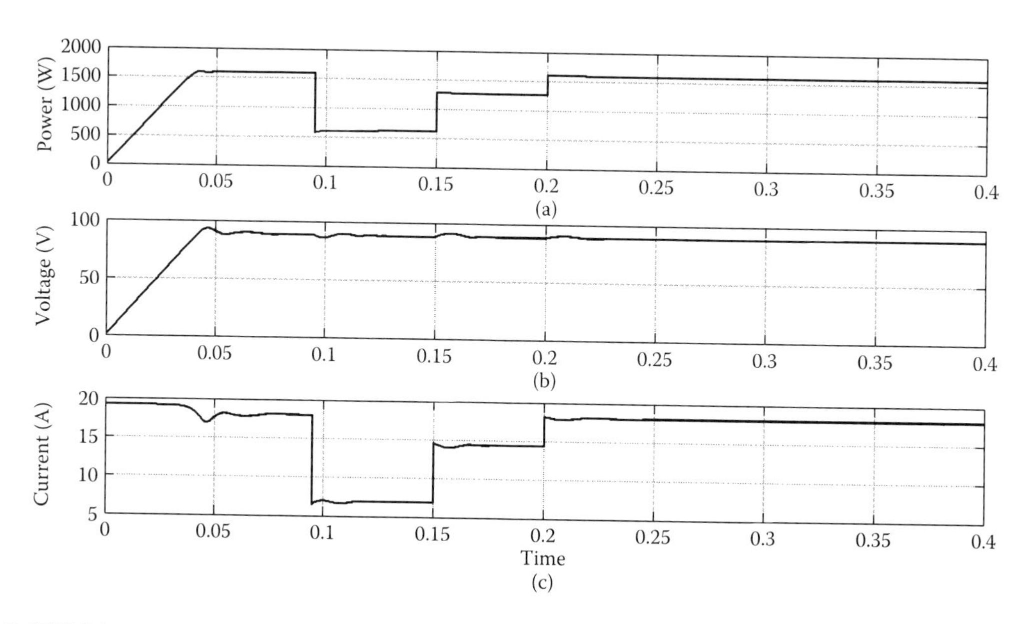

FIGURE 5.14
Performance of PV array: (a) power, (b) voltage, and (c) current at varying irradiance and constant temperature using constant voltage MPPT technique.

temperature, respectively. The battery charging characteristics are presented in Figures 5.12, 5.14, and 5.16 using P&O, incremental conductance, and constant voltage approaches.

Performance analysis of these MPPT techniques is carried out at a fixed temperature of 25°C and varying solar irradiance. The solar irradiance is suddenly varied at 300 W/m^2

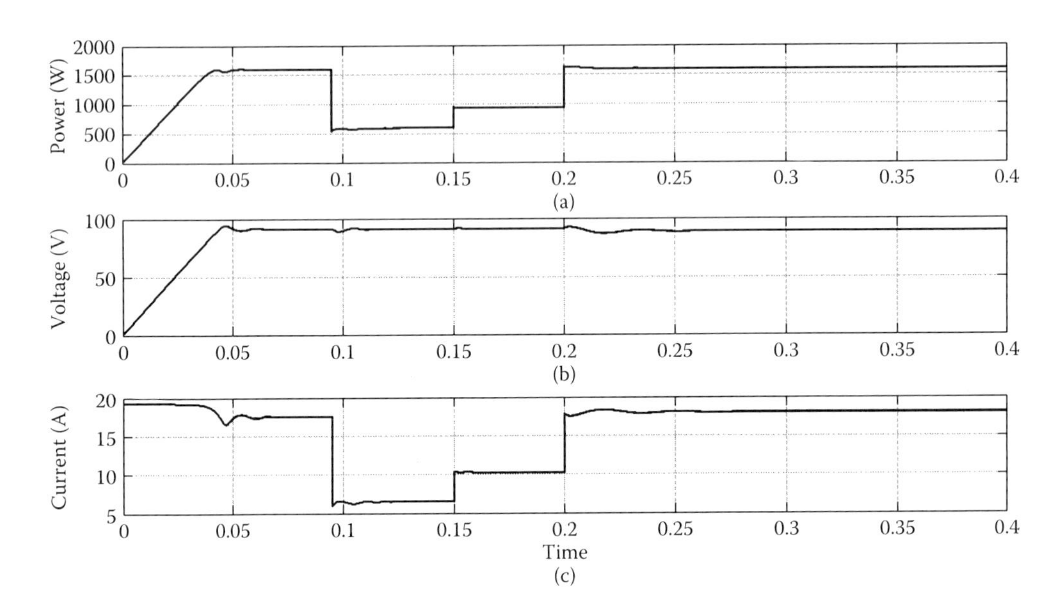

FIGURE 5.15
Performance of PV array: (a) power, (b) voltage, and (c) current at varying irradiance and constant temperature using incremental conductance MPPT technique.

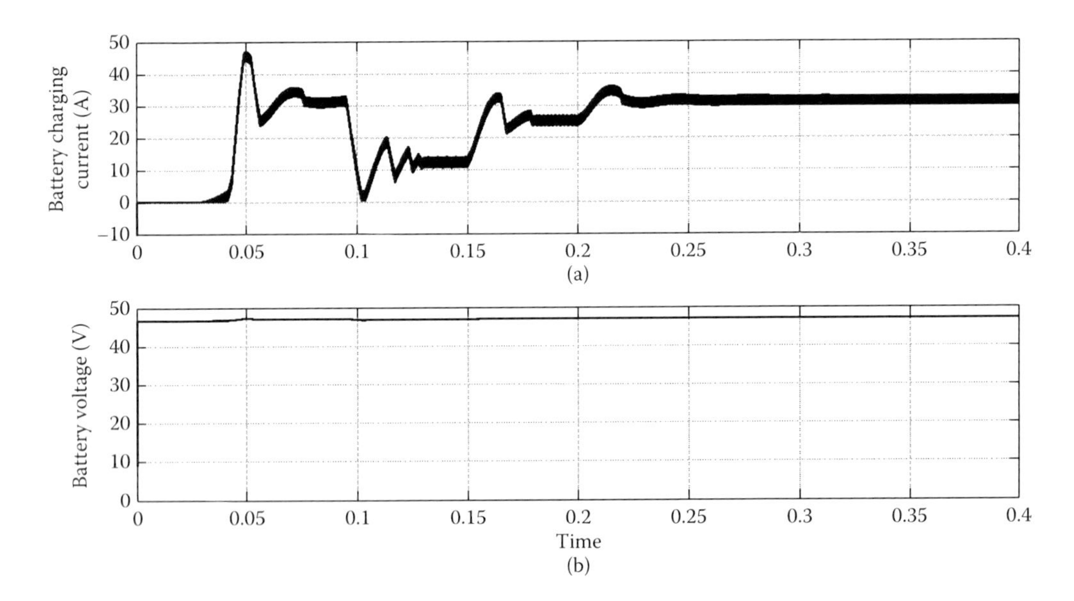

FIGURE 5.16
Performance of PV array: (a) battery charging current and (b) battery charging voltage at varying irradiance and constant temperature using constant voltage MPPT technique.

and 700 W/m^2 from STC (i.e., 1000 W/m^2). In this case, it is observed that the energy extracted from the PV system is less compared to the constant solar irradiance and temperature. From Table 5.2, we can see that the performance of the constant voltage technique is better compared to the other technique in terms of power ripple, voltage fluctuations,

TABLE 5.2

Performance of P&O, Incremental Conductance, and Constant Voltage MPPT Techniques

Parameters	MPPT Techniques		
	P&O	Incremental Conductance	Constant Voltage
Power ripple (W)	25	0.8	1.1
Voltage fluctuations (V)	4.7	2.7	1.6
Response time (s)	0.26	0.27	0.23
Battery charging voltage (V)	47.64	47.78	47.9
Battery charging current (A)	Not constantly oscillating between 10 and 50 A throughout the period	32	31

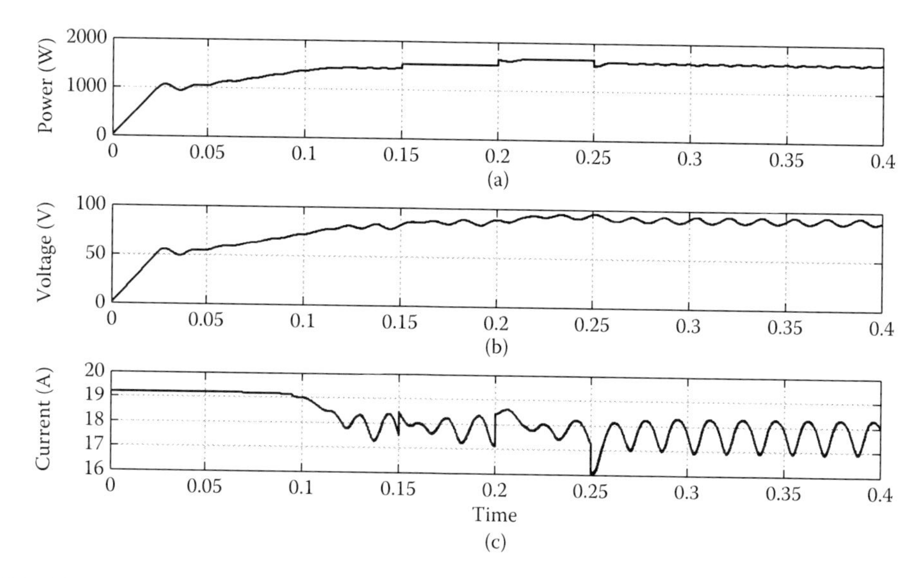

FIGURE 5.17
Performance of PV array: (a) power, (b) voltage, and (c) current at constant irradiance and varying temperature using P&O MPPT technique.

response time, and battery charging time. However, this technique is not suitable for suddenly changing metrological parameters like solar irradiance and temperature. Therefore, the incremental conductance technique is better as compared to other techniques, as mentioned earlier regarding dynamic conditions.

5.4.3 Performance Analysis for Constant Solar Irradiance and Varying Temperature

It is also observed that solar cell temperature has a direct impact on the output of the solar cell, and there is decrease in voltage of about 0.023 V per degree rise in temperature. Therefore, the study of the impact of temperature on a PV system is vitally important. The performance of the PV system at a solar irradiance of 1000 W/m^2 and different temperatures (i.e., 20°C, 37°C, and 50°C) is studied to understand the impact of cell temperature on PV output and is presented in Figures 5.17 through 5.22 for three MPPT techniques. The analytical results are

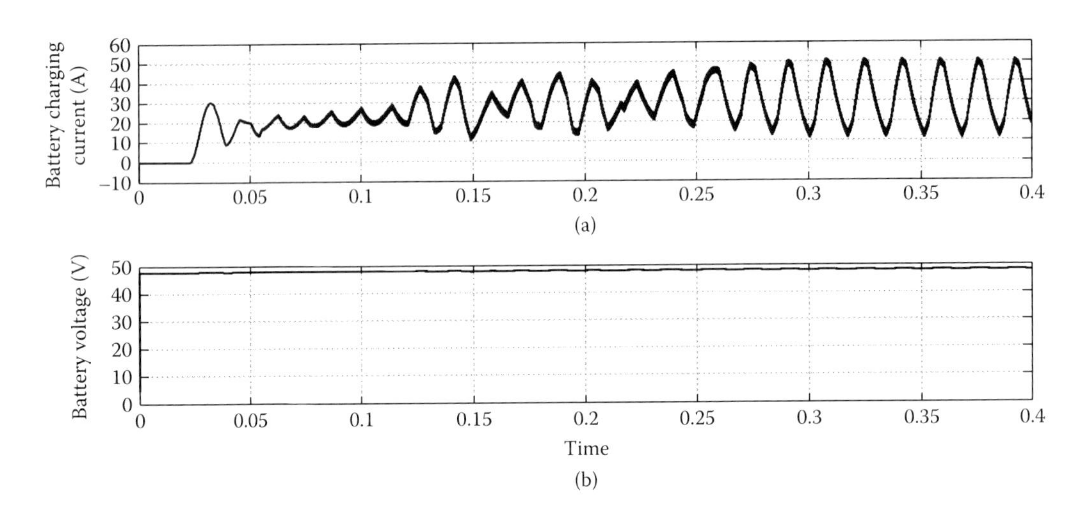

FIGURE 5.18
Performance of PV array: (a) battery charging current and (b) battery charging voltage at varying irradiance and constant temperature using P&O MPPT technique.

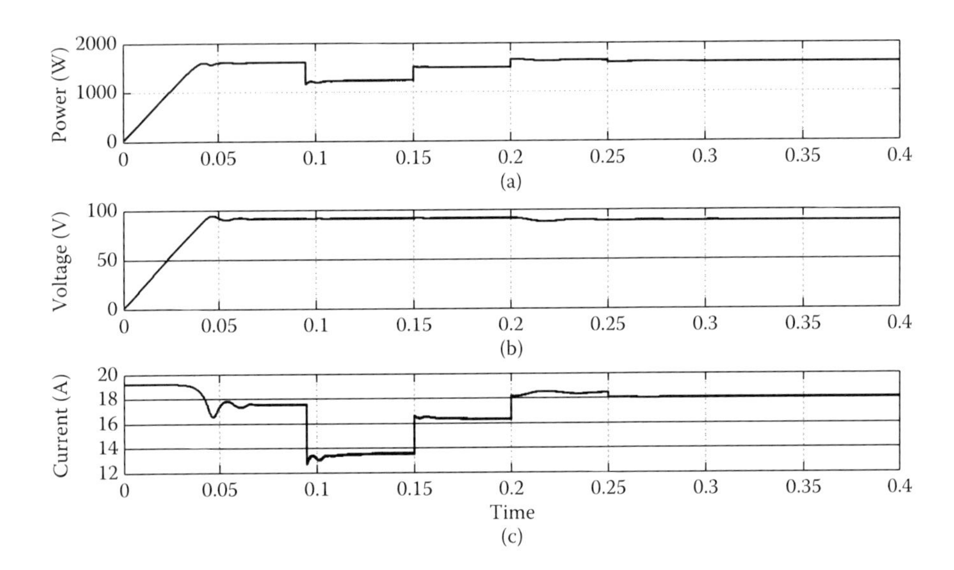

FIGURE 5.19
Performance of PV array: (a) power, (b) voltage, and (c) current at constant irradiance and varying temperature using incremental conductance MPPT technique.

also presented in Table 5.3. From this analysis, it is observed that the performance of the constant voltage technique is better compared to the P&O and incremental conductance methods. As mentioned earlier, the constant voltage approach cannot be employed for practical cases; hence, the incremental conductance technique is best for this case. The performance of P&O, Incremental Conductance and Constant Voltage algorithms is presented in Table 5.4.

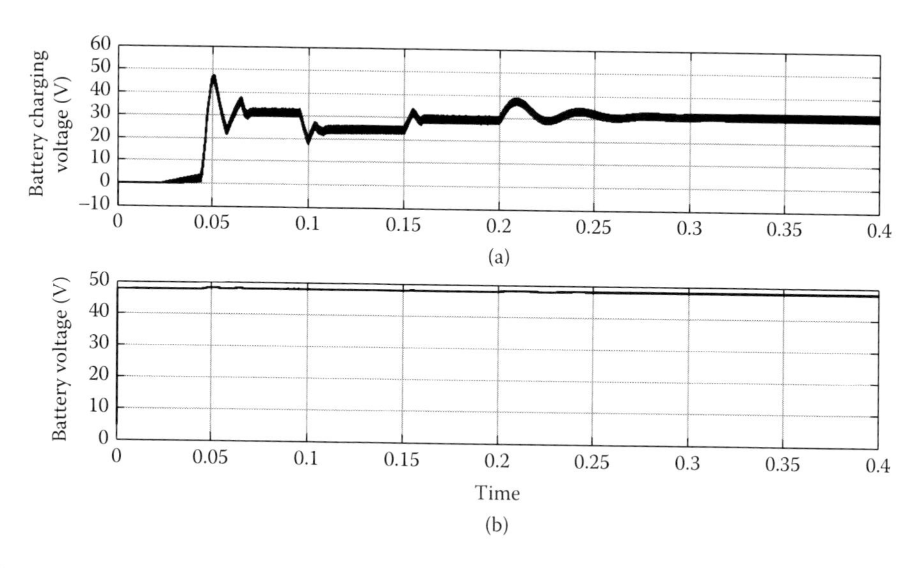

FIGURE 5.20
Performance of PV array: (a) battery charging current and (b) battery charging voltage at constant irradiance and varying temperature using incremental conductance MPPT technique.

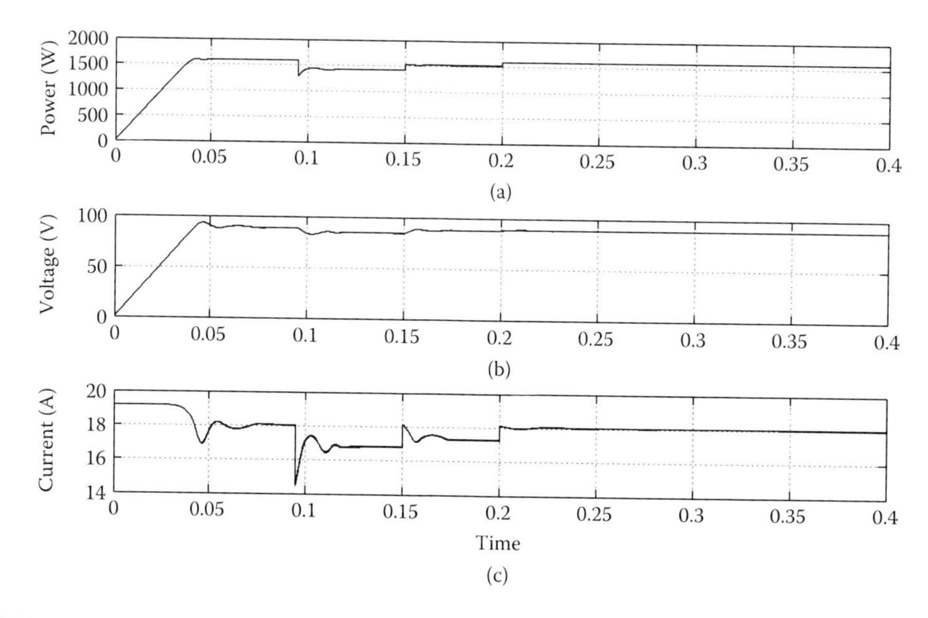

FIGURE 5.21
Performance of PV array: (a) power, (b) voltage, and (c) current at constant irradiance and varying temperature using constant voltage MPPT technique.

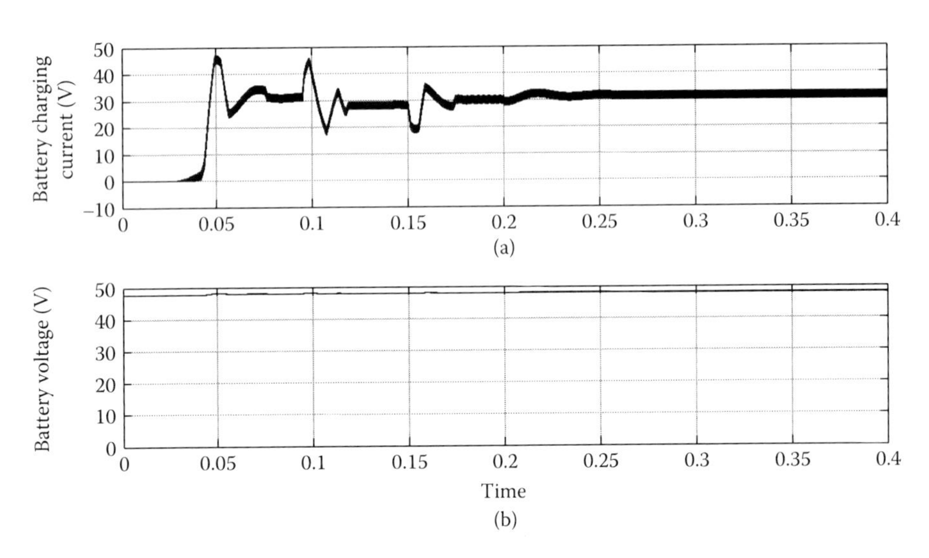

FIGURE 5.22
Performance of PV array: (a) battery charging current and (b) battery charging voltage at constant irradiance and varying temperature using constant voltage MPPT technique.

TABLE 5.3
Performance of P&O, Incremental Conductance, and Constant Voltage MPPT Techniques

Parameters	MPPT Techniques		
	P&O	Incremental Conductance	Constant Voltage
Power ripple (W)	20	4	1.7
Voltage fluctuations (V)	3.5	2.9	1.5
Response time (s)	0.28	0.28	0.24
Battery charging voltage (V)	47.68	47.75	47.89
Battery charging current (A)	Not constantly oscillating between 0 and 59 A throughout the period	31.96	31.89

TABLE 5.4
Performance of P&O, Incremental Conductance, and Constant Voltage MPPT Techniques

Parameters	MPPT Techniques		
	P&O	Incremental Conductance	Constant Voltage
Power ripple (W)	22	3.5	1.5
Voltage fluctuations (V)	3.2	1.8	1.3
Response time (s)	0.24	0.28	0.22
Battery charging voltage (V)	47.89	47.94	48.1
Battery charging current (A)	Not constantly oscillating between 0 and 45 A throughout the period	26.97	28.1

5.5 Performance Analysis for Varying Solar Irradiance and Temperature

It is clearly observed that solar irradiance and temperature have a significant impact on the PV power generation. Therefore, it is important to study the impact of these parameters simultaneously. The performance analysis of various MPPT techniques under varying irradiance and temperature is carried out and presented in Figures 5.23 through 5.28 using P&O, incremental conductance, and constant voltage MPPT techniques, respectively.

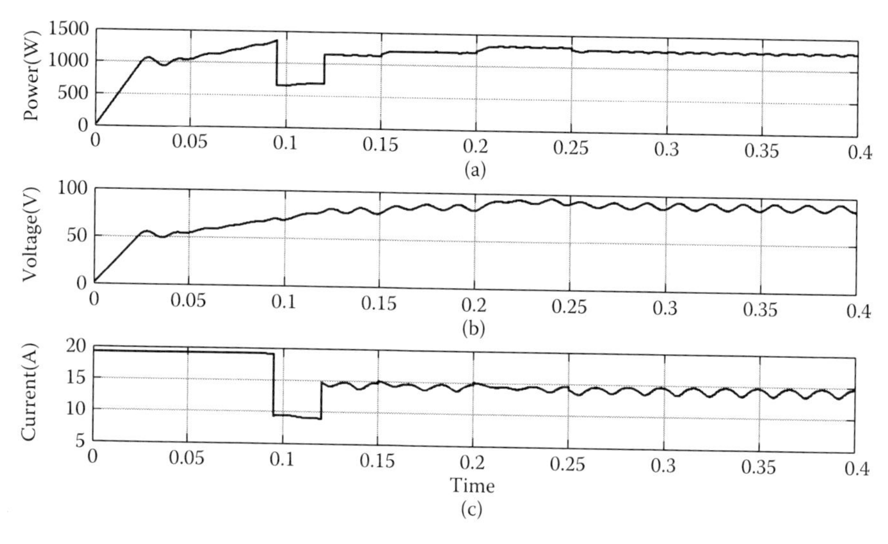

FIGURE 5.23
Performance of PV array: (a) power, (b) voltage, and (c) current at varying irradiance and temperature using P&O MPPT technique.

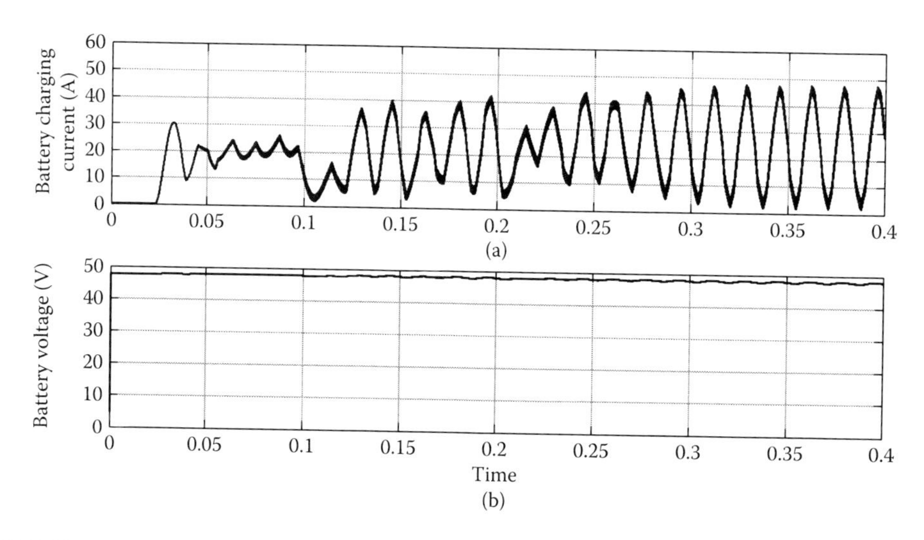

FIGURE 5.24
Performance of PV array: (a) battery charging current and (b) battery charging voltage at varying irradiance and temperature using P&O MPPT technique.

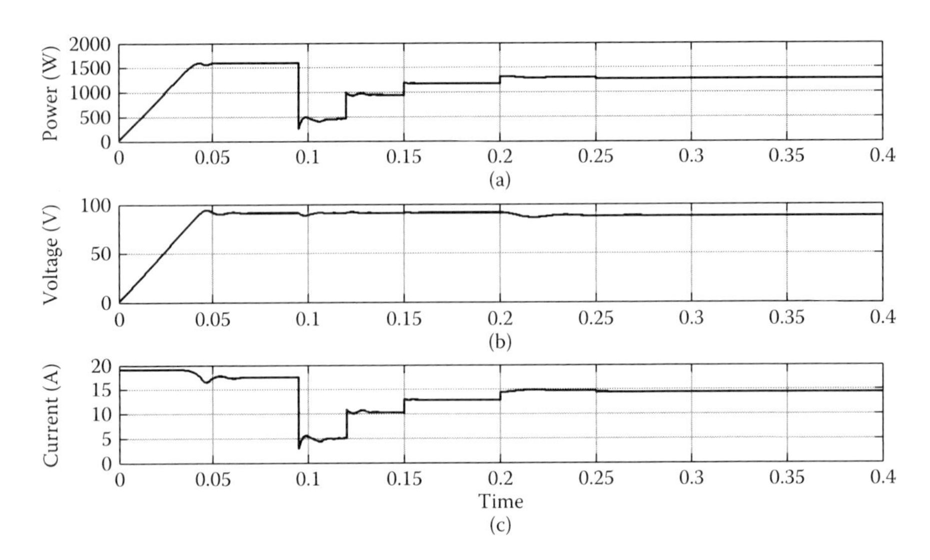

FIGURE 5.25
Performance of PV array: (a) power, (b) voltage, and (c) current at varying irradiance and temperature using incremental conductance MPPT technique.

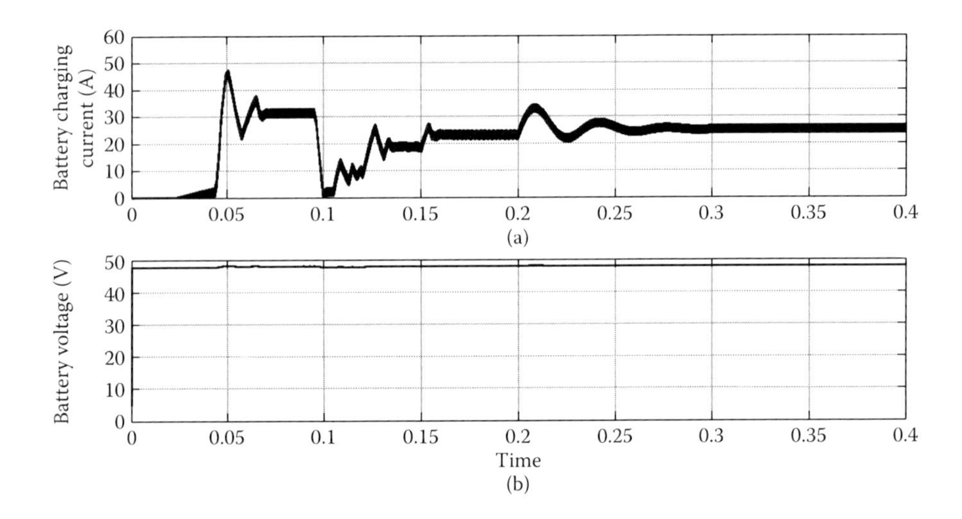

FIGURE 5.26
Performance of PV array: (a) battery charging current and (b) battery charging voltage at varying irradiance and temperature using incremental conductance MPPT technique.

In this case, the P&O MPPT technique contains more oscillations then in varying temperature and constant irradiance conditions, and the battery charging current is also oscillatory. The incremental conductance technique performs pretty well in this situation, but it has a large response time. The constant voltage MPPT technique again gives a higher amount of extracted energy than the other two techniques.

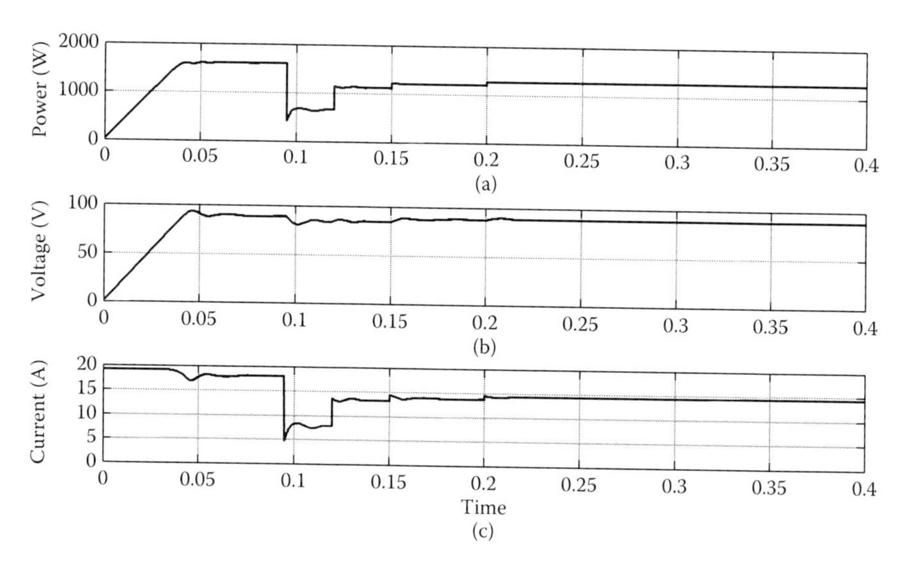

FIGURE 5.27
Performance of PV array: (a) power, (b) voltage, and (c) current at varying irradiance and temperature using constant voltage MPPT technique.

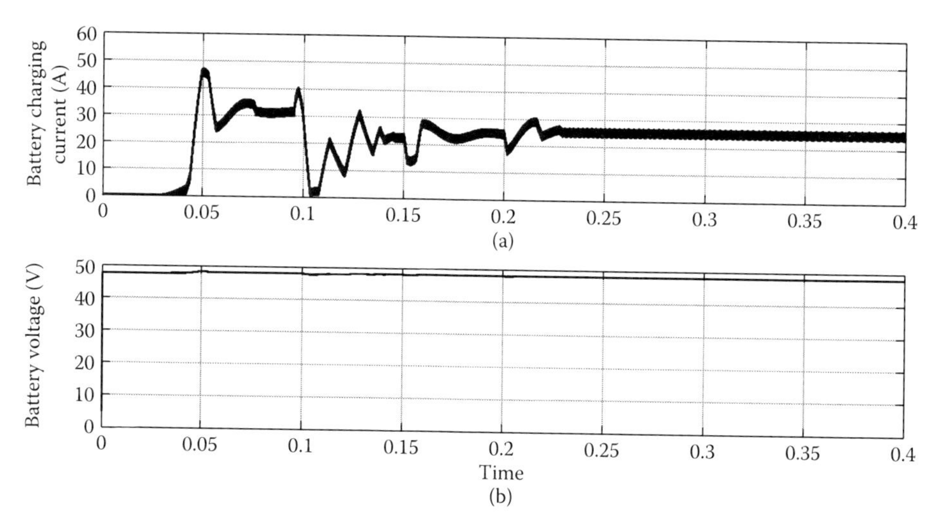

FIGURE 5.28
Performance of PV array: (a) battery charging current (b) battery charging voltage at varying irradiance and temperature using constant voltage MPPT technique.

5.6 Comparison of Various MPPT Techniques

A comparison of various MPPT techniques is presented in Table 5.5. From the earlier discussion we can conclude that P&O and incremental conductance (IC) are the basic maximum power point algorithms, and the speed of these algorithms varies, meaning

TABLE 5.5

Comparison of Various MPPT Techniques

MPPT Technique	PV Array Dependent	True MPPT	Analog and Digital	Speed	Complexity	Sensed Parameter	Periodically Tuned
P&O/Hill climbing	No	Yes	Both	Varies	Low	V, I	No
IC	No	Yes	Digital	Varies	Medium	V, I	No
Fractional voltage	Yes	No	Both	Medium	Low	V	Yes
Fractional current	Yes	No	Both	Medium	Medium	I	Yes
Fuzzy logic	Yes	Yes	Digital	Fast	High	Varies	Yes
dP/dV and dP/dI Feedback control	No	Yes	Digital	Fast	Medium	V, I	No
Neural network	Yes	Yes	Digital	Fast	High	Varies	Yes

the speed is dependent upon the step size used. Fractional voltage and fractional current schemes are not true MPPT algorithms, as in these techniques the sensor requirement is less compared to other techniques. The fuzzy logic and the neural network technique are used with digital systems and their speed is fast, but their complexity is greater compared to other hardware.

5.7 Charge Controllers and MPPT Algorithms

In this section, the design and simulation of the charge controller and MPPT techniques for standalone solar photovoltaic systems are discussed. A depiction of a standalone PV system with converter and charge controllers is presented in Figures 5.29 and 5.30, respectively.

In this standalone system we get the current and voltage from the PV array system; these parameters are given to the control algorithm of the duty cycle of the converter to obtain the maximum power from the photovoltaic system. Different MPPT techniques (P&O or INC) generate duty cycles for the converter to provide the maximum power to the battery bank for charging. The battery is connected in between the converter and the inverter that supply the power to the DC and AC load.

In the standalone system using the charge controller technique, we implement the P&O technique to obtain the maximum power from the photovoltaic system. It will charge the battery up to a certain limit and after that it will stop.

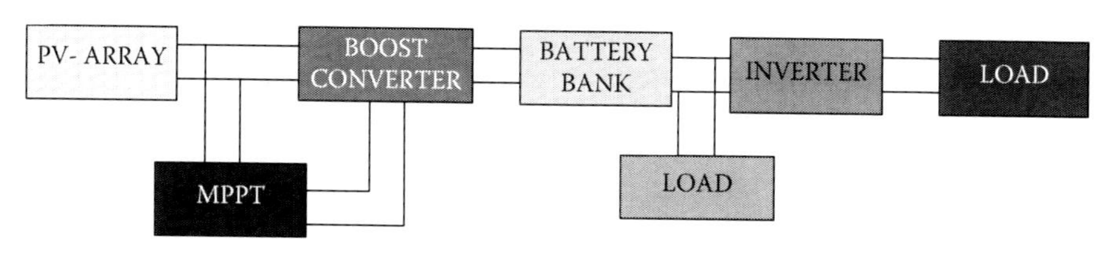

FIGURE 5.29
Block diagram of standalone system with converter.

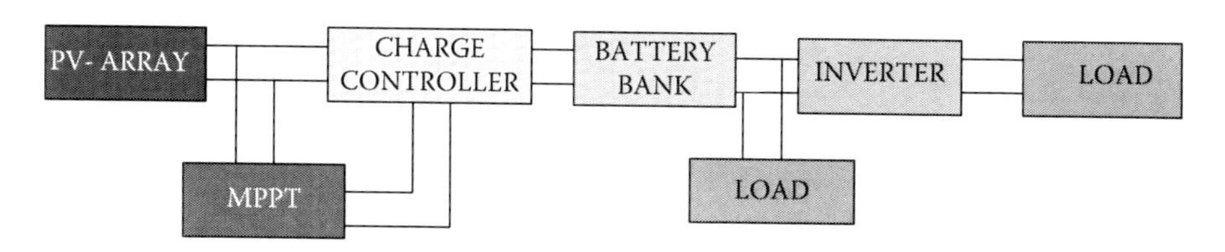

FIGURE 5.30
Block diagram of standalone system with charge controller.

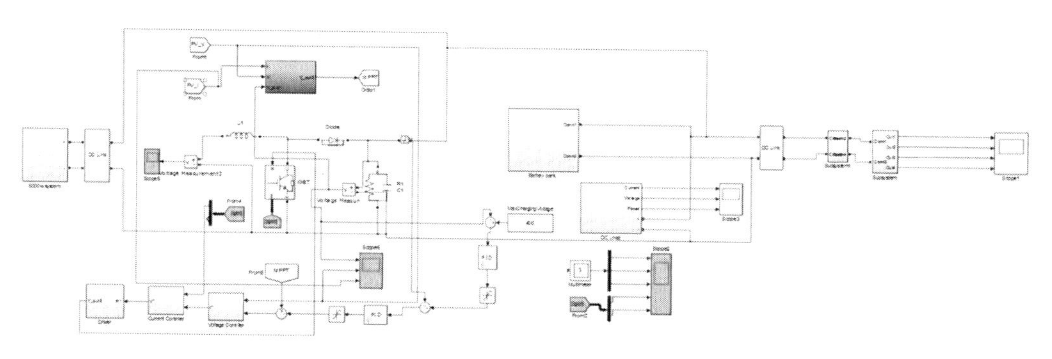

FIGURE 5.31
Simulation model of standalone system with charge controller.

5.8 MATLAB® Simulation Model of PV System with Charge Controller

The Simulink model of the standalone system with charge controllers and for a battery is presented in Figures 5.31 and 5.32, respectively.

In the charge controller we will use the maximum power point technique for charging and discharging the battery. The MPP at the P-V curve is used to obtain the maximum power output to charge the battery. In this circuit diagram an RC network is connected in parallel to the battery for the charging and discharging time constant. From the battery terminal we get the voltage; this is compared with the maximum charging voltage. Then the error signal is generated, which will be provided to the PI controller. The

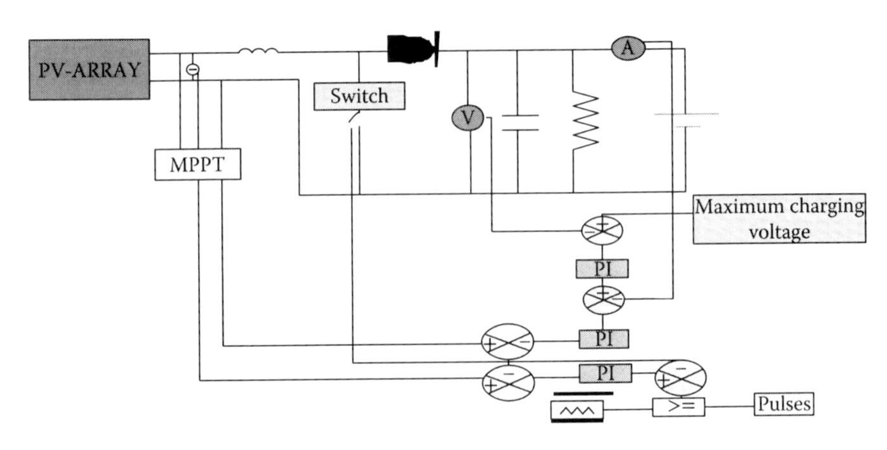

FIGURE 5.32
Charging circuit block diagram for the battery.

MATLAB® models of the P&O and incremental conductance techniques are provided in Figures 5.33 and 5.34, respectively. Also the MATLAB® coding is provided for the incremental conductance technique.

MATLAB® Code:

```
function D = mppt_INC (dV, dV, I, V, y)
%#codegen
N=1e-5; %max
Step=N*abs (dI/dV);
If (dV==0) || (dI=0)
D=y;
else if dI>0
D=y step;
Else
D=y+Step;
End
If dV ~ =0
If dI/dV>= -I/V
D=y;
end
If dI/dV>= -I/V
D=y-Step;
else if dI/dV>= -I/V
D=y+Step;
end
end
% if dV==0
% D=y-step
% end
%y=D;
end
```

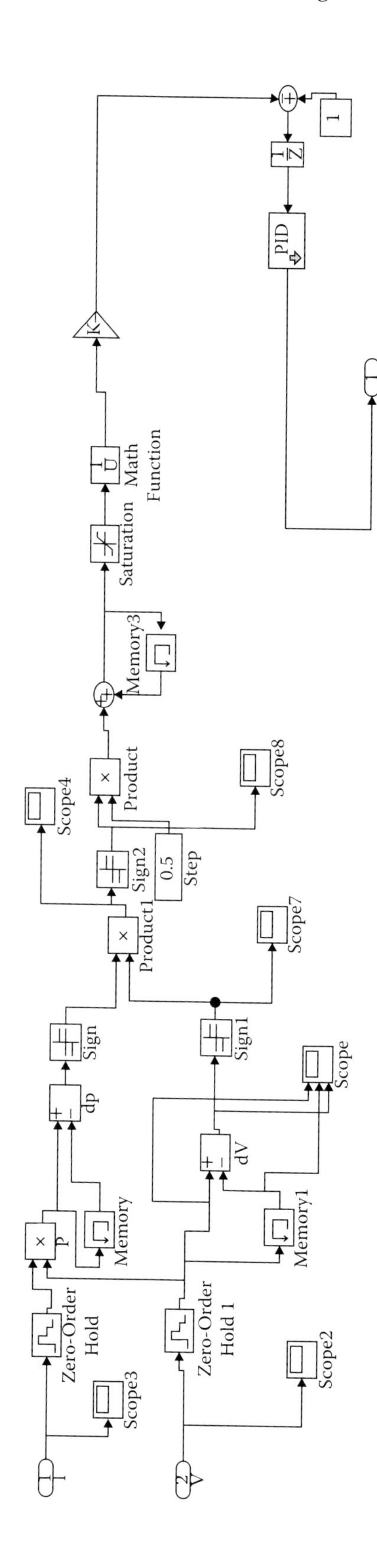

FIGURE 5.33
MATLAB® simulation of the P&O technique.

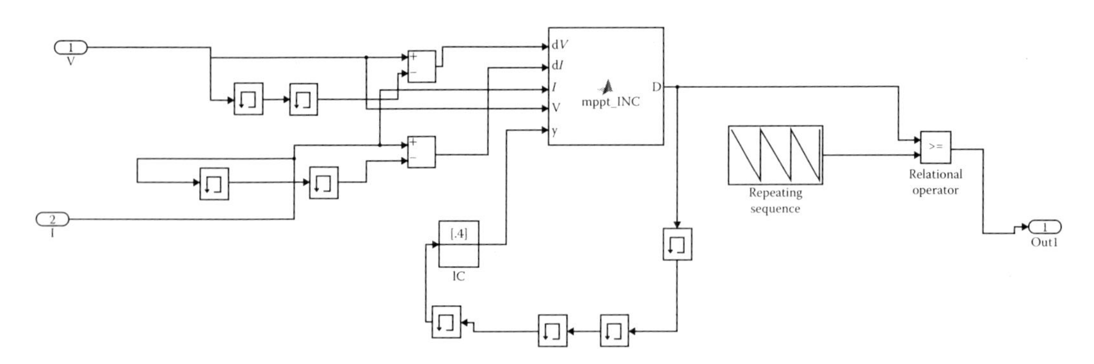

FIGURE 5.34
MATLAB® simulated modeling of INC technique.

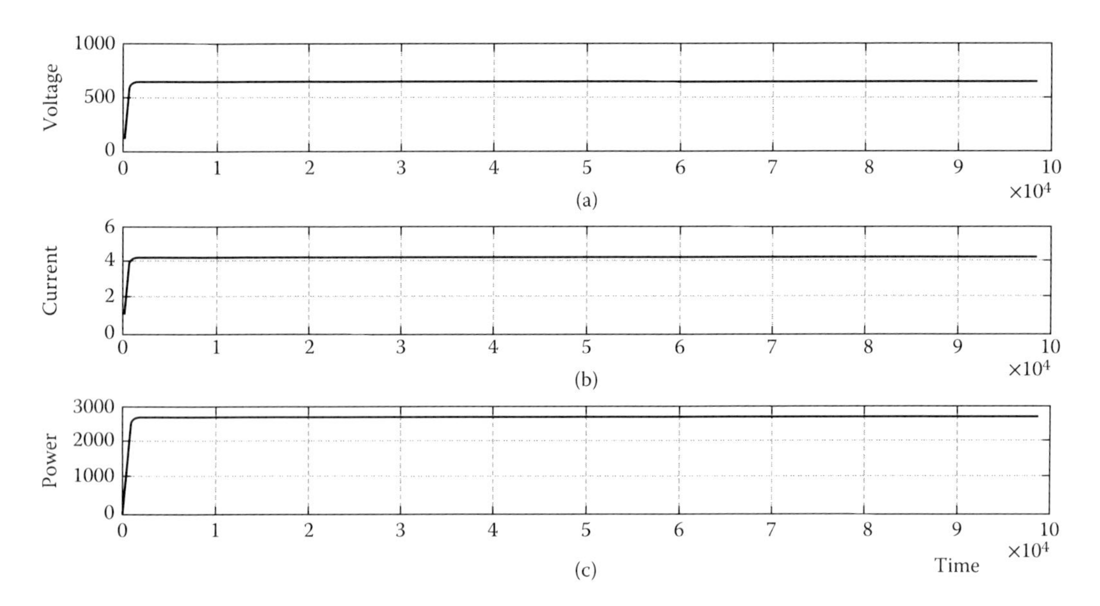

FIGURE 5.35
(a) Voltage, (b) current, and (c) power at the converter terminal with P&O.

In the standalone system, to step up the DC voltage, the boost converter has been used. To get the maximum power from the array, the MPPT technique, which will change the duty cycle of the converter according to the maximum power point, has been implemented. The results obtained are presented in Figures 5.35 and 5.36 and in Table 5.6.

The effect of partial shading is also presented in Figures 5.37 and 5.38.

In the partial shading condition, the INC response is good compared to the P&O approach. To show the partial shading condition, the signal generator has been used to provide different solar irradiances, like 1000, 800, and $400\,\text{W/m}^2$, at different time intervals. The results are presented in Figure 5.39.

The MATLAB® Simulink model for the P&O approach is shown in Figure 5.40, and the results in different scenarios are presented in Figures 5.41 through 5.43. The performance

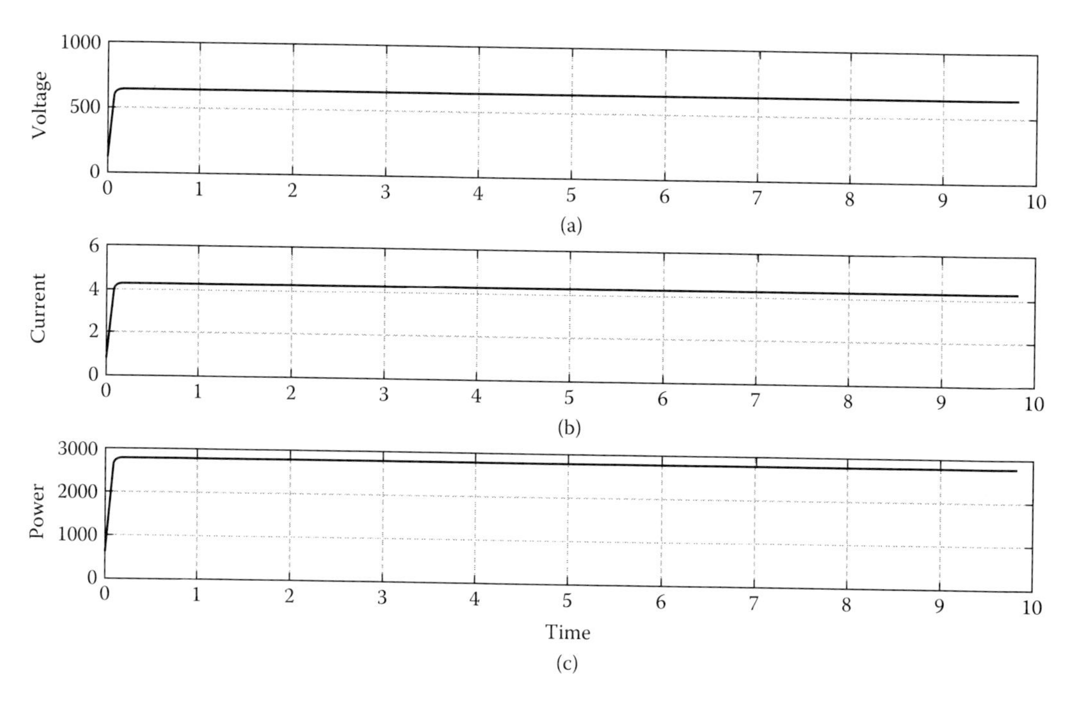

FIGURE 5.36
(a) Voltage, (b) current, and (c) power at boost converter terminal with INC.

TABLE 5.6
Parameter Competition at Converter Terminal with MPPT

S. No.	Technique	Power (W)	Current (A)	Voltage (V)	Load
1	P&O	2752	4.28	642.2	$R = 150$
2	IC	2790	4.31	646.2	$R = 150$

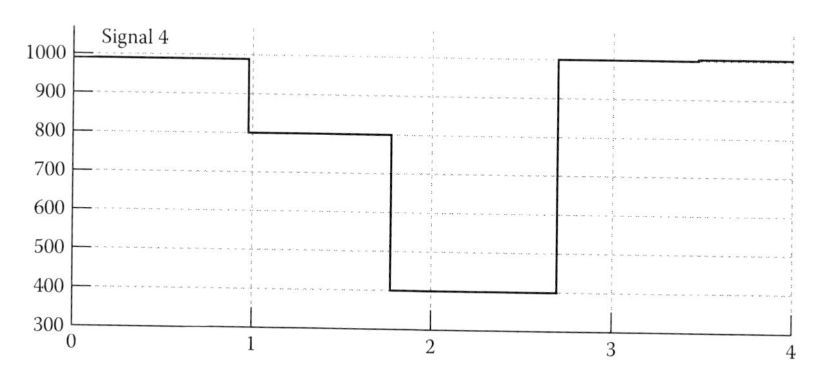

FIGURE 5.37
Signal generator with different radiation levels.

in partial shading conditions is presented in Figures 5.44 through 5.46 at AC load terminal, DC load terminal, and battery terminal, respectively.

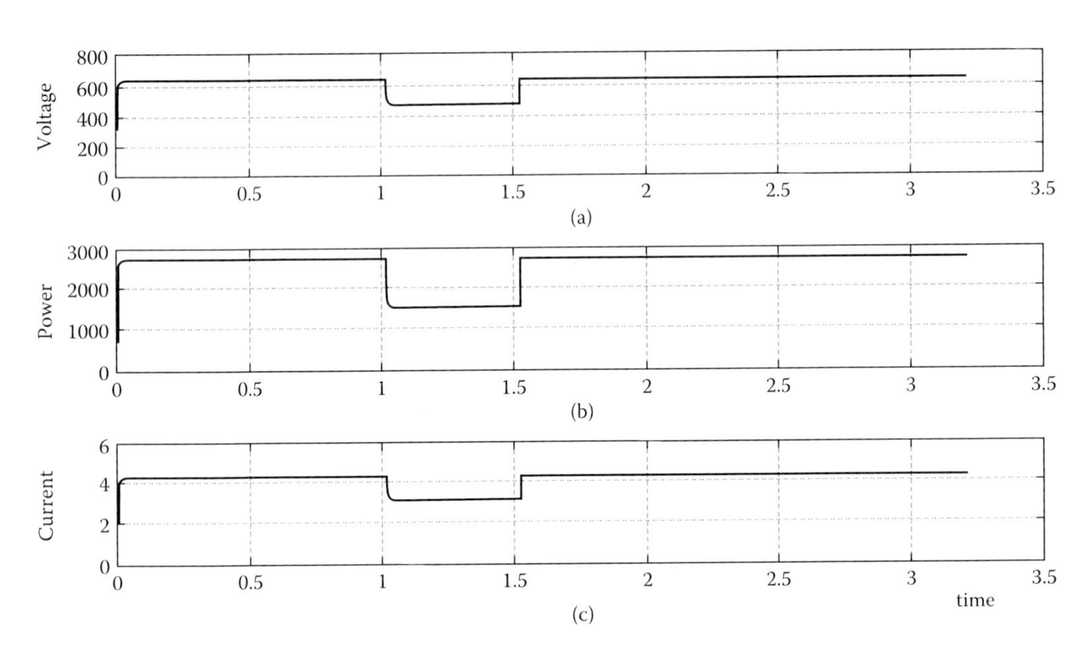

FIGURE 5.38
(a) Voltage, (b) power, and (c) current under partial shading with P&O.

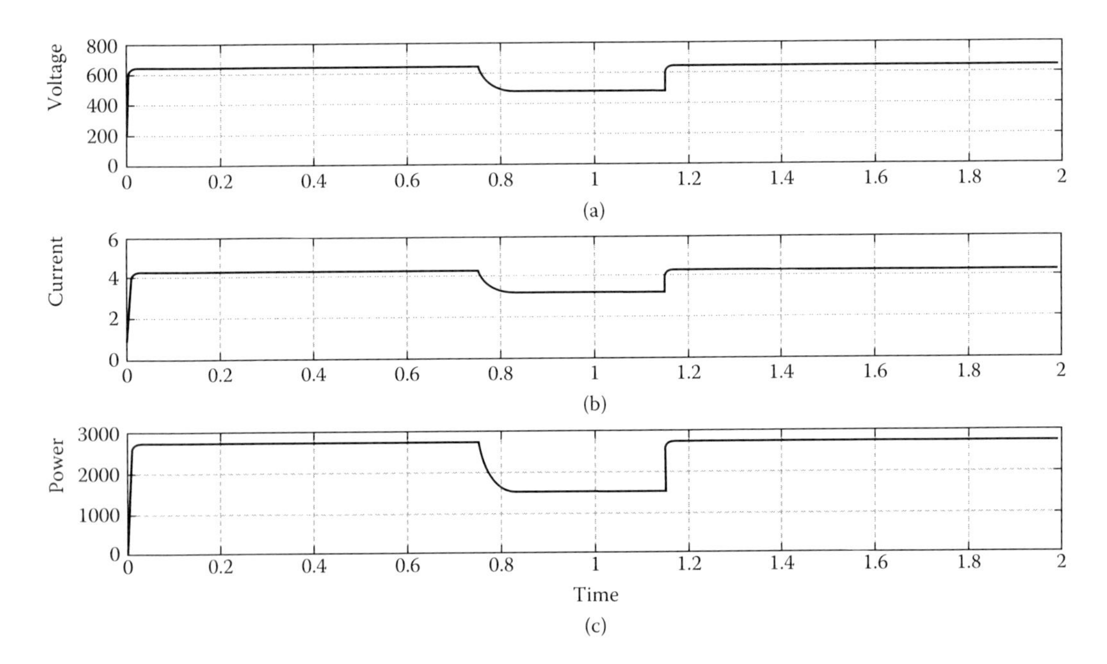

FIGURE 5.39
(a) Voltage, (b) current, and (c) power under partial shading with INC.

The MATLAB® Simulink model for the INC approach is shown in Figure 5.46, and the results in different scenarios are presented in Figures 5.47 through 5.49. The performance under partial shading conditions is presented in Figures 5.50 through 5.52 at AC load terminal, DC load terminal, and battery terminal, respectively.

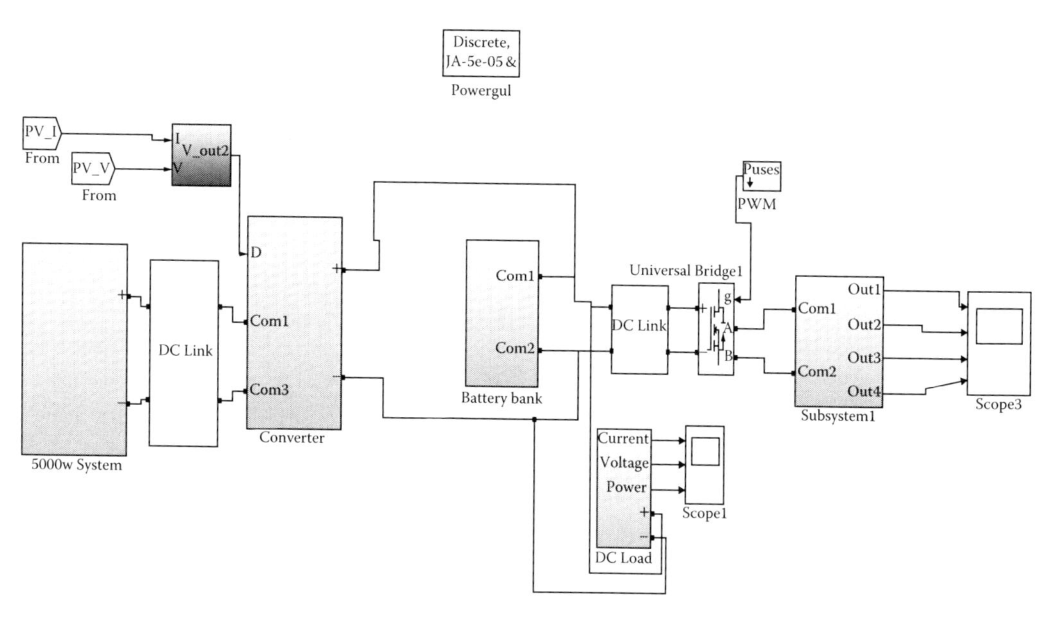

FIGURE 5.40
Simulated standalone system with P&O technique.

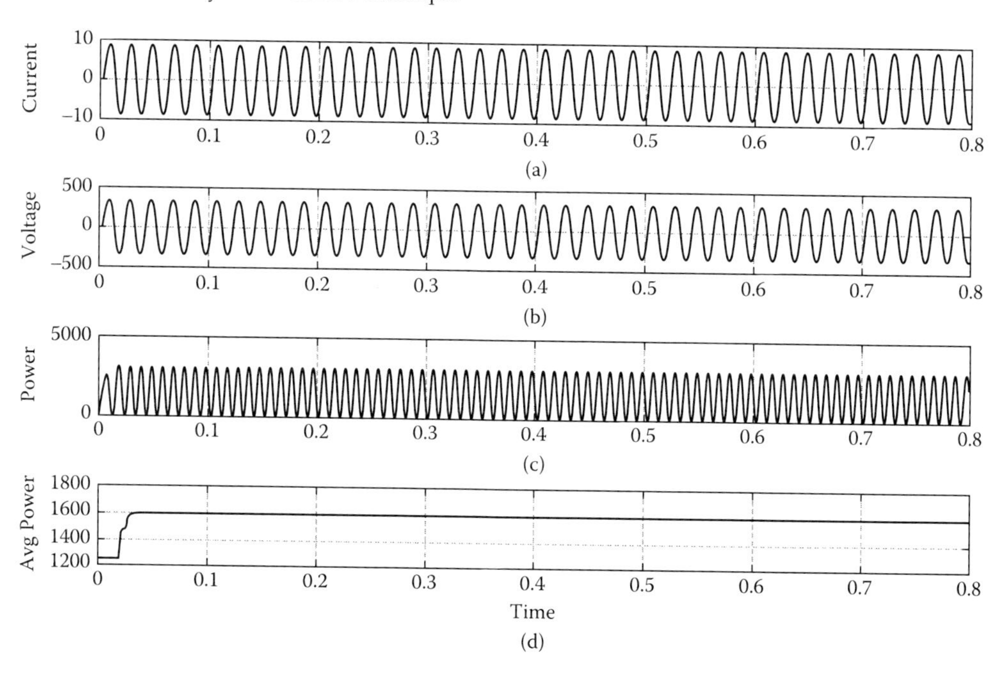

FIGURE 5.41
(a) Current, (b) voltage, (c) power, and (d) average power for AC load.

The results with the charge controller technique are presented in Figures 5.53 and 5.54 and in Table 5.7. In this work, the performance analysis of the P&O and incremental conductance techniques is carried out and presented. The performance is evaluated on different meteorological conditions like constant irradiance and temperature, varying irradiance, and constant temperature. The comparison of various parameters with different MPPTs are presented in Table 5.7.

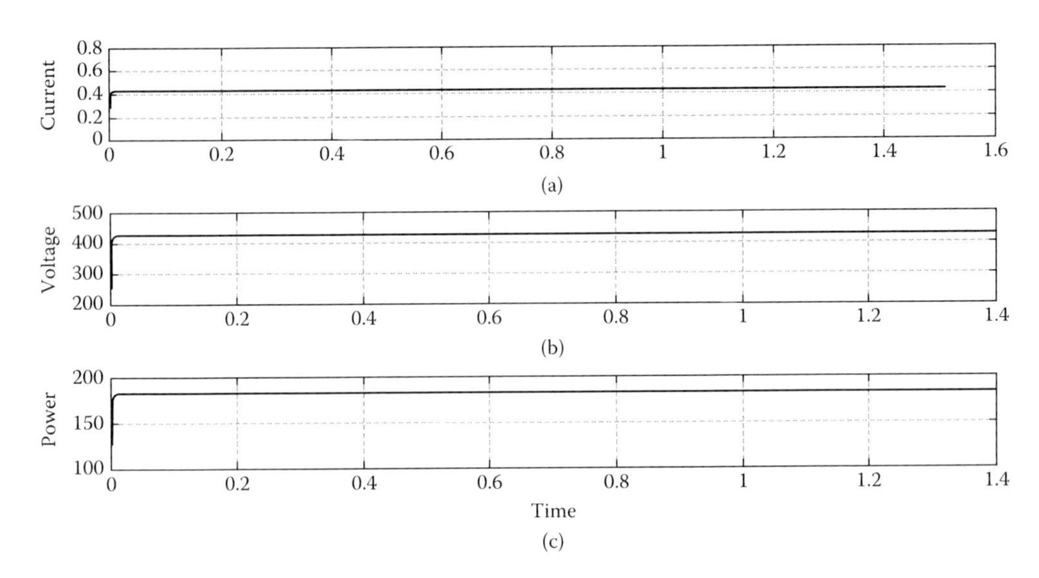

FIGURE 5.42
(a) Voltage, (b) current, and (c) power for DC load.

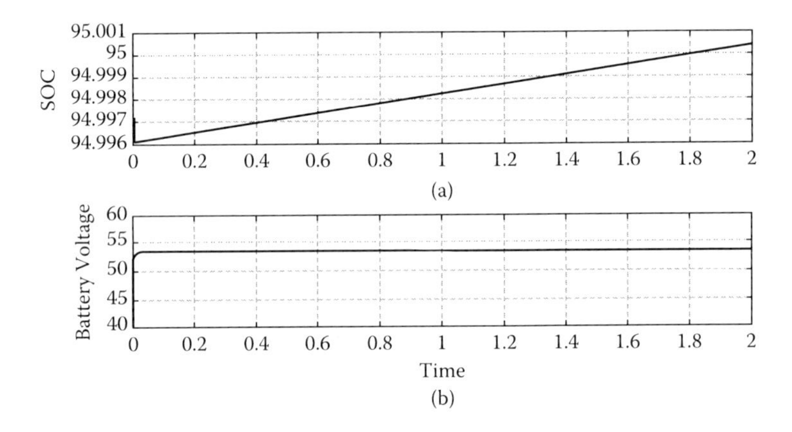

FIGURE 5.43
(a) State of charge (SOC), and (b) battery voltage.

In addition, battery charging characteristics are evaluated on the basis of battery voltage and the battery charging current. We can conclude that the performance of the constant voltage technique is better compared to others at constant irradiance and temperature, whereas the P&O approach is better at constant irradiance and varying temperature. The control algorithm of the charge controller executes the P&O technique to allow the module to operate at the maximum power point according to solar irradiation and match the load with the source impedance to provide maximum power. This MPPT model is more suitable because it costs less, has an easier circuit design, and the efficiency of the circuit

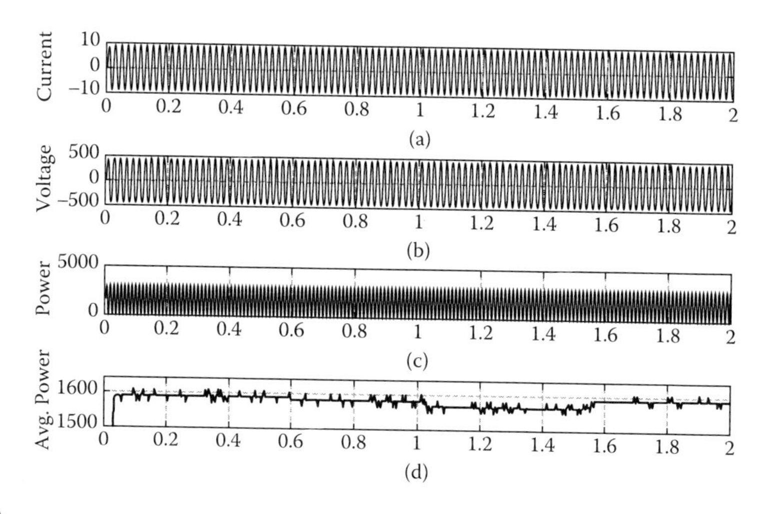

FIGURE 5.44A
(a) Current, (b) voltage, (c) power, and (d) average power for AC load w.r.t time.

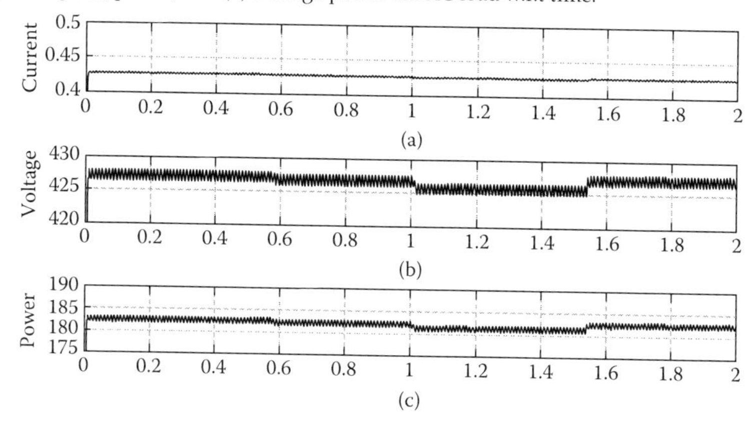

FIGURE 5.44B
(a) Current, (b) voltage, and (c) power for DC load w.r.t time.

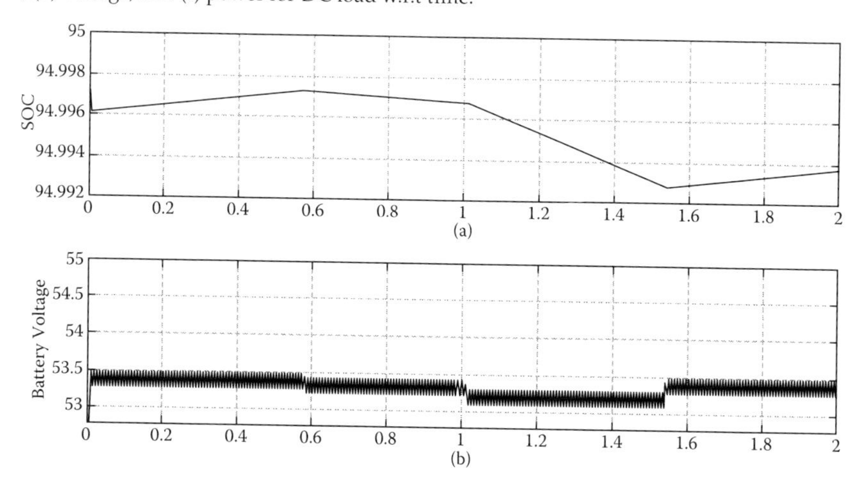

FIGURE 5.45
(a) SOC and (b) battery voltage w.r.t time.

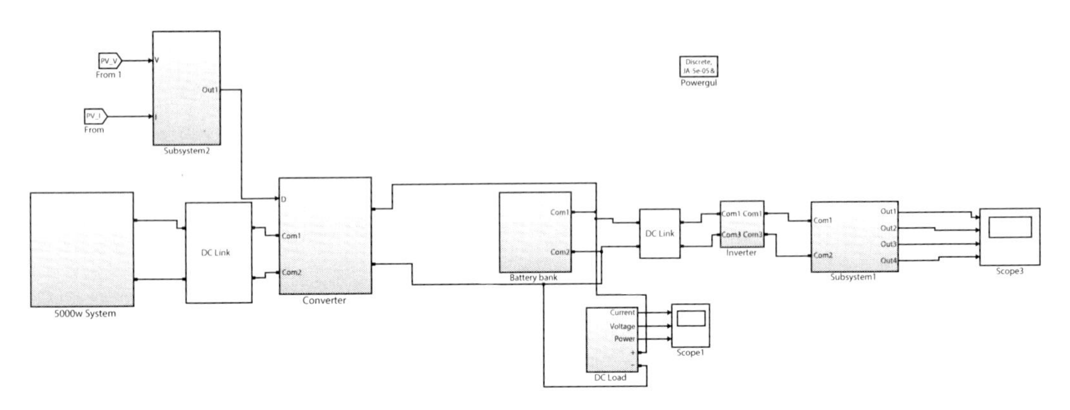

FIGURE 5.46
Simulated model of INC with standalone system.

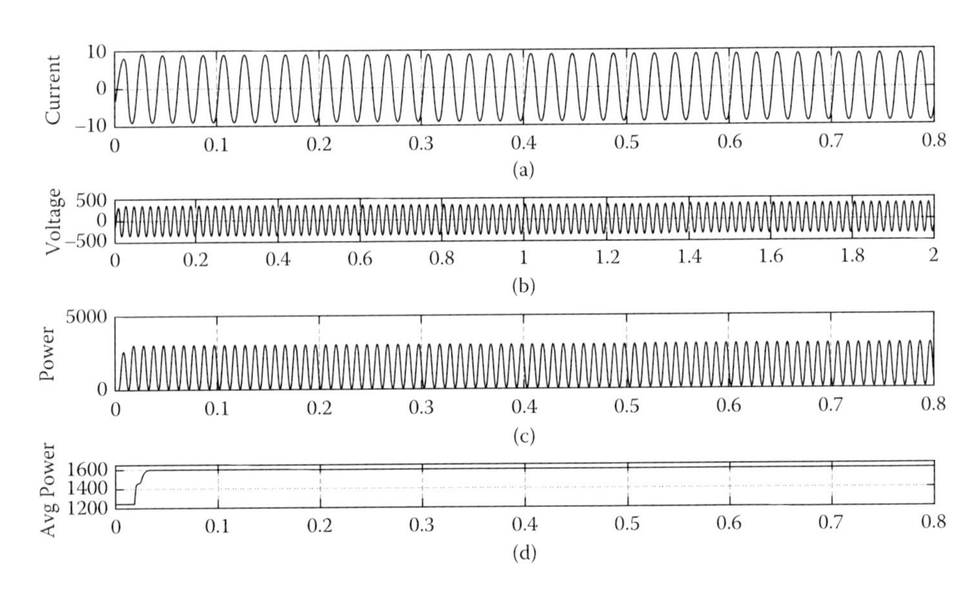

FIGURE 5.47
(a) Current, (b) voltage, (c) power, and (d) average power for AC load w.r.t time.

is increased by 20% to 25% in the case of an MPPT solar charge controller compared to a circuit with MPPT. However, the performance of the incremental conductance approach is better compared to perturb and observe at varying irradiance and constant temperature. Therefore, the incremental conductance technique is better for evaluating the performance of a PV system under meteorological parameters.

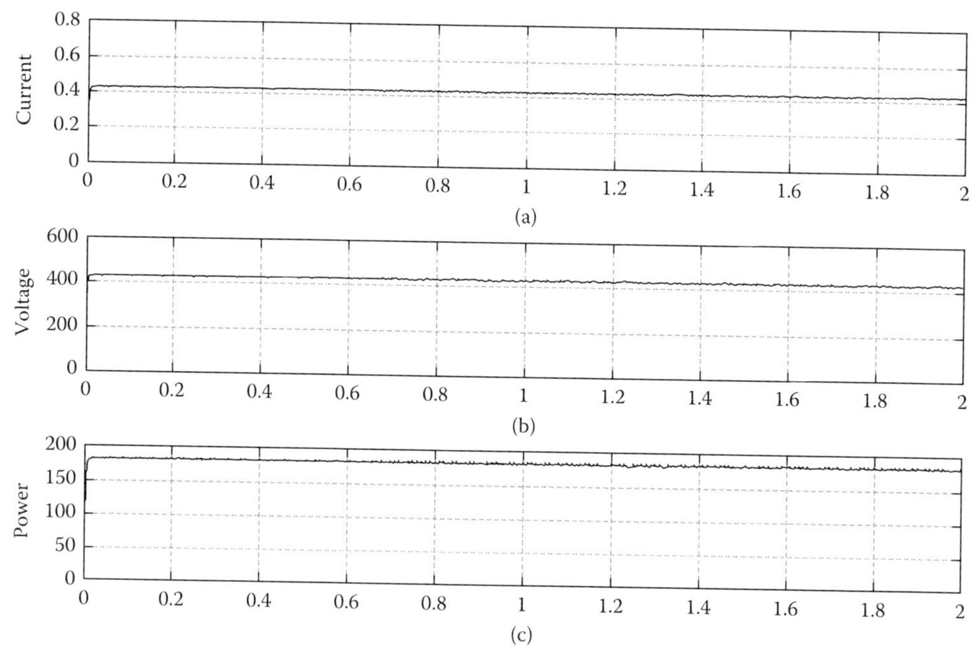

FIGURE 5.48
(a) Current, (b) voltage, and (c) power for DC load w.r.t time.

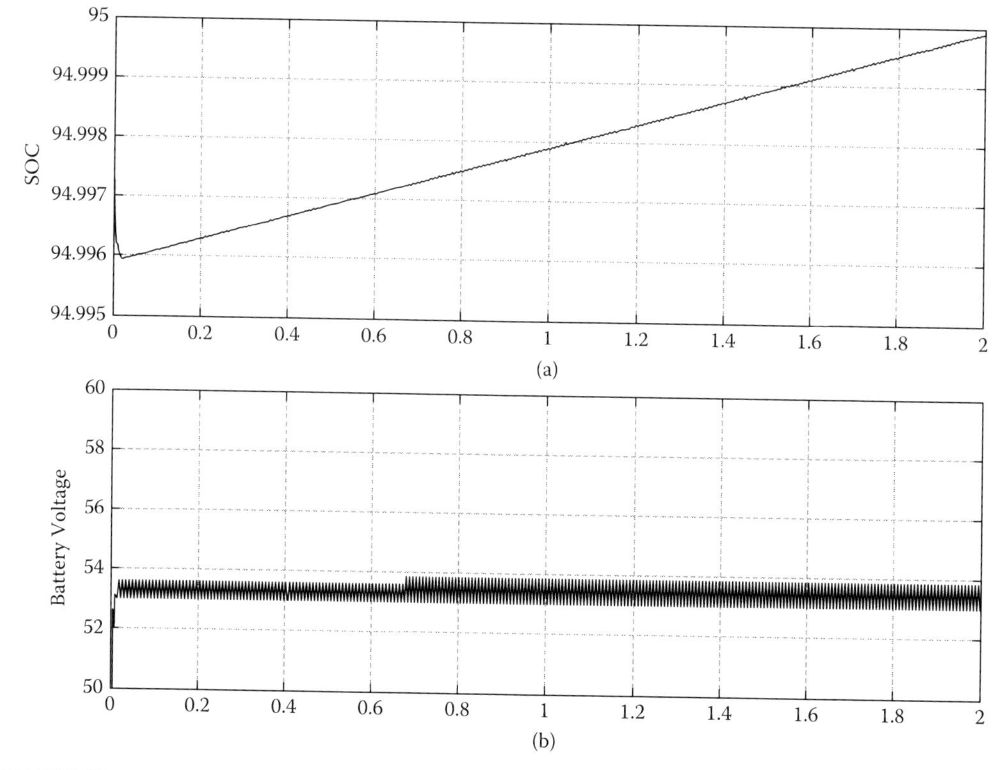

FIGURE 5.49
(a) SOC and (b) battery voltage w.r.t time.

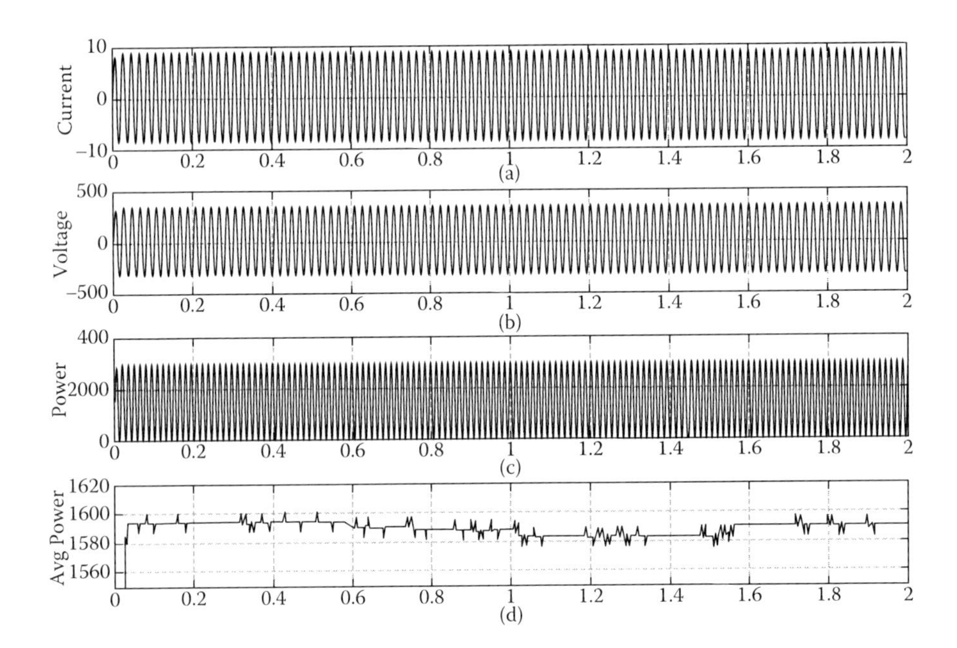

FIGURE 5.50
(a) Current, (b) voltage, (c) power, and (d) average power for AC load w.r.t time.

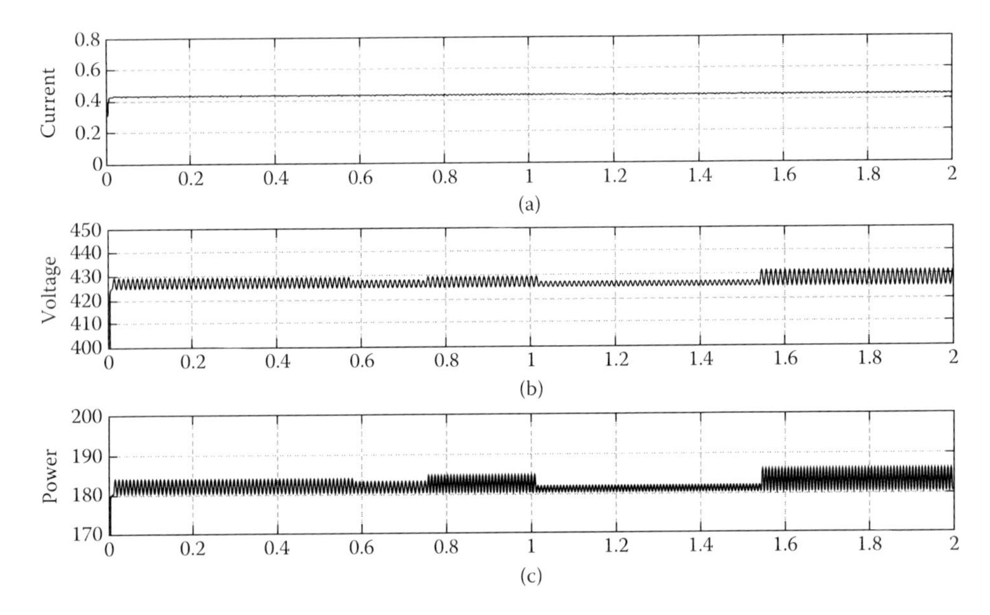

FIGURE 5.51
(a) Current, (b) voltage, and (c) power for DC load w.r.t time.

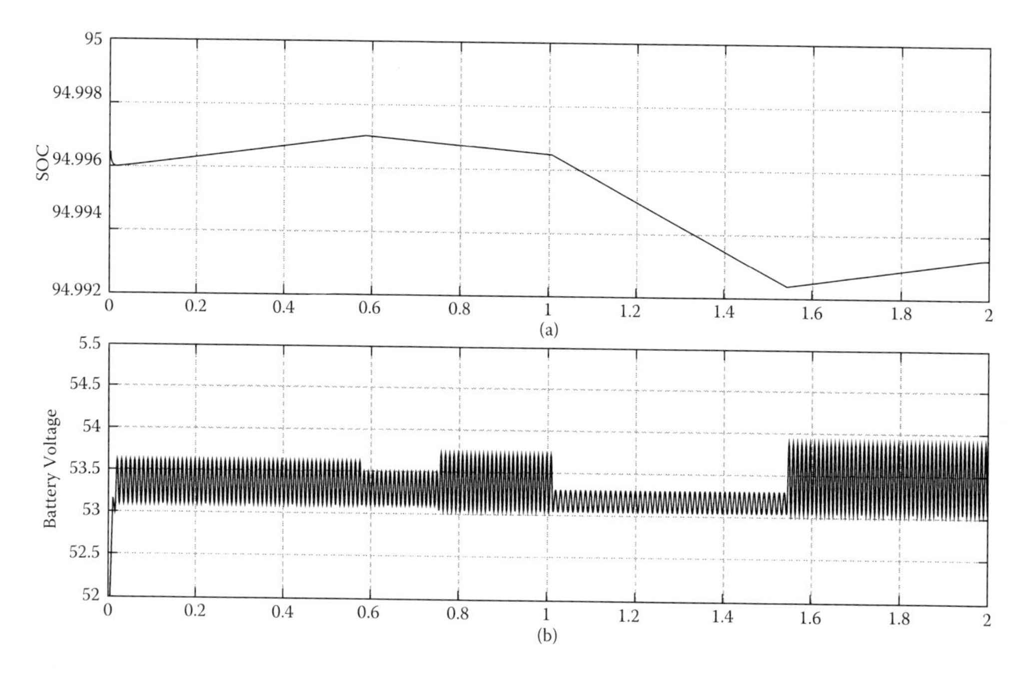

FIGURE 5.52
(a) SOC and (b) battery voltage w.r.t time.

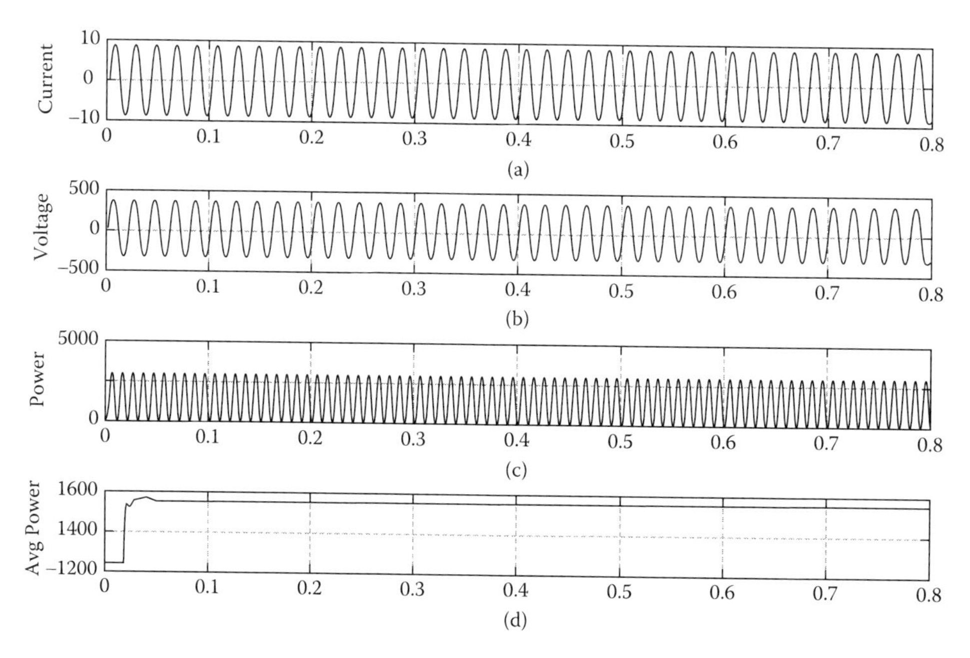

FIGURE 5.53
(a) Current, (b) voltage, (c) power, and (d) average power for AC load w.r.t time.

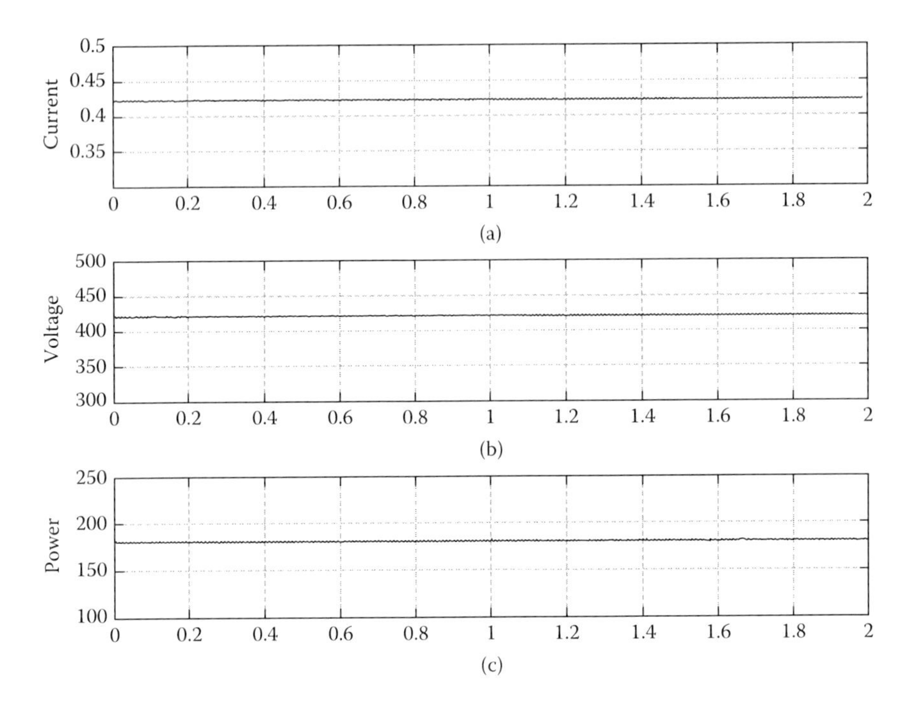

FIGURE 5.54
(a) Current, (b) voltage, and (c) power for DC load w.r.t time.

TABLE 5.7
Comparison of Various Parameters with Different MPPTs

S. No.	Parameter	Average Power		Voltage		Current	
		AC	DC	AC	DC	AC	DC
1	**P&O**	1592W	195.5W	339(p-p)	442V	9.21(p-p)	0.4425A
2	**IC**	1594W	180.5W	338(p-p)	424.5V	8.85(p-p)	0.425A
3	**Charge controller**	1550W	179W	335(p-p)	423V	8.7A(p-p)	0.423A

6

Converter Design

6.1 DC to DC Converters

The literal meaning of the word "DC" in the world of electricity and electronics is direct current. So, by definition, a DC to DC converter is an electronic circuit or an electromechanical device that converts a source of direct current (DC) from one voltage level to another. It is basically a type of electric power converter. This voltage conversion can be from a higher voltage level to a lower voltage level or vice versa. The power levels can range from very small power ratings, like that of small batteries used in day-to-day life to very high power ratings encountered in high-voltage power transmission.

6.1.1 Brief History

Earlier, when the power semiconductors and allied technologies had not yet been developed, one way to convert the voltage of a DC supply to a higher voltage for low-power applications was to convert it to AC (alternating current) by using a vibrator, followed by a step-up transformer and a rectifier. For higher power, an electric motor was used to drive a generator of the desired voltage. These were relatively inefficient and expensive procedures used only when there were no alternatives to power a car radio, etc. Although it was possible to derive a lower voltage from a higher voltage with a linear electronic circuit, or even a resistor, these methods dissipated the excess energy as heat. Energy-efficient conversion only became possible with solid-state switch mode circuits. The development of power semiconductors and integrated circuits made it economically viable to use technologies implementing semiconductor devices for DC to DC conversions, be it step-up or step-down.

6.2 Classification of DC to DC Converters

Based on the direction in which it allows the conduction to take place, and hence the conversion, the converters are classified as follows:

- **Unidirectional converters**: These converters allow the flow of charges in one direction only. They generally cater to various onboard loads such as sensors, controls, utility, entertainment, and safety equipment and appliances.
- **Bidirectional converters**: The converters allow the flow of charges in both directions. They are used in places where battery charging and regenerative braking

are required. The power flow in a bidirectional converter is usually from a low-voltage end, such as a battery or a super capacitor, to a high-voltage side. This operation is referred as the boost mode operation. During the operation of regenerative braking, the power flows back to the low voltage bus to recharge the batteries, referred to as the buck mode operation.

6.2.1 Classification on the Basis of Voltage Stepping Operation

On the basis of the voltage stepping operation, the DC to DC converters can be classified as follows:

- **Buck converter:** A buck converter is a step-down converter that produces a lower average output voltage than the DC input voltage.
- **Boost converter:** A boost converter is a step-up converter that produces a higher average output than the DC input voltage.
- **Buck–boost converter:** A buck–boost converter is a step-up as well as a step-down converter that produces a higher or a lower value of output voltage compared to the applied input voltage, depending upon the duty cycle of operation.

6.2.2 Buck Converter

A buck converter (DC to DC) comprises an inductor, a diode, and a switching device that belongs to the transistor family. Also a diode (called a free-wheeling diode) is used to allow the load current to flow through it when the switch (i.e., a device) is turned off. The load is an inductive (*R-L*) one. In some cases, a battery or back electromotive force (back-EMF) is connected in series with the load (inductive). Due to the load inductance, the load current must be allowed a path, which is provided by the diode; otherwise—for example, in the absence of the aforementioned diode—the EMF of the inductance, which tends to increase as the load current decreases, may cause some damage to the switching device. If the switching device used is a thyristor, this circuit is called a step-down chopper, as the output voltage is normally lower than the input voltage.

Normally, due to the turn-on delay of the device used, the duty ratio k is not zero, but has some positive value. Similarly, due to the required turn-off time of the device, the duty ratio k is less than 1.0. So the range of duty ratio is reduced. It may be noted that the output voltage is lower than the input voltage. Also, the average output voltage increases as the duty ratio is increased. So, a variable DC output voltage is obtained from a constant DC input voltage. The load current is assumed to be continuous. The load current increases in the ON period, as the input voltage appears across the load, and the load current decreases in the OFF period, as it flows in the diode, but is positive at the end of the time period, T.

6.2.3 Boost Converter

A boost converter employs an inductor, a switch, and a diode. When the switch is turned ON the inductor is short-circuited and hence stores a large amount of charge, which is then boosted through the diode to the load as the switch is turned off. A capacitor is often connected in parallel to the load to supply a continuous current through it. The switching device used can be a metal-oxide-semiconductor field-effect transistor

(MOSFET) or any other semiconductor switching device, depending on the type of application. This is called a boost converter (DC to DC), when a self-commutated device is used as a switch. Instead, if a thyristor is used in its place, this is called a step-up chopper. The variation (range) of the output voltage can be easily computed.

6.2.4 Buck–Boost Converter

A buck–boost converter (DC to DC) consists of a diode in series with the load. The inductor, L, is connected in parallel after the switch and before the diode. A capacitor, C, is connected in parallel with the load.

The polarity of the output voltage is opposite to that of the input voltage. When the switch is turned ON, the supply current flows through the path, switch, and inductor during the given time interval. The currents through both source and inductor increase and are same, with (di/dt) being positive. The polarity of the induced voltage is same as that of the input voltage.

When the switch, S, is put OFF, the inductor current tends to decrease, with the polarity of the induced EMF reversing. (di/dL) is negative now, with the polarity of the output voltage being opposite to that of the input voltage. The path of the current is through L; a parallel combination of load, C; and diode, D, during the time interval, T_{off}. The output voltage remains nearly constant as the capacitor is connected across the load.

6.2.5 Uses

DC to DC converters are widely used in portable electronic devices such as cellular phones and laptop computers, which are supplied with power from batteries primarily. Such electronic devices often contain several subcircuits, each with its own voltage level requirement that is different from that of the supplied voltage level and hence employ DC to DC converters. They are used in high-efficiency LED power sources. DC to DC converters are also developed to maximize the energy harvest for photovoltaic systems and wind turbines, and they are called power optimizers in this case.

6.3 DC to AC Converters

History of inverters:

- David C. Prince coined the term inverter in 1925: Until the mid-20th century, gas-filled tubes and vacuum tubes were used as switches in inverters; Thyratron was the most widely used gas tube–controlled switch.
- In 1957, silicon-controlled switches were first used in inverter operation.

Disadvantages of gas and vacuum switches inverters:

- Very high voltage drop: Slow response, inefficient, very costly
- Large size and heavy weight

Why the inverter market is growing:

- Gap between power demand and supply: Decreasing tolerance among consumers for frequent power cuts, growth of power generation using renewable energy sources
- Grid tie up

6.4 Classification of Inverters

Classification based on output voltage:

- Square wave inverters: Modified square wave inverters
- Pure sine wave inverters

Voltage source inverter:

- Direct control over output AC voltage. Input to inverter is a stiff DC voltage source, classified as half bridge and full bridge
- Can be of two or more levels (i.e., multilevel)

Current source inverter:

- Current in the output AC waveform is controlled
- Current in the input side is kept constant

Multilevel inverter:

- Numerous industrial applications require higher power apparatus in recent years: A multilevel inverter (MLI) is introduced for high-power applications. It has overshadowed the conventional two-level inverter
- It can be easily interfaced to the renewable energy sources

What is level? The number of values that the output voltage can take is defined as the level of the inverter. Types of multilevel inverters:

- Diode clamped
- Flying capacitor

6.4.1 Diode Clamped Multilevel Inverter (DCMLI)

A neutral point is created when two capacitors are connected in between a DC voltage source.

The diodes are connected through that neutral point, which is where the name "diode clamped" comes from.

An "N" level DCMLI consists of:

- N–1 capacitors on the DC bus: (N–1)(N–2) clamping diodes
- 2(N–1) switching devices

Advantages of DCMLI:

- More voltage levels, which means lesser harmonic content: High inverter efficiency
- Control method is simple

Disadvantages of DCMLI: More levels, which means more clamping diodes

6.4.2 Flying Capacitor Multilevel Inverter (FCMLI)

There is a ladder structure of DC side capacitors present in its structure.

An N–level flying capacitor MLI consists of:

- N levels of phase voltage
- N–1 capacitors

Advantages of flying capacitor MLI:

- Phase redundancies are available
- Large number of capacitors enables it to work in the case of short duration outages and deep voltage sags

Disadvantages of flying capacitor MLI:

- Precharging all the capacitors to the same voltage level is difficult
- More capacitors means a more expensive and bulky circuit

6.4.3 Cascade Multilevel Inverter (Cascade MLI)

A cascade multilevel inverter consists of:

- A series of H-bridges
- Uses separate DC sources for each H-brid
- Output phase voltage level = m = 2Ns+1

where Ns = number of voltage sources

Advantages of Cascade MLI:

- Compared to DCMLI and FCMLI, it requires fewer components to achieve the same number of voltage levels.
- Overall weight and price in comparison to the other two types is less
- Reduced switching losses and device stresses
- Can be used in car batteries and braking systems

Disadvantages of Cascade MLI:

- Requirements of separate DC sources limits its uses for real power conversion

Multilevel converter applications:

- Energy and power system
- Wind energy conversion
- Production
- Conveyor motor drive
- Transportation
- High-speed train traction

6.5 Applications

Wind energy conversion:

- A back-to-back converter provides the indirect AC to DC to AC connection required
- In variable windmills, there is independent rotational speed of the blades and the frequency of the load

Advantages:

- Voltage and frequency control of the local grid: Improvement of the power quality
- Better integration of wind energy to the electrical energy

Conveyor belt motor drive: General requirements

- High availability to permit a continuous operation of the process: High reliability to avoid interrupts in production
- Each AC to DC to AC drive is composed of a three-level active front-end neutral point clamped (NPC) inverter connected in a back-to-back configuration with a three-level inverter

Useful in high speed train traction: Operation

- The main reason is that high speed trains are driven by higher fundamental frequencies (up to 400 Hz)
- Demand high dynamic performance and efficiency
- Multilevel converters reduce the transformer by operating at higher voltages
- Deliver good power quality for the catenaries and motors (less need of filters)
- Can operate at lower switching frequencies while keeping dynamic performance

6.6 Photovoltaic (PV) Inverter

One of the most incredible things about photovoltaic power is its simplicity. It is almost completely solid state, from the photovoltaic cell to the electricity delivered to the consumer.

Whether the application is a solar calculator with a PV array of less than 1 W or a 100-MW grid-connected PV power generation plant, all that is required between the solar array and the load are electronic and electrical components. Compared to other sources of energy humankind has harnessed to make electricity, PV is the most scalable and modular. Larger PV systems require more electrical bussing, fusing, and wiring, but the most complex component between the solar array and the load is the electronic component that converts and processes the electricity: the inverter.

Photovoltaic power consists of DC voltage and current, but the majority of appliances are AC, so to use this power with AC appliances, it must be converted into AC. A PV inverter is a device that converts the DC voltage output taken from the PV array into the AC voltage that can be used in our home and exported back to the grid. The method that is used to convert DC power extracted from a PV array into AC power is known as inversion.

PV inverters have very good efficiency—often between 93 and 97 percent—because it is a semiconductor-based, solid-state device and its power consumption is very low. The main features of any PV inverter are that it must extract maximum power from the PV array and convert it into a suitable pure AC voltage.

6.6.1 Maximum Power Point Tracking

The method an inverter uses to remain on the ever-moving maximum power point (MPP) of a PV array is called maximum power point tracking (MPPT). PV modules have a characteristic I-V curve that includes a short-circuit current value (Isc) at 0 V DC, an open-circuit voltage (Voc) value at 0 A, and a "knee" at the point the MPP is found—the location on the I-V curve where the voltage multiplied by the current yields the highest value: the maximum power. The MPP for a module at full sun in a variety of temperature conditions is discussed further. As cell temperature increases, voltage decreases. Module performance is also irradiance dependent. When the sun is brighter, module current is higher, and when there is less light, module current is lower. Because sunlight intensity and cell temperature vary substantially throughout the day and the year, array MPP current and voltage also move significantly, greatly affecting inverter and system design. The knee, or maximum power point, of the I-V curve varies dramatically according to the effects of both cell temperature, as shown here, and irradiance.

Classification: Based on the connection, the inverters are mainly classified as follows:

Standalone inverter: It is used in isolated systems where the inverter draws its DC energy from batteries charged by photovoltaic (PV) arrays. Many standalone inverters also incorporate integral battery chargers to replenish the battery from an AC source, when available. Normally these do not interface in any way with the utility grid, and as such, are not required to have anti-islanding protection.

6.7 Grid Tied Inverter

A grid tie system is comparatively simple but best for an area where utility service is reliable and well maintained. Grid-connected systems can supply solar power to your load and use utility power as a backup. As long as there is enough electricity flowing in from your PV system, no electricity will flow in from the utility company. If your system is generating more power than you are using, the excess will flow back into the grid, turning your meter backwards.

Grid tie inverters are mainly of three types: string inverters, solar microinverters, and centralized inverters (discussed in Section 6.9).

6.7.1 String Inverter

A string inverter is the type most commonly used in home and commercial solar power systems. It is a large box that is often situated some distance from the solar array. Depending on the size of the installation, more than one string inverter may be present.

Its advantages:

- Allows for high design flexibility
- High efficiency
- Robust
- Three phase variations available
- Low cost
- Well supported (if buying trusted brands)
- Remote system monitoring capabilities

Its disadvantages:

- No panel-level MPPT
- No panel-level monitoring
- High voltage levels present a potential safety hazard

6.7.2 Solar Microinverter

A solar microinverter is a device used in photovoltaic systems. It converts the output generated by a single PV module into the alternating current. And the output of all the microinverters is combined and often fed to the electrical grid. Microinverters have several advantages over conventional inverters. The main advantage is that small amounts of shading, debris, or snow lines on any one solar module prevent a complete module failure and do not disproportionately reduce the output of the entire array. Each microinverter harvests optimum power by performing maximum power. Simplicity in system design, simplified stock management, and added safety are other factors introduced with the microinverter solution.

Advantages:

- No need to calculate string lengths, thus systems are simpler to design
- Ability to use different makes/models of modules in one system, particularly when repairing or updating older systems
- Panel-level MPPT
- Panel-level monitoring
- Lower DC voltage, which increases safety (no need for ~600 V DC cabling requiring conduits
- Allows for increased design flexibility; modules can be oriented in different directions

- Increased yield from sites that suffer from overshadowing, as one shadowed module doesn't drag down a whole string

Disadvantages:

- Higher costs in terms of dollars per watt, currently up to double the cost compared to string inverters
- Increased complexity in installation
- Given their positioning in an installation, some microinverters may have issues in extreme heat
- Increased maintenance costs due to the presence of multiple units in an array

6.8 Hybrid Inverter with Batteries and Grid-Connected System

If utility power is reliable and well maintained in the area and energy storage is not a priority, a grid tie system can work. But if the utility power goes down, even if there is solar, the PV system will be off for the safety of the utility workers. The availability of solar power is dependent on the grid source and only in the daytime. Systems with a battery can supply power 100% of the time.

6.9 Inverter Topologies

6.9.1 Solar Topologies: String Inverter

Modular inverter approach:

- Solar panels are connected serially (strings)
- Multiple inverters are used in the plant (e.g., one per string)
- Efficiency is good because every string can be operated in its maximum power point and no string diodes are used
- Nominal power in the range of 1 to 10 kWp

6.9.2 Solar Topologies: Centralized

Centralized inverter:

- Several strings (serially coupled solar panels) are connected in parallel
- Only a single inverter is used for the solar plant
- Nominal power in the range of several MWp
- Special diodes are required to allow different string voltages (diodes cause losses)
- Not all solar panels can be driven in their MPP

6.9.3 Centralized Multistring

Centralized inverter with multiple DC to DC modules:

- Each string of solar panels is connected to its own DC to DC converter.
- Only a single DC to AC inverter is used for the solar plant (cost efficient)
- Efficiency is good because every string can be operated in its maximum power point and no string diodes are used
- Nominal power of the inverter in the range of 5 to 100 kWp

6.9.4 Solar Topologies: Microinverters

Integrated inverters:

- In each solar panel a separate inverter and MPP tracker are integrated
- No DC cabling, but extensive AC cabling is necessary
- Less efficient than string inverters
- Only economical for small systems
- Nominal power of the inverters is 50 to 400 kWp

7

Energy Storage for PV Applications

7.1 Introduction to Batteries

Electrochemical devices that can covert chemical energy into electrical energy by means of oxidation reduction technique are called cells, and a combination of two or more cells connected in series is a battery. When an appliance connects to an electric source, current starts flowing between its two terminals and provides power to the device. Batteries are essential for electronic devices and are extensively used for operating devices when no main power supply is available. In addition to general applications, batteries are used in nonconventional energy resources for storing energy so that it can be used whenever it is needed. Mobile phones, laptops, and computers, which have become a part of our daily life, cannot work without a battery. Batteries have been used for more than 200 years and are found almost everywhere in industrial and consumer products, such as surgical instruments, toll collection devices, metering devices, clocks, security alarms, and cameras.

Batteries can be classified into two main categories:

i. Primary batteries, which cannot be recharged once they have been discharged (i.e., these are disposable and need to be discarded). These are also called **dry batteries** as sometimes the electrolyte of the battery is absorbed in the separator. These are cheap and lightweight and are mostly used in portable electronic devices. Examples of primary batteries include lithium, alkaline, zinc-carbon, and zinc-chloride.

ii. Secondary batteries, which can be recharged several times as per requirements. These are called **storage batteries** as they are used for electrical energy storage purposes. These are more expensive compared to primary batteries. Such batteries are used in cell phones, laptops, computers, and solar PV systems. Examples of secondary batteries include lithium-ion, nickel-cadmium, and lead-acid. Recently it has become more important to design more compact batteries with a longer lifespan and greater operating capacity.

For small, high-power applications we use 1.5 V batteries. If we need a battery with a longer lifespan, then use a battery with a rating 3 to 3.6 V. In the 1960s silver oxide batteries were introduced and are still used for watches. In the 1970s the manganese battery with a rating of 1.5 V was replaced by the alkaline battery. Lithium-ion batteries, which were introduced in 1970s, have an even longer life span. Some lithium ion batteries can last as long as 10 years before needing to be replaced and are still extensively used in digital cameras, watches, or in computer clocks. For high-current applications lithium batteries can be used, but they are very expensive.

To conduct electric current, a material needs to have charged particles in it. Electrons are the charge carriers in solid conductors like copper, aluminum, iron, lead, etc. With liquids and gases, ions are the charge particles that carry electric current. Ions can have a positive or negative charge, for example, H^+ hydrogen ion, Cu^{++} copper ion, and OH^- hydroxyl ion. Distilled water is an example of poor conductor of electricity, as it has no ions, and salt water is a good conductor as it has ions to carry the electric current.

An electrolyte is a material that can be decomposed by electrolysis. Salt water, sulfuric acid, and copper sulfate are examples of electrolytes. Decomposition of a liquid by passing electric current through it is called electrolysis. If we have a beaker with a salt water solution with two electrodes of copper in it and connect it to a battery that has an electrode with a positive connection (called an anode) and a negative connection (called a cathode), current will flow through the solution, and it will be decomposed into hydrogen and oxygen and air bubbles will appear around the electrodes. If we use copper sulfate as an electrolyte in place of salt water, we will see that the cathode is gaining copper, whereas the anode is losing copper. Electrolysis is used to extract aluminum from its ore, metal electroplating, and copper refining.

Electroplating is the process of coating a thin layer of one metal upon another metal by using electrolysis. Some examples are silver plating of nickel alloys, tin plating of steel, and chromium plating of steel.

7.2 The Simple Cell

In a battery the fundamental unit that is used to generate electrical energy from chemical energy is called a cell. A cell comprises four components: anode, cathode, electrolyte, and separator. The electrode pair (anode and cathode) is usually made up of copper and zinc. Other electrode pairs are zinc-iron and zinc-lead. The potential measured between the electrodes may vary among different pairs of electrode materials. By comparing the EMF of each electrode to a standard electrode material, the EMF of pairs of electrodes can be determined. The standard electrode used to make the comparison is hydrogen. A list of materials according to their electric potential (shown in Table 7.1) makes the electrochemical series.

A simple cell has two types of faults—one due to polarization and other due to local action.

7.2.1 Polarization

When a simple cell is connected for long time, its current starts to decease rapidly because of the formation of a thin layer of hydrogen bubbles around the copper anode. This is

TABLE 7.1

Electrochemical Series Elements in Order of Electric Potential

Potassium	Lead
Sodium	Hydrogen
Aluminum	Copper
Zinc	Silver
Iron	Carbon

called polarization of a cell. The current decreases as the hydrogen is formed and prevents full contact with the electrode and electrolyte and thus increases cell resistance. To reduce the effect of polarization, we need to use a chemical depolarizing agent. For example, potassium dichromate can be used as a depolarizer, which removes the hydrogen bubbles when they form and so allows the delivery of steady current.

7.2.2 Local Action

If commercial zinc is put into a dilute solution of sulfuric acid, the zinc dissolves and hydrogen gas is released. The reason behind this is impurities in the zinc—for example if traces of iron are present in the zinc, then it forms small primary cells, which are short-circuited by the electrolyte solution so localized current flows and causes corrosion. This is known as local action in a cell. To prevent this event, the zinc surface should be rubbed with a small amount of mercury, which forms a protective layer over the surface of the electrode.

In a simple cell, the electromechanical series is used to predict the behavior of the cell when it is made up of two metals:

i. The metal with a higher position in the series acts as a negative electrode and vice versa. For example, in a cell zinc is the negative electrode, as it is in higher position than copper, and copper is the positive electrode.
ii. According to their position in series, if separation is higher between the two metals, then the EMF produced by the cell will be higher. So the electromechanical series represents the metals' order of reactivity and their compounds.
iii. Materials with a higher position in series will have a higher oxidation property than those lower in series.
iv. In cells with two metal electrodes, the one that is higher in series tends to dissolve in the electrolyte.

7.2.3 Corrosion

Gradual destruction of a metal in a damp atmosphere by means of cell action is called corrosion. In addition to the requirement of air and moisture for the formation of rust, an anode, cathode, and electrolyte are necessary for corrosion. So corrosion will take place if the metals that are widely spaced in electromechanical series are used in contact with each other in an electrolyte.

Corrosion badly affects the materials by weakening the structure and wasting material, which leads to a reduction in the life of the materials and the added expenses of replacement. To reduce the adverse effect of corrosion, metals should coated by plastic coats, grease, paints, enamels, or painting with tin or chromium. To prevent iron from corrosion, make it galvanized (i.e., zinc plating).

7.2.4 EMF and Internal Resistance of a Cell

Electromotive force (EMF), denoted as E, is the potential difference between two terminals of a cell when there is no load. EMF is measured by a voltmeter connected in parallel with the cell. The voltmeter should have high resistance so that no current or at least a very negligible amount of current passes through it so that the cell can be in a no-load condition. For example, if the voltmeter resistance is 1MΩ and the cell has resistance of 1Ω, then the total resistance of the circuit will be nearby 1MΩ and the cell will remain at no load.

When a load is connected at the cell terminals, the voltage measured will be reduced, as there is an increase in the internal resistance of the cell that opposes the current flow, and this resistance acts in series with other resistances in the circuit.

For example, let us take a cell with EMF E volts, internal resistance R Ω, and current I amperes.

The potential difference at no-load condition is V.

Now, $V = E$

When load R is connected to the cell terminals, current I starts to flow, causing the voltage drop: $I*R$

So, now the potential difference at the cell terminal is less than the total EMF and is given by:

$$V = E - I*R$$

A battery with an EMF of 12 V and internal resistance 0.01 Ω delivers a current of 100 A; the terminal potential difference:

$$\begin{aligned} V &= 12 - (100)(0.01) \\ &= 12 - 1 = 11 \text{ V} \end{aligned}$$

The internal resistance can be calculated from:

$$R = \frac{E - V}{I}$$

7.2.5 Battery

Two or more cells connected in series or in parallel forms a battery.

i. Cells connected in series:
   Total EMF of battery = sum of all series-connected cells' EMFs
   Total internal resistance of battery = sum of all series-connected cells' internal resistances
ii. Cells connected in parallel:
   If each cell of the battery has an equal EMF and internal resistance:
   Total EMF of battery = EMF of any one cell
   Total internal resistance of n cells $= \frac{1}{n} \times$ internal resistance of one cell

7.2.6 Lead-Acid Cell

The most common storage cell is a lead-acid cell.

A lead-acid cell has the following parts:

i. Hardened rubber, glass, ebonite, or plastic container
ii. Lead oxide (PbO_2) anode
iii. Spongy lead cathode
iv. Glass, wood, or celluloid separator
v. Sulfuric acid and distilled water solution as an electrolyte

To measure the specific gravity of the lead-acid cell, a hydrometer is used, and its value varies depending upon charged and discharged conditions. It is about 1.26 V when the cell is in a fully charged condition and 1.19 V when the cell is completely discharged. The potential difference of a lead-acid cell is 2.12 V.

7.2.7 Discharging

A cell is discharged when supplying a load.

During discharge:

i. The anode plate lead oxide and cathode plate spongy lead are converted into lead sulfate.
ii. The oxygen of the lead oxide combines with hydrogen in electrolyte and forms water.

So the relative density of the electrolyte falls and it becomes weakened. When fully discharged, the potential difference of a lead-acid cell is 1.8 V.

7.2.8 Charging

To charge a cell, it needs to connect to a DC supply by connecting the positive terminal of the cell to the positive terminal of the supply, and the negative terminal of the cell to the negative terminal of the supply. The charging current flows in an opposite direction to that of the discharging current, and so the chemical reaction is also reversed.

During charging:

i. Lead sulfate will be converted back into lead oxide on the anode plate and into spongy lead on the cathode plate.
ii. Oxygen released from the electrolyte combines with the lead anode plate so the water content reduces in terms of electrolytes and the relative density also improves.

The anode plate turns dark brown in color when fully charged and light brown during discharging. The cathode plate turns grey when fully charged and light grey when discharged.

The merits of lead-acid batteries include the following:

- It is easy to manufacture and economic.
- Its self-discharging rate is lowest among all other batteries.
- It can discharge at a very high rate.

The disadvantages of lead-acid batteries include:

- Short lifespan.
- Low energy density.
- It cannot be stored in a discharged condition.
- Due to the presence of lead, which is toxic in nature, the lead-acid battery is not environment friendly.

7.2.9 Nickel Cadmium (Ni-Cd) and Nickel Metal Cells

In addition to the lead-acid battery, other types of batteries have been developed that have a lower cost per unit energy density. Among them nickel-cadmium, which is prepared from nickel hydroxide (NiOH), has wide acceptance. In nickel-cadmium and nickel metal cells, the anode plate is made up of nickel hydroxide enclosed in finely perforated steel tubes; the addition of nickel and graphite reduces its resistance. The steel tubes are assembled into nickel steel plates. The cathode plate is made up of iron oxide of the nickel cell or nickel metal cell, and the resistance is reduced by adding a little mercuric oxide. In a nickel-cadmium cell, the cathode plate is made up of cadmium. The electrolyte used in both types is a potassium hydroxide solution. This does not undergo any chemical reactions, so very little is used. The electrodes are insulated from each other by insulating rods assembled in steel containers. These containers are insulated from each other by placing them in a nonmetallic crate. An alkaline cell has an average potential difference of about 1.2 V.

The merits of nickel cadmium cells over nickel iron and lead-acid cells include:

- Robust construction
- Has a longer lifespan
- Lighter in weight for a given capacity
- It can withstand heavy charging and discharging current without damage
- It will not damage if left indefinitely in any state of charging or discharging
- It does not self-discharge

The disadvantages of nickel cadmium and nickel metal cells over lead-acid cells include:

- More costly
- It needs more cells to connect for a given EMF
- Internal resistance is higher
- It has to be kept in the sealed position
- It has lower efficiency

Nickel cells are often used in heavy work conditions, like in very high temperature areas, areas with vibration, areas with long duty periods, and areas with high discharging currents, for example, marine and traction work, lighting system in railway carriages, and portable radios used in the military for diesel or petrol engine start.

7.3 Battery Storage System

Batteries are also used as an energy storage system in renewable energy resources. These are a secondary type of batteries, or rechargeable batteries. In PV systems the batteries are used to store energy during the day and are used as a supply whenever needed in night, low light, or cloudy weather conditions. Batteries are also used in a PV system so that it can operate near its maximum power condition and supply stable voltage and surge current to electrical loads and inverters. To protect the battery from overcharge or discharge, a battery charge controller is used.

In a standalone solar system, the power generated by the solar panels are stored by using lead-acid batteries, as it has many advantages over other types of batteries and so is a popular choice. A battery is made up of a number of cells. For a 12 V battery, six lead-acid cells needs to connect, as each cell has a voltage rating of about 2 V.

In order to properly select the batteries to be used in a standalone PV system, the designer should have good knowledge of design specifications, performance characteristics, and operational requirements. The following information is given as a review of basic battery specifications and characteristics commonly used in the design and application of PV systems.

7.4 Functions Performed by Storage Batteries in a PV System

- **Storage of energy**

  Batteries are used in a PV system to store the energy produced by PV array during the day and to supply energy to loads in night time, winter, less light conditions, or in cloudy weather.

- **Voltage stabilization**

  It is used to supply loads at a stable voltage by suppressing the voltage fluctuations and so protecting the loads from damage.

- **Supply of surge current**

  Batteries are used to supply a high starting current to electrical loads like motors and inductive loads. Performance of a PV system with battery storage depends on system parameters and the design of the battery as per load requirements; otherwise, it will fail to work.

- **Capacity**

  The capacity of a battery normally is measured in ampere-hour (Ah). It is defined as the maximum Ah a fully charged battery can deliver under certain specified conditions.

7.4.1 Efficiency

Energy efficiency of a battery =

Energy in watt hours (Wh) discharged/Energy in Wh required for complete recharge

The charge efficiency, or Ah efficiency, = Ah discharged/Ah required for complete recharge

7.5 Types of Batteries

7.5.1 Lead-Acid Batteries

The most commonly used energy storage device for a PV system is the lead-acid battery. It is available in 6 V and 12 V ratings and is often in a tough plastic container.

Specifications of lead-acid batteries:

Specific energy: 25–35 Wh/kg

Lifetime: 25–750 cycles

Advantages: low cost, high efficiency, simple operation

Disadvantages: relatively short lifetime

Lead-acid batteries come in the following types:

1. **Flooded cell battery**
   This type of battery is extensively used in renewable energy resources. There are two types of flooded cell batteries: one is flat and the other tubular. In this type of battery electrodes are completely submerged in the electrolyte, and during its fully charged state, oxygen and hydrogen are produced from water due to a chemical reaction at the anode and cathode, and this comes out of the battery through vents. That is why periodically adding water to the battery is necessary.
2. **Gelled batteries**
   Silicon oxide is added to the electrolyte, and it forms a warm liquid. This is then added to the battery, and it becomes a gel after cooling. Here the oxygen and hydrogen that are produced during charging are transported between the anode and cathode via the voids and cracks in the gelled electrolyte.
3. **Absorbed Gas Mat [AGM] batteries**
   In AGM batteries mats made up of glass are sandwiched between anode and cathode plates. The electrolyte is absorbed by the glass plates. Oxygen from the anode plate recombines with hydrogen at the cathode plate by moving through the electrolyte in glass mats. Both gel and AGM batteries need controlled charging. To reduce water loss and gassing, electrodes are made up of lead calcium. Voltage and current should be controlled below the C/20 rate.
4. **Captive electrolyte lead-acid batteries**
   In this type of battery, the electrolyte is in an immobilized form. This type of maintenance-free, sealed lead-acid battery is also called captive electrolyte or valve-regulated lead-acid (VRLA) batteries. A sealed lead-acid battery is basically a combination of gelled electrolyte and absorbed glass mat batteries. In this type of battery, the electrolyte cannot replenish, so they are intolerant of overcharge.

These batteries have fewer electrolyte freeing problems than the flooded type. Here also hydrogen and oxygen gases are formed from water during charging due to the chemical reaction at the anode and cathode plates, but these gases again recombine to form water. Thus, there is no need to add water like in the flooded type. One more advantage of the captive design over the flooded type is that is it less prone to freezing.

According to different designs, charge regulation voltage varies for captive type batteries, so it is necessary to follow the recommendations given by manufacturers. If no information is available, then use a charge regulation voltage of not more than 14.2 V for a 12 V battery at 25°C.

This type of lead-acid battery is preferred for PV systems due to the following points:

- Easy to transport
- It requires less maintenance so it can be used in remote application areas
- Addition of extra water not needed.

Problems:

- Excessive overcharge and loss of electrolyte (mainly in warm climates)
- This battery is very sensitive to temperature, charging process, and voltage regulation

To reduce gassing, lead calcium batteries are used in captive electrolyte systems, or you can also use lead antimony in hybrid grids.

- **Lead-Calcium Batteries**

  One more type of lead-acid battery is the lead calcium battery, where the primary alloying element used is calcium (Ca) with lead in plate grids. It has following advantages and disadvantages:

  Advantages:

  - It has greater mechanical strength than the pure lead grid
  - Reduced gassing, so water loss is reduced
  - Less maintenance
  - Self-discharging rate is low

  Disadvantages:

  - Battery lifespan decreases when operating at high temperature and if discharged repeatedly to greater than 25% of discharge depth
  - Acceptance of charge is poor after deep discharge

7.5.2 Nickel-Cadmium (Ni-Cd) Batteries

In this type of battery, the anode is made up of cadmium and the cathode is made up of nickel hydroxide. As separators, nylon is used, and potassium hydroxide is used as the electrolyte, and the whole arrangement is placed in a stainless-steel casing. Its life span and temperature tolerance are better as compared to the lead-acid battery. Today, due to environmental regulatory rules cadmium has been replaced by metal hydrides. If the battery is kept idle for long, then its capacity is reduced due to the memory effect.

In a battery, the memory effect means the process of remembering its depth of discharge in the past. For example, if it is discharged repeatedly to 25%, it will remember this and when discharged to greater than 25%, the cell voltage will drop. To recover the battery's full capacity, it should be reconditioned by fully discharging it and then again fully charging it.

7.5.3 Nickel-Metal Hydride (Ni-MH) Batteries

This is a more advanced version of Ni-Cd batteries that has a high energy density. For the anode, metal hydride is used instead of cadmium. It has high peak power and less of a memory effect than Ni-Cd batteries, but it is expensive. If overcharged, the battery can get damaged easily.

Specifications of nickel-metal hydride batteries:

Specific energy: 65–75 Wh/kg
Lifetime: 700 cycles
Advantages: high specific energy, good deep discharge, environmentally friendly
Disadvantages: costlier, self-discharging rate is higher

7.5.4 Lithium Ion Batteries

Lithium ion batteries have an energy density three times that of lead-acid batteries. The voltage rating is 3.5 V, which can be achieved by connecting a few cells in series. In this type of battery, the electrodes used are thicker so as to compensate for a passivation film generated by the reaction of lithium electrodes and electrolytes during every time it charges and discharges. Because of the thicker electrode, lithium ion batteries are costlier than Ni-Cd batteries. Overcharging can damage the battery.

7.5.5 Lithium Polymer Batteries

This type of battery has an electrolyte and separator, both of which are made up of a solid polymer. The reaction between the lithium electrodes and electrolyte is less now.

Specifications of lithium batteries:

Specific energy: 100–150 Wh/kg
Lifetime: 1,000 cycles
Advantages: Specific energy is higher, longer life span
Disadvantages: Costlier, not as safe

7.5.6 Calcium-Calcium

A recent development in batteries is the calcium-calcium. In this type both anode and cathode electrodes are made up of calcium instead of antimony. The obvious benefits are 80% less fluid loss than antimony batteries, and its self-discharging rate is lower than antimony batteries, so it can remain unused for a longer time without losing much of its charge.

They have a disadvantage, in that they demand more during charging if they are overdischarged. However, they avoid gassing because the bubbles that are produced during charging cause the acid to become thoroughly mixed. If the bubbles are not there, the acid can stratify at different densities and weights, which is a common phenomenon. We want an even acid weight of 1.28, but get an acid weight of 1.35 at the bottom and 1.17 at the top—this can affect the battery by sulfating it and also grid corrosion even if the battery is fully charged.

7.5.7 Battery Parameters

Important battery parameters are:

Battery capacity: The battery or cell capacity is the maximum capacity of charge stored in it, and it is measured in ampere-hours (Ah). With Ah, battery voltage is also specified, and so the total energy storage capacity can be obtained as follows: Ah × V = watt-hour. Rechargeable batteries are available at different ratings—3 Ah,

7.5 Ah, 12 Ah, 20 Ah, 50 Ah, 150 Ah, and so on up to 1200 Ah. As the Ah increases, so does the need for active material to store the charge, and the size of the battery also increases. If a fully charged 50 Ah battery is rated at 10 hours, then it will discharge at a steady current rating of 5 A for 10 hours, but if the load current is increased to 10 A, the battery will discharge in 3 to 4 hours, as the increase in discharging current will reduce the effective capacity of the battery.

Battery voltage: The terminal potential difference of a battery during operating conditions is known as the nominal voltage, or working voltage, of the battery. Batteries will different voltage ratings are available, which are specified by the manufacturers—for example, 3 V, 5 V, 6 V, 12 V, 24 V, etc.

Depth of discharge (DoD): This is the percentage withdrawal of energy from a battery of its full capacity. State of charge is full charge (i.e., 100% minus DoD). For example, if a state of charge of a battery is given as 75%, it means that it DoD is 25%. As we continue to discharge the battery, its DoD increases. If we discharge a large battery adversely, it may affect its life cycle. If DoD is restricted to a limit where the life cycle of a battery can be improved, then the user can get more life out of the battery.

Battery life cycle: The number of charge and discharge cycles that a battery can complete until its nominal capacity falls below 80% of the initial rated capacity is the battery life cycle. The value ranges from 500 to 1,500 cycles. As a battery ages, its capacity reduces, but the battery can still work for a number of cycles with lesser capacity.

Discharge or charge rate or C-rate: The discharge or charge rating of a battery can be represented as the C-rating. To find the C-rating divide the battery capacity by the time required in hours for complete charge or discharge of the battery. Let the number of hours required to fully charge or discharge be denoted by X and capacity by C. Then the C-rating of the battery is C/X. If $X = 10$ hours, then the C-rating is C/10 or 0.1C. So the higher the charge or discharge time, the lower the C-rating.

The charge or discharge current can be calculated from the C-rating if we know the number of hours it is operating and divide its Ah capacity by this number of hours. For example, if 10 Ah is the capacity and the C-rating is 0.1C, this means it takes 10 hours to fully charge or discharge, so the charging or discharging current is 10/10 = 1A.

Self-discharging of a battery: The electrical capacity of the battery is lost when the battery is not in use and is sitting on the shelf. This is due to internal electrochemical reactions in the cell, and it is equivalent to applying a small load across the battery terminals. The self-discharging of batteries should be as low as possible. As temperature rises, the self-discharging increases in battery, so it is recommended that batteries be kept at a lower temperature. Lithium ion batteries lose 8% capacity in the first month. If stored at 30°C a lead-acid battery will lose 50% of its stored charge in three to four months, whereas Ni-Cd will lose 50% in just six weeks.

7.6 Selection of Batteries

Batteries are mostly chosen based on economy (i.e., by lowest price). But we don't want to select an improper and inadequate battery for our systems that reduces its reliability and life. There are many factors by which the battery for the PV system can be chosen, such as the following:

i. Life cycle of battery
ii. Performance at extreme temperatures
iii. Self-discharging rate
iv. Maximum voltage and current drain capacity in ampere-hours
v. Watt-hour per volume and weight of battery
vi. Cost per watt-hour
vii. Maintenance requirements

Lead-acid automotive batteries are widely used in PV systems. Today, they have had pasted or tubular plates and grids with low or high antimony or pure lead or calcium alloys added to them. The gelled type battery or recombination type can also be used. Ni-Cd and a few other alkaline-based batteries are also comparable with the lead-acid battery in terms of longer service life and maintenance-free operation in PV system applications.

7.6.1 Batteries Used in PV Applications

As mentioned, the most common battery in a PV system is the lead-acid type. Alkaline batteries are also suitable, but at present Ni-Cd is the only type of battery that has the desired performance and life cycle cost for PV applications. The requirement of maintenance-free batteries is increasing day by day. Gelled electrolyte batteries are automatic, stationary, and traction and maintenance free, and they are also used for PV applications. For daily applications like street lighting, use a PV system with a shallow DoD. SLI (i.e., starting, lighting, and ignition) batteries are used though they have smaller life span of two to three years and poor cycling ability.

Stationary batteries have deep discharge capabilities and are used in telecommunication, emergency lights, UPSs, and in navigational systems. In electric vehicles, rechargeable motive power batteries or traction batteries are used with a PV array.

Research is ongoing into sealed lead-acid batteries for PV applications, and recently a tubular type battery was developed with acid immobilization by silica gel and antimony-free lead grids with thicker electrode plates. Lead plates strengthened by calcium and a small amount of antimony make it cheaper and good to use for remote applications, but antimony also increases battery self-discharging. So according to some PV companies, batteries with low antimony are the best choice.

7.7 Installation, Operation, and Maintenance of Batteries

The Commission of the European Community (CEC) started concerted efforts toward the investigation of "Battery charge control and management in PV systems" in 1987. The objective was to identify the problems of operation in PV plants. The problems are:

i. Improper maintenance and operations
ii. Improper sizing of batteries
iii. No routine test or documentation regarding the condition of the system

In various PV plants it has been seen that batteries are damaged due to aging, deep discharge, and failure of the cell casing structure. Whereas in some other cases, overcharging and a large number of operating cycles are observed for five years. Also explosions due to hydrogen building up in cells have been observed. In most of the cases, the CEC found that operation and maintenance of the plant is not proper and is not documented. Also no routine test of voltage, temperature, or specific gravity has been done. In some cases, if proper maintenance and operation were adopted, this problem can be detected and avoided. Peripheral components for processing need to be installed, battery parameters need to be acquired, and battery management measures need to be adopted. Batteries in PV systems are subjected to two cycles as follows:

a. Daily cycle: This is the varying profile and amplitude of PV energy supplied and electrical energy given to the load.
b. Seasonal cycle: This depends on variations in insolation during the year, which can cause stress and aging in the battery.

The most common problems are as follows:

1. Due to overcharging-based corrosion of anode plates, excessive gassing takes place in the battery, which results in a loosening of active material. This loosened material deposits at the bottom of the battery as sediment. Overcharging leads to a high temperature rise in the cell, which can destroy it.
2. The battery starts gradually running down if it is undercharged consistently, and this can be identified by the decreasing specific gravity and electrodes becoming lighter in color. This also causes white lead sulfate sedimentation in the battery. Due to the strain of lead sulfate on the electrodes, it takes more space than the original active materials and results in buckling.
3. When nonconducting materials are present in the battery, it creates a layer between the battery terminals and connectors, which increases its resistance in the path of a large current through the load. Corroded terminals will not interfere with charging or discharging of batteries at low currents.
4. Short circuits may take place in the batteries due to excessive sedimentation and separator breakdown, which is called "treeing," as a tree-like structure forms from the from anode to the cathode. Treeing is caused by the presence of certain materials in the grid, like cadmium, or it can be due to a phenomenon called "mossing" in which sediments are brought by gas to the surface of the electrolyte and form bridges over the separator.
5. If a battery is operated for several days at partial state of charge (SOC) without equalization or is left in an unused condition for long time as fully or partially discharged, then a large amount of lead sulfate is deposited and forms large crystals instead of normal tiny ones on the plates, which is called sulfation. This may also take place due to temperature variations. These crystals will increase the internal resistance of the cell and reduce discharge and increase the charge voltage.
6. After the battery is fully charged, an increase in plate potential beyond the cutoff level causes decomposition of water into hydrogen and oxygen so there is water loss. The quantity of this gas formation depends on the excess current that is not absorbed by the battery.

Batteries made for PV applications should be placed in a separate room with good ventilation and moderate temperature to avoid accidents and should be placed on wooden or plastic planks. They also can be placed on a steel step with acid-resistant paint. A sealed maintenance-free battery can be placed with normal ventilation in the usual area or on iron racks. But we should follow the installation guidelines given by the manufacturers for different types of batteries, including battery room designs.

For installation of lead-acid and Ni-Cd type batteries for PV applications, all guidelines on installation and maintenance are available from the Institute of Electrical and Electronics Engineers (IEEE) in European standard form, with safety precautions and design criteria.

7.8 System Design and Selection Criteria for Batteries

When designing a battery system selecting a battery the criteria include many decisions and tradeoffs. Which type of battery should be used in PV applications depends on various factors. There is no specific battery for all PV applications, so the decision needs to be made based on common sense and a careful review of the different battery specifications we have discussed so far and considering the particular type of application.

Depending on the physical properties, selecting a battery may be easy, but other decisions may be more difficult and tradeoffs between desirable and undesirable battery specifications need to be considered. With good knowledge, experience, and familiarization, designers should make a difference in various types of batteries and their application areas.

7.9 Comparison of Various Batteries

It can be difficult to choose batteries for commercial as well as domestic energy storage, as we need to consider various options to find the ideal one for our application. The various factors that we should keep in mind are different Ah ratings, operating voltages, battery storage capacity, sizes, weights, brands, and battery chemistries.

The most important factor is to confirm that the chosen battery can supply electricity whenever needed in the site as a regular supply or backup supply at time intervals and also at the best price.

Different parameters for comparison are:

- **Life cycle or number of cycles of the battery:** If the life cycle of a battery lasts only one cycle, that means each kWh that we are extracting from the battery will have a higher cost. The life cycle gives the number of cycles needed by the battery for its true battery storage capacity to reach 80% of nominal capacity at full charge condition.
- **Depth of discharge (DoD):** This is the percentage of how much battery has been used. For example, if a battery with a nominal capacity of 100 kWh has 50 kWh energy stored, that means its current DoD is 50%, and when 20 kWh is left, this

means its DoD is 80%. Batteries cannot drain off all the stored energy, as this will damage the components. So there should be a maximum specified DoD for the nominal cycle lifetime of the battery, and when DoD increases, the battery life cycle decreases.

- **Round-trip efficiency:** This is the percentage of how much energy can be extracted from a battery from the amount of energy it took to store it. For example, if a battery charged to 1 kWh and the amount of energy we can extract from it is 800 Wh, that means it has round-trip efficiency of 80%. The loss is due to various factors like heat or other inefficiencies in the system. The actual round-trip efficiency of battery depends on its DoD; however, it is just taken as the average.

Rs./kWh is nowhere sufficient to fairly contrast different battery models and types. So we need a more appropriate matric COS (i.e., cost of stored energy), which is a closer approximation of true battery cost.

7.9.1 The Effect of DoD on COS

The lifetime of a battery depends strongly on the DoD for all those cycles—in general a higher average DoD per cycle can significantly decrease the lifetime of the battery. The expected battery cycle lifetime and corresponding COS for the lead-acid battery option is given as a function of DoD. The trend outlines the fact that there is an optimal DoD operating point at which the COS—and thus the true cost of the battery—is minimized. This should dictate the way a system is operated (i.e., shallow or deep cycling) and sized. The DoD for this particular battery should be maintained at around 30% to prolong its lifetime energy production or, conversely, the system should be designed such that the average DoD is kept as close as possible to 30%.

7.9.2 Safe Disposal of Batteries

Battery disposal is a very important topic to discuss as it is related to environmental and human health. It has become a topical subject in the UK, as every year over 300 million batteries are deposited into landfill sites, weighing 2,000 tons. Some batteries have such substances that can harm humans, wildlife, and the environment or carry a fire risk, and some can be recycled for the metal content.

Discarded batteries are a source of toxic materials like cadmium, mercury, and lead, which can pollute soil and ground-level water if not disposed of correctly. They have many adverse effects on humans and wildlife. For example, prolonged exposure to cadmium can cause cancer and disease of the lungs and liver. Lead-acid batteries have sulfuric acid, which can burn the skin or cause irritation. Mercury can damage the human brain, kidney, spinal cord, and liver. So it is very important to correctly dispose of batteries.

Factors affecting the battery life and performance in PV systems:

- Manufacturing faults: A reliable and trustworthy manufacturer should be chosen.
- User abuse: Provide documentation, proper supervision, or training for commissioning
- Accidents: Be careful to avoid dropping the spanner across the battery terminals
- Improper design of PV system: Avoid sulfation (in the case of lead-acid batteries), stratification, and freezing

Possible remedies:

- Full charging must be carried out (at least periodically)
- Should be restricted to specified DoD
- Provide as rapid a recharge as possible after a deep discharge This can be achieved by:
    - Choosing an appropriate battery
    - Sizing it properly
    - Providing an appropriate method of charge control

7.10 Super Capacitors

The key point of an ultimate energy storage device is that it should have high energy density, which can be released rapidly. Batteries with a high energy level have been introduced for single-use or in rechargeable systems, but they can release rapidly, and it takes minutes to hours to discharge, not seconds. Standard capacitors with high power can discharge rapidly, but they do not have high energy density. A super capacitor known as the Cooper Bussmann PowerStor is lead-free, environmentally friendly, compliant with the Restriction of Hazardous Substances (RoHS) Directive, and does not have disposal issues. The European Waste Electrical and Electronic Equipment Directive (WEEE) directive needs companies to recycle and reuse lithium batteries used for back-up, which is also followed by Japan and China. Super capacitors have both high energy density, which is nearly 100 times that of electrolyte capacitors, and high power—10 to 100 times the power of a battery.

Super capacitors are having following advantages over a battery:

- Do not need replacing as it has a longer life.
- It is charged rapidly, so it can be used as a back-up in seconds.
- It can provide higher peak power as it has low ESR.
- It does not have disposal issues at the end of its life, as it does not have any heavy metals so it is environmentally friendly.
- It has higher energy density than a battery, and customized packaging is available.

Batteries have far more energy than required and the following drawbacks:

- Longer charging time.
- Cannot hold fully charged condition.
- Due to memory effect, it degrades with shallow discharge.
- Have a shorter life, so replacement and maintenance are needed.

Rechargeable batteries have memory effect, life cycle, replacement, maintenance, and cost issues compared to super capacitors. Super capacitors are extensively used in toys, solar traffic lighting systems, transmitters, and remote monitoring systems. Following are two examples of the use of PowerStor super capacitors in place of rechargeable batteries:

1. Restaurant pagers. The pagers run for two hours while a patron is waiting for a table. When returned to the host/hostess, the pager requires only 10 seconds to be

ready for the next customer. In contrast, of a Ni-Cd battery is used, then because of its memory effect, it will perform poorly with shallow DoD. So by using super capacitors the life of the product increases, and it does not require any replacement, so there are no added cost issues and they are used for main power.

2. A data storage system used in network server called RAID that provides high-integrity, faster data storage can use super capacitors instead of expensive lithium batteries in case of a main failures. Super capacitors provide a low-cost green alternative that doesn't require maintenance or replacements and can be recharged in seconds after a main failure. So, here use of super capacitors will provide a maintenance-free product with a longer life, increasing reliability, and at less expense as it eliminates the requirement of maintenance.

7.11 Fuel Cells

Fuel cells are similar to a battery in that it is also an electrochemical energy conversion device, but there is a major difference in that fuel cells are designed for continuous replenishment of the reactants consumed, which means it needs an external source such as oxygen and fuel to produce electricity, unlike a battery, which has limited energy storage capacity. Also the electrodes of fuel cells are catalytic, which means it will never discharge permanently and are stable, whereas in batteries, electrodes are changed during charging or discharging reactions.

In fuel cells, generally hydrogen is used on the anode side and oxygen on the cathode side, and reactants flow in and reaction products flow out. Here, continuous operation is possible if the flows can be maintained.

Fuel cells are becoming very popular as opposed to other modern fuels such as methane or natural gas, which produce carbon dioxide as by-product, because of its higher efficiency and its by-product, which is water vapor when using pure hydrogen as a fuel. Fuel cells with hydrogen as a fuel are lighter in weight, compact in size, and do not have any moving parts.

The only disadvantage is that fuel cells are an expensive alternative to internal combustion engines. However, research is continuing to make fuel cells available at reasonable prices in few years. Fuel cells are useful in remote application areas such as weather stations, spacecraft, and military applications.

8

Mounting Structure

8.1 Introduction

The PV arrays are the major components of the photovoltaic systems. These arrays experience loadings from wind and ice, in addition to their own load. Also the location of the PV array affects the performance of the PV systems. The mounting structure of the PV system should have the optimal structural platforms, lightweight array frames, innovative deployment systems, and higher-efficiency photovoltaic components. There are many ways to install the mounting structures, such as ground, roof, or integrated with the building itself. In this chapter, different mounting structures are discussed. Also the impact of wind loading on the PV array is assessed.

8.2 Assessment of Wind Loading on PV Array

To reduce the heating effect of the solar PV module, a certain minimum gap has to be provided from the ground level or any other obstacle, specifically when the module is placed at a location with high temperatures. If this gap is not provided, the efficiency and energy generation of the solar PV module will be reduced. The effect of temperature may also be seen and analyzed through the I–V curve of the solar module as shown in Figure 8.1.

From this figure, it can be observed that the power output reduces with the increase in temperature. Therefore, the solar PV array should be designed in such a way that the air circulates properly. But on the other hand, this also leads to the problem of uplifting the array due to the high speed of wind, depending on geographic location.

This wind loading is entirely dependent upon on the location of the site. If the site of installation of the solar PV is near a coastal area, the wind speed is high compared to the site of installations at the flat lands. This should be taken care of when designing the module mounting structure. The module site of installation (i.e., on a rooftop, or ground mounted, or a tilted surface, etc.) at a given site should also be considered. Accordingly, the structural engineer calculates the specific wind loading, considering the material used for the mounting structure and local climate conditions.

The PV array structure should have the following essential considerations:

1. The geographical location of the site of installation
2. The wind zone map or loading details of the site
3. The type/place of installation (i.e., rooftop, ground mounted, tilted roof, building walls, etc.)

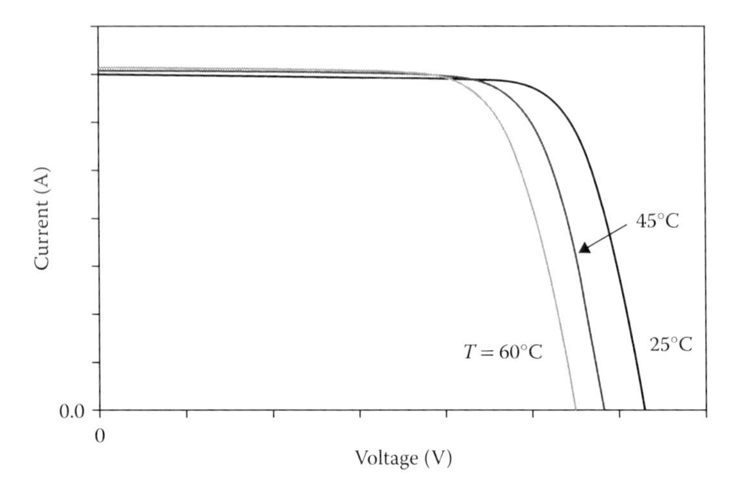

FIGURE 8.1
I–V characteristics of PV panel.

4. Follows the relevant standards made by government agencies, for example, India follows the 15875.3 code of practice for designing loads (wind loads) for buildings and structure
5. The maximum number of PV modules
6. Should be vetted by a structural engineer

8.3 Types of Module Mounting Systems

The mounting structures are classified into three major groups:

1. Rooftop Module Mounting Structure
   a. Flat rooftop
   b. Tilt rooftop
      i. One-sided
      ii. Two-sided
   c. Building integrated
2. Ground Module Mounting Structure
   a. Fixed
   b. Tracking
      i. Seasonal tracking
      ii. Nonseasonal tracking
      iii. Single axis
      iv. Dual axis
3. Building-Integrated Photovoltaic (BIPV)

Rooftop Module Mounting Structure: Today, popularity of rooftop solar PV systems is growing rapidly because the space on rooftops of houses, office premises, parking lots, school/college buildings, factory or manufacturing units, etc., can be properly utilized for these purposes. This helps in the following ways:

- No extra space arrangement is required (within the existing infrastructure, the solar PV system can be installed).
- The unused space can be utilized for power generation.
- The current electricity bill paid to the utility can be reduced by adding up solar PV power.
- Fulfillment of government agencies obligations.

The module mounting structure can be designed based on the structure of the roof. The PV structure can be installed on any type of roof, including concrete, tin, fiber, tiles, polycarbonate, wood, slate, asbestos, metal, etc. An example of a PV array structure is presented in Figure 8.2.

Ground-Mounted PV System: The majority of home solar installations are rooftop-mounted panels. Roofs offer a practical and ready location to install solar cells on most occasions. But on some occasions the rooftop is just not suitable or optimal. In this situation the alternative is to have the solar PV system mounted on the ground. This type of system is more suitable for rural areas where lot of land is freely available. A ground-mounted structure is shown in Figure 8.3.

FIGURE 8.2
Flat rooftop PV system.

FIGURE 8.3
Ground mounted PV array.

The following are the advantages of ground-mounted structures:

- Typically, there is much more space on the ground than on the roof.
- Panels are easier to clean and maintain.
- More scope and lower cost to install a sun tracking system.
- Ability to operate a manual seasonal tilt adjusting system.

Disadvantages of ground-mounted solar systems include:

- More expensive to install due to cost of frame, foundation, and footings.
- Urban settings often do not have the available land space.

Building-Integrated Photovoltaics: A building-integrated photovoltaic system consists of integrating photovoltaic modules into the building structure, such as the roof or the façade. By simultaneously serving as building structure material and power generator, BIPV systems can provide savings in material and electricity costs and reduce the use of fossil fuels and greenhouse gas emissions. Figure 8.4 shows the BIPV technology employed at CIS Tower, Manchester, England.

FIGURE 8.4
The CIS Tower in Manchester, England.

8.4 PV Array Row Spacing

When designing a PV system that is tilted or ground-mounted, determining the appropriate spacing between each row is important. Further, it is also important to do it right the first time to avoid accidental shading from the modules that are ahead of each row. When designing a solar system, there is often the need to understand how long a shadow will be so that the row spacing between solar modules can be properly planned. Most locations for solar projects tend to get around 5 to 6 net sun-hours per day, so anything that obstructs that sunlight needs to be avoided at all costs. Shading just one corner of a module can cut production in half. This is mainly an issue on ground mounts and some flat roof mounts, where rows of solar panels need to be optimally spaced to best use the available space. With limited solar resources and steep penalties for failure, properly determining the correct shade spacing is a critical calculation in solar system design. The calculation of inter-row spacing can be understood with the help of Figures 8.5 and 8.6.

The first step in calculating the inter-row spacing for modules is to calculate the height difference from the back of the module to the surface. This can be accomplished with the following procedure.

Height difference = sin(tilt angle) × module width

In this case, a solar module has a width of 39.41 inches at a tilt angle of 15 degrees

Height difference = sin(15) × 39.41

Height difference = 10.2 rounded down to 10

In this case, a window during the winter for the worst-case scenario has been considered. From the chart it can be seen that we have highlighted this window and drawn a horizontal line out to the left of the chart to narrow in on the solar elevation angle at those times. Then an estimation at the 17-degree angle has been done using the following formula:

Module row spacing = height difference/tan(17)

Module row spacing = 10/tan(17)

Module row spacing = 32.7 rounded up to 33

The inter-row spacing between the trailing edge of the first row of the module and the leading edge of the next row needs to be 33.

In the next step, the azimuth angle will be calculated. Take a look again at Figure 8.6. You will see we have drawn two vertical reference lines down from each time reference. The difference between south going in either direction turns out to be 44 degrees, and we will use this in the following formula to determine the minimum module row spacing.

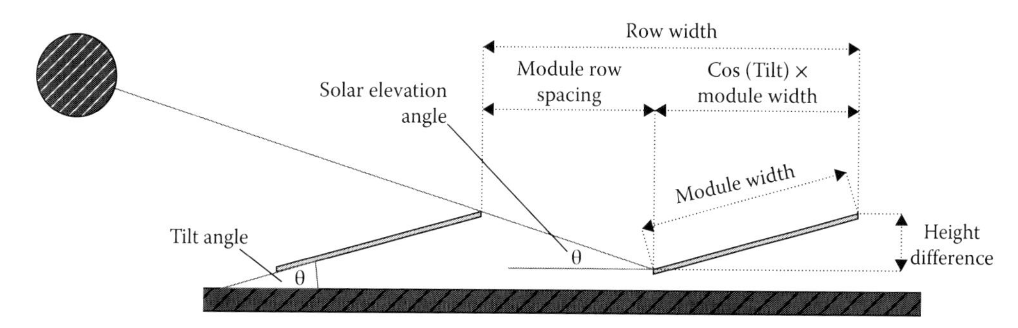

FIGURE 8.5
Inter-row spacing for PV modules.

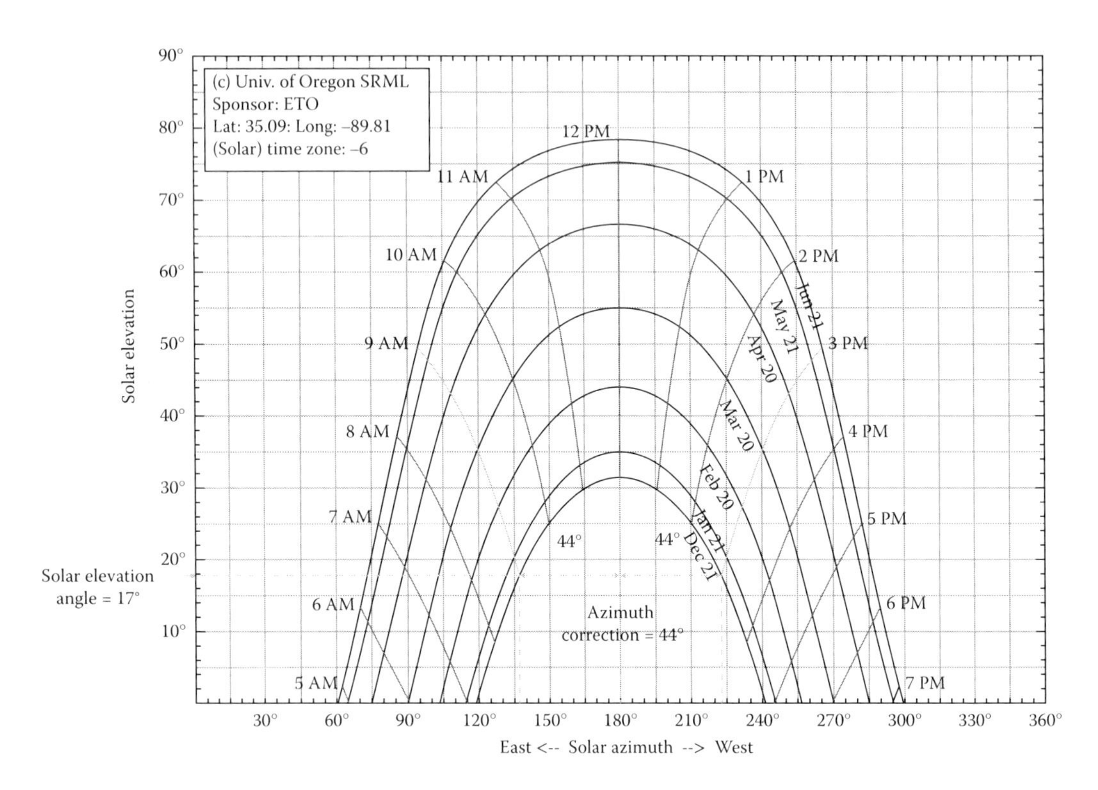

FIGURE 8.6
Chart for inter-row spacing for PV modules.

Minimum module row spacing = module row spacing × cos(azimuth correction angle)
Minimum module row spacing = 33 × cos(44)
Minimum module row spacing = 23.7 rounded up to 24

The following formula gives you the distance from the trailing edge of one row to the trailing edge of the subsequent row or your row width:

Row width = minimum module row spacing + cos (tilt angle) × module width
Row width = 24+ cos(15) × 39.41
Row width = 62

8.5 Standards for Mounting Structures

In a given country and for a particular type of installation, some of the standards specified by some or all of these bodies may apply, subject to the national regulations. Any desired standards documents can be purchased from the body directly or from a third party.

All operators, installation contractors, and maintenance people that work with PV mounting structures have to perform according to the standards.

Some of the major applicable standards are:

1. UL2703, standard for mounting systems, mounting devices, clamping and ground lugs for use with flat plate PV modules and panels.
2. ICC-AC428, acceptance criterion for modular framing systems used to support PV modules.
3. UL3703, outline of investigation for solar trackers.
4. UL3730, defines PV junction boxes intended to be attached to PV modules and panels.
5. UL9703, defines outline of investigation for distributed generation wiring harnesses.
6. UL6703, outline of investigation for connectors for use in photovoltaic systems.

9

Balance of Systems

9.1 Cabling

A cable interconnects solar panels and other electrical components of a PV system. These cables are designed to be UV resistant and weather resistant. Also, it can be used within a large temperature range and is generally laid outside.

One common factor of the photovoltaic system is the outdoor use, characterized by high temperature and high UV radiation. Single-core cables have a maximum permissible DC voltage of 1.8 kV U_{max}. The phase-to-ground DC voltage rating must be 1.5 kV DC and the temperature range must be −40°C to +90°C ambient, 120°C on the conductor, for 25 years' service life to protect against thermal aging. Ambient temperature and conductor temperature are derived from the Arrhenius law for aging of polymers. Aging of polymers doubles for every 10°C rise. Two types of cables are used in PV applications:

1. DC cables
2. AC cables

9.1.1 DC Cables

Individual modules are connected using cables to form the PV array. The module cables are connected into a string, which leads into the array junction box, and a main DC cable connects the array junction box to the inverter. The size of these DC cables depends on the current ratings of the module and further strings or arrays. In order to eliminate the risk of ground fault and short circuits, the positive and negative cables, each with double insulation, are laid separately.

For DC cables, it is preferable to use the copper type compared to the aluminum type because of the decreased transmission loss. But copper cables are costlier compared to aluminum. Hence, the designer needs to create the system in such a manner that the quantity of DC cable required is small.

9.1.2 AC Cables

Figure 9.1 illustrates a typical AC cable. Many AC modules use flexible, sunlight-rated, multiconductor cables with pin and socket type connectors. Typically, AC modules are electrically connected to their supply circuit in parallel with other AC modules. They are used from the inverter output up to the power evacuation.

FIGURE 9.1
A typical four-core AC power cable.

The calculation of the current-carrying capacity of the cable is:

$$C_{cc} > 1.25 \times I_{sc} \times N$$

C_{CC} = Current carrying capacity
I_{sc} = Short circuit current of each module
N = Number of strings

For AC cables, generally aluminum is used because of the various advantages like flexibility, less costly compared to copper, and high current-carrying capabilities.

Generally, cross-linked polyethylene (XLPE) cables are used as per Indian Standard IS: 7098 part 1. They are ideal for the transmission and distribution of power because of their high corrosion resistance and they have fewer atmospheric pollutants.

Advantages of XLPE over PVC cables:

- Longer life
- Small dielectric losses
- Higher current-carrying capacity
- Higher short-circuit rating of 250°C as opposed to 160°C for PVC cables
- 100 times more moisture resistance capacity compared to PVC cables
- Have higher emergency overload capacity (up to 60%)
- Low installation cost due to the light weight

9.2 Fuses

A fuse is an electrical device with a characteristic of low resistance that is placed in the circuit to protect the load. In solar PV systems, the fuses are placed inside the array junction boxes (AJBs) to protect the inverter from any overcurrent coming from the modules. The ratings of these fuses are decided based on the current rating of the entire string or array connected to AJB. In electrical engineering, IEC 60269 is a set of technical standards for low-voltage power fuses. The standard is in four volumes, which describes general requirements, fuses for industrial and commercial applications, fuses for residential applications, and fuses to protect semiconductor devices. The IEC standard unifies several national standards, thereby improving the interchangeability of fuses in international trade.

1. IEC 60269-1-low voltage fuses-part 1: General requirements
2. IEC 60269-2-low voltage fuses—part 2: Supplementary requirements for fuses for use by authorized persons (fuses mainly for industrial applications)
3. IEC 60269-3-low voltage fuses—part 3: Supplementary requirements for fuses for use by unskilled persons (fuses mainly for household applications)
4. IEC 60269-4-low voltage fuses—part 4: Supplementary requirements for fuse links for protection of semiconductor devices
5. IEC 60269-5-low voltage fuses—part 5: Guidance for the application if low voltage fuses
6. IEC 60269-6-low voltage fuses—part 6: Supplementary requirement for fuse links for protection of solar PV energy systems.

9.3 Lightning Arrestor

Lightning is a common cause of failure in PV and wind energy conversion systems. A damaging surge can occur from lightning that strikes a long distance from the system or even between clouds, but most lightning damage is preventable. Some cost-effective techniques are generally accepted by power system engineers based on decades of experience.

Lightning arrestors are designed to absorb voltage spikes caused by electrical storms and effectively allow the surge to bypass the power wiring and equipment. Surge protectors should be installed at both ends of any long wire run that is connected to any part of your system, including AC lines from an inverter. Arrestors are made for various voltages for both AC and DC. Be sure to use the appropriate arrestors for your particular application. Many system installers routinely use delta surge arrestors, which are inexpensive and offer some protection where the threat of lightning is moderate. PolyPhaser and Transtector arrestors are high-quality products for lightning-prone sites and larger installations. These durable units offer robust protection and compatibility with a wide variety of system voltages. Some devices also have indicators to display failure modes. A lightning arrester is shown in Figure 9.2.

9.4 Data Loggers

A data logger is an electronic device used to automatically monitor and record environment parameters over time, allowing conditions to be measured, documented, analyzed, and validated. The data logger contains a sensor to receive the information and a computer chip to store it. Then the information stored in the data logger is transferred to a computer for analysis. A data logger has the following advantages:

- Measurements are always taken at the right time. Unlike a human, the computer will not forget to take a reading or take a reading late or too early.
- Graphs and tables of results can be produced automatically by the data logging software.

FIGURE 9.2
Lightning arrestor.

The hardware is compact but offers high functionality and complies with or exceeds the following minimum requirements:

- 1 MB of user memory
- 2 ports of RS-485 serial interface with Modbus RTU
- 4 analog inputs of 4–20 mA
- 4 RTD inputs
- 16 digital inputs
- 4 digital outputs

The unit should be configurable with suitable programming software complying with IEC 61131-3. The user program should have a removable data card of any industry standard format and battery backup for data buffering.

The data logger hardware shall be housed in a polycarbonate enclosure of IP 22. The enclosure shall have provision for cable entry and have a power strip with six switched sockets of 5 A and a suitable ventilation fan with a removable filter module. The block diagram of the setup is shown in Figure 9.3.

9.5 Junction Box

An electrical junction box is a container for electrical connections. They are usually intended to conceal them from sight and deter tampering. A small metal or plastic junction

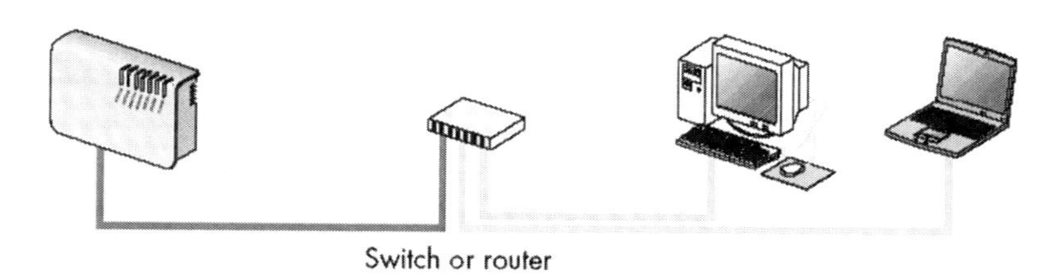

FIGURE 9.3
Data logger setup.

FIGURE 9.4
Junction box.

box may form part of an electrical conduit or thermoplastic-sheathed cable wiring system in a building. If designed for surface mounting, it is used mostly in ceilings, under floors, or concealed behind an access panel—particularly in domestic or commercial buildings. An appropriate type may be buried in the plaster of a wall or cast into concrete with only the cover visible. It sometimes includes built-in terminals for the joining of wires. A similar, usually wall-mounted, container called a pattress accommodates switches, circuits, and the associated connecting wiring. The term junction box may also be used for larger items, such as piece of street furniture.

Junction boxes form an integral part of a circuit protection system where circuit integrity has to be provided, as for emergency lighting. In such an installation, the fireproofing around the incoming or outgoing cables must also be extended to cover the junction box to prevent short circuits inside the box during an accidental fire.

These junction boxes are compliant with IP65/IP54, depending upon location and environment. This will be helpful in fulfilling requirements like dust free, vermin free, and weatherproof. The junction box could be made from mild steel (MS) sheet metal, aluminium die cast, or thermoplastics. Low-diameter DC cables with cable glands from strings will be fed into the input side and higher-diameter DC cables from the output side of junction boxes. The junction box may be AC or DC depending upon requirements. A typical junction box is shown in Figure 9.4.

10

Site Selection and Assessment

10.1 Introduction

With the rapid depletion of fossil fuel reserves, it is feared that the world will soon run out of its energy resources. This is a matter of concern for developing countries whose economy heavily leans on its use of energy. Under the circumstances, it is highly desirable that renewable energy resources be utilized with maximum conversion efficiency to cope with the ever-increasing energy demand. Furthermore, the global economic and political conditions that tend to make countries more dependent on their own energy resources have caused growing interest in the development and use of renewable energy-based technologies. In terms of environmental advantages, renewable energy sources generate electricity with insignificant contribution of carbon dioxide (CO_2) or other greenhouse gases (GHG) to the atmosphere, and they produce no pollutant discharge into water or soil, and hence power generation from renewable energy is important. Major types of renewable energy sources (RESs) include solar, wind, hydro, and biomass, all of which have huge potential to meet future energy challenges. Based on the availability of the resources in a particular area, a proper site is identified for the installation of an RES-based power-generating system. Further, the site assessment is done and proper sizing of equipment is carried out for an optimal and cost-effective power system.

10.2 Site Location

The first step to design and develop a renewable energy-based system at a particular site is to perform the feasibility analysis on the basis of availability of resources and their ability to generate efficient and reliable power. Therefore, the data regarding the potential of major available renewable energy sources at a proposed site is collected, and accordingly the site is selected for the proposed power systems. As a case study, the following example explains the importance of site location.

10.3 Site Assessment

To design the techno-economic viable system based on solar energy, the data on hourly global solar radiation are essential so that the photovoltaic systems could be financially

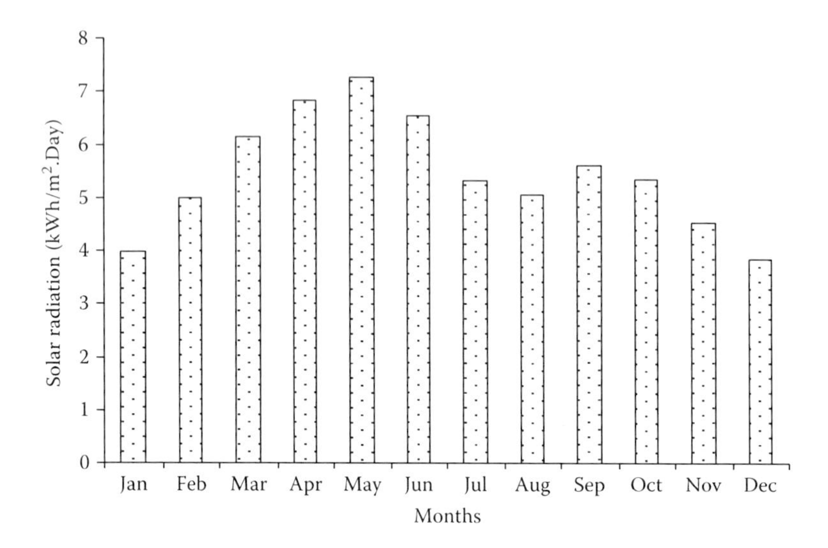

FIGURE 10.1
Monthly average daily global radiation on a horizontal surface.

attractive compared to other options for providing access to electricity. In addition, such data are equally important in the design of energy-efficient buildings. As the accurate measurement of solar radiation involves cost and man-power, this information is hardly available in many developing countries, including India, but such measurements are necessary to meet energy needs with this environmentally friendly and clean energy source. However, very few stations in India measure solar irradiance, and the common practice is to calculate solar irradiance from easily measurable metrological parameters like sunshine duration or temperature at stations where no measured data of solar irradiance are available. Accurate modeling depends upon the quality and quantity of the measured data used, and where measurements have not been done, it is a better approach for estimating the solar irradiance of location. It is necessary to collect the meteorological or environmental data for the site location under consideration to predict the performance of a PV system at that site. Figure 10.1 shows the monthly average daily solar radiation data for a horizontal plane of the site. It can be clearly seen in Figure 10.1 that solar energy incidence in the considered site is very high, especially during the summer months, where it reaches $7\ \text{kWh/m}^2/\text{day}$ on the horizontal plane.

10.4 Selection of PV Module and Inverter

The proper sizing of the PV module and inverter is important in the design of solar energy–based power systems. Ideally, a larger size and higher rating for a PV module is always desirable. The size of a PV module and inverter can be determined based on the current and voltage requirements at the particular site. In this chapter a case study of a household has been considered to explain the sizing of the PV module and inverter. Table 10.1 shows the household data at a particular site.

TABLE 10.1

The Household Load Data

Electrical Load	No. of Units	Operating Hours Per Day	Wattage Per Unit Used	Total Units Consumed Per Day
Lighting lamp	5	4 lamps from 18 to 24 and 1 lamp from 0 to 6	60	1.8
Washing machine	1	From 11 to 13	250	0.5
Refrigerator	1	From 0 to 24	180	2.4
Water pump	1	For three hours daily	120	0.36
TV	1	From 17 to 24	80	0.56
Fan	3	From 0 to 24	60	1.44
Electric iron	1	For 1 hour in 2 days	1500	0.75
AC 1.5 ton	2	6 hours each daily	2000	24
Total			6530	31.81

- **Sizing of the PV Array**

  The peak power of the PV generator (P_{PV}) is obtained as follows from Equation 10.1:

$$P_{PV} = \frac{E_L}{\eta_V \eta_R PSH} S_F \quad (10.1)$$

where E_L (daily energy consumption) = 8.44 kWh/day and (PSH) peak sun hours = 5. The efficiencies of the system components ($\eta_R = 0.92$, $\eta_v = 0.9$) and the safety factor to compensate for resistive losses and PV-cell temperature losses $S_f = 1.15$.

Substituting these values in Equation 10.1, we obtain the peak power of the PV generator:

$$P_{PV} = 8.836kW_p \quad (10.2)$$

To install this power, a monocrystalline PV module type MBPV-CAAP (Moser Baer model) of a gross area of $A_{PV} = 1.65\ m^2$ and peak power of $P_{MPP} = 200\ W_P$ is selected. The number of the necessary PV modules (N_{PV}) is obtained from Equation 10.3:

$$N_{PV} = \frac{P_{PV}}{P_{MPP}} \quad (10.3)$$

V_{MP} for MBPV200 watt peak is 28.81 volts.

No. of modules in series $= \dfrac{220 \times 1.414}{28.81} = 11$

Hence the number of modules in parallel = 4

For MBPV CAAP 200 W_p module

V_{OC} (open-circuit voltage) = 35.99 volts and

I_{SC} (short-circuit current) = 7.65 amperes

Now, we obtain an open-circuit voltage and short-circuit current for this SPV array as:

$V_{OC} = 35.99 \times 11 = 395.89\ V$

$I_{SC} = 7.65 \times 4 = 30.6$ A

- **Sizing of the Battery Block:** The storage capacity of the battery block for such systems is considerably large. Therefore, special lead-acid battery cells (block type) with a long lifetime, high cycling stability rate, and capability of withstanding very deep discharge should be selected. Such battery types are available, but at a much higher price than regular batteries. The ampere-hour capacity (C_{AH}) and watt-hour capacity (C_{WH}) of the battery block necessary to cover the load demands for a period of two days without sun is obtained as follows from Equation 10.4 and Equation 10.5:

$$C_{AH} = \frac{1.5xE_L}{V_B xDoDx\eta_B x\eta_V} \tag{10.4}$$

$$C_{WH} = C_{AH} V_B \tag{10.5}$$

where V_B and η_B are the voltage and efficiency of the battery block, and DoD is the permissible depth of discharge rate of a cell. Assuming realistic values for $\eta_B = 0.85$, DoD = 0.75, and $V_B = 220V$, we obtain:

$$C_{AH} = 500\ AH$$

$$C_{WH} = 110\ KWH$$

- **Design of the Battery Charge Controller:** The battery charge controller is required to safely charge the batteries and to maintain a longer lifetime for them. It has to be capable of carrying the short-circuit current of the PV array. Thus, in this case, it can be chosen to handle 31 A.
- **Design of the Inverter:** The inverter must be able to handle the maximum expected power of AC loads. Therefore, it can be selected at 20% higher than the rated power of the total AC loads. The rated power of the inverter then becomes 7836 W. The specifications of the required inverter will be 7836 W, 220 VAC, and 50 Hz.

11

Grid Integration of PV Systems

11.1 Grid-Connected PV Power Systems

Photovoltaic power systems are broadly classified into standalone systems and grid-connected systems. In this section, the grid-connected system is discussed. The block diagram of a grid-connected PV system is presented in Figure 11.1. The grid-connected system consists of the PV array, DC–DC converter, inverter, and associated controllers.

The penetration of grid-connected PV systems will increase drastically in the near future. However, because the grid was not particularly designed for large-scale distributed generation, utility companies may be concerned about the implications of variable solar generation on the power quality, its impact on the low-tension (LT) distribution grid, and the safety of its workforce. Therefore, there are lot of challenges and issues in integrating distributed energy sources like solar photovoltaics. In this chapter, the issues and challenges and their possible solutions are discussed.

Distributed kW scale solar PV plants (mainly rooftop models) are only now beginning to be deployed with the help of supporting policies and regulations. Hence, most of the utility companies have limited experience in dealing with distributed solar PV systems connected primarily to the LT distribution grid. They have some valid concerns regarding distributed solar PV, as it is their responsibility to maintain a reliable grid supply:

- **Power quality:** Distribution companies (DISCOMs) are apprehensive about the quality of the power being injected into their distribution grids. This has mainly to do with flicker, harmonics, and DC injection.
- **Safety:** Utility companies are rightly concerned about the safety of their personnel, especially while working around the possible formation of an unintentional island due to the operation of the distributed solar PV systems.
- **Low voltage distribution grid:** They are also concerned about the impact on the LV distribution grid (voltage levels, power factor, higher wear and tear of equipment, etc.) due to the high penetration of a large number of distributed solar generators.
- **Transaction costs:** Another logistical worry for utility companies is the significantly higher transaction effort in terms of metering, inspection, and certifications.

In addition other issues related to PV integration to distribution grids include:

- **Inverter and power quality:** The inverter is at the heart of the solar PV system, ensuring power quality and grid integration. The three important technical parameters that can affect the quality of the power being injected into the grid are

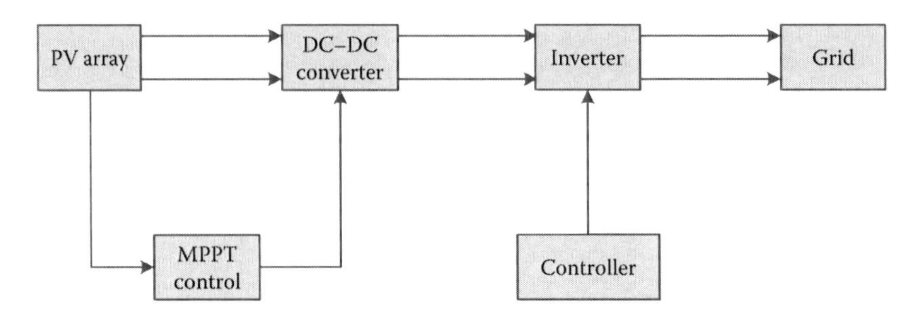

FIGURE 11.1
Block diagram of grid-connected PV power system.

harmonics, flicker, and DC injection. The issues may be taken care of with associated Institute of Electrical and Electronics Engineers (IEEE) standards.

- **Central Electricity Authority (CEA)**
    - Harmonics: IEEE 519, wherein the total harmonic distribution (THD) level is less than 5%;
    - Flicker: International Electrotechnical Commission (IEC) 61000;
    - DC injection: IEEE 1547, wherein the maximum permissible level is 0.5% for the full-rated output at the interconnection point.
- **Inverter functioning range:** Apart from the three basic parameters mentioned earlier, the solar system is allowed to function only within a certain range of voltage and frequency and is thus subject to the quality of the grid. These ranges vary from country to country. The CEA (for India) presently permits the system to function within a range of 80% to 110% of the nominal voltage and between the 47.5 and 51.5 Hz frequency bands.
- **Inverter support to the grid:** Today's inverters are capable not only of reliably integrating with the distribution grid and maintaining the power quality mandated by it, but can also provide additional support to improve some grid characteristics and assist in fault recovery. Through the low/high voltage ride-through (LHVRT) and low/high frequency ride-through (LHFRT) functions, inverters can stay connected to the grid in times of momentary grid faults/failures. Inverters can provide grid-supporting functions like power frequency droop and reactive power support to help maintain local grid parameters within their normative limits—at times in a more cost-effective manner than the centralized options. Such smart functionality in inverters comes practically at no extra cost in most cases and can be activated and incorporated without serious technical challenges. A significant share of the string inverter market in India is already capable of providing most of these functionalities.

IEEE 1547, the reference guideline for distributed generation (DG), is being revised, and many of the functions mentioned earlier are likely to be included. A draft version (IEEE 1547a) already allows some of these functions. More advanced functions, some of which may require communication capability between inverter and utility/energy markets, are being considered in some countries. Examples of such functions are limiting maximum active power upon instruction from the utility company, supporting instructions to connect/disconnect, ability to update default settings in response to changing grid conditions, etc.

- **Inverter and safety:** Islanding refers to the condition in which a distributed solar system continues to energize the circuit even when the grid power from the utility company is unavailable. Islanding can be dangerous to utility workers, who may not realize that a circuit is still powered when working on repairs or maintenance. To ensure safety, most countries, including India, mandate the anti-islanding functionality, which requires the PV system to stop energizing the grid as soon as grid power is unavailable. An emerging function is "intentional islanding," which allows distributed solar PV to continue to power certain loads when the grid is down, with adequate safety considerations. This is of utmost relevance in India, where load shedding or power cuts are common in most DISCOM service areas.

Research into these issues is presented in the next sections.

11.2 Inverter Control Algorithms

The classic configuration of a grid-connected solar PV system consists basically of two parts. The first stage is the energy supply represented by a photovoltaic array. In the output stage, an inverter is used to convert and adapt the energy accumulated in an intermediate-storage element (DC-link capacitor), to be consistent with the voltage and power quality requirements of the utility grid. For low-power photovoltaic systems, the classical two-level inverter is typically employed as the interface between the DC-link and grid. For high-power, medium-voltage applications, a multilevel converter is widely used.

Several control techniques have been proposed for grid-connected applications in the last few decades. The controllers attempt to achieve stability, low harmonic content, fast dynamic response, and simplicity. However, trade-offs are usually required. The classical hysteresis current controller, for example, has an inherent stability and fast and robust dynamic response, but is affected by variable switching frequency and interference between phases in isolated neutral three-phase systems. These weaknesses are overcome with the voltage-oriented control method, which achieves constant switching frequency, and consequently a well-defined harmonic spectrum, by controlling the load currents, which are split into direct quadrature (d-q) components, according to the synchronous reference frame theory. On the other hand, its stability and dynamic performance are highly dependent on load parameters. The fast development of powerful microprocessors has been opening space for a new class of control techniques, due to advantages of flexibility and the ability to process complex online calculations. The predictive control technique makes use of these benefits to calculate the optimum control action among all the possibilities to fulfill certain predefined criteria. Nevertheless, the performance is susceptible to variations in load parameters. Another emerging control method, known as direct power control, utilizes online estimations of control variables and a switching table to select the proper switching state. This technique has been shown to be very suitable for grid-connected applications, because it regulates directly the instantaneous active and reactive powers.

11.2.1 Synchronous Reference Frame–Based Current Controller

The DC–AC stage must only inject the active component of grid current. So for this, the steady-state current error between the actual and desired grid current should remain zero at any grid frequency. The phase locked loop (PLL) tracks the phase of the input voltage signal and generates unit voltage templates (sine and cosine components). The d-q components of currents pass through a filter, which filters out high-frequency harmonic components. Then the d-q frame is again transformed back to one-phase components. This current is then compared with the source current and any error between them is fed to the hysteresis-based pulse width modulation (PWM) signal generator to produce the final switching signals, which are the pulses for the inverter. The Reference current extraction using synchronous reference frame (SRF) theory and Inverter control using PI controller is presented in Figure 11.2 and Figure 11.3 respectively.

System terminal voltages are given as,

$$v_\alpha = V_m \cos(\omega t + \Phi) \tag{11.1}$$

And current is given as:

$$i_\alpha = I_m \cos(\omega t + \theta_n) \tag{11.2}$$

The following equation gives:

$$\begin{bmatrix} i_d \\ i_q \end{bmatrix} = \begin{bmatrix} \cos\theta & \sin\theta \\ -\sin\theta & \cos\theta \end{bmatrix} \begin{bmatrix} i_\alpha \\ i_\beta \end{bmatrix} \tag{11.3}$$

i_d = In-phase component of grid current

i_q = Reactive component of grid current

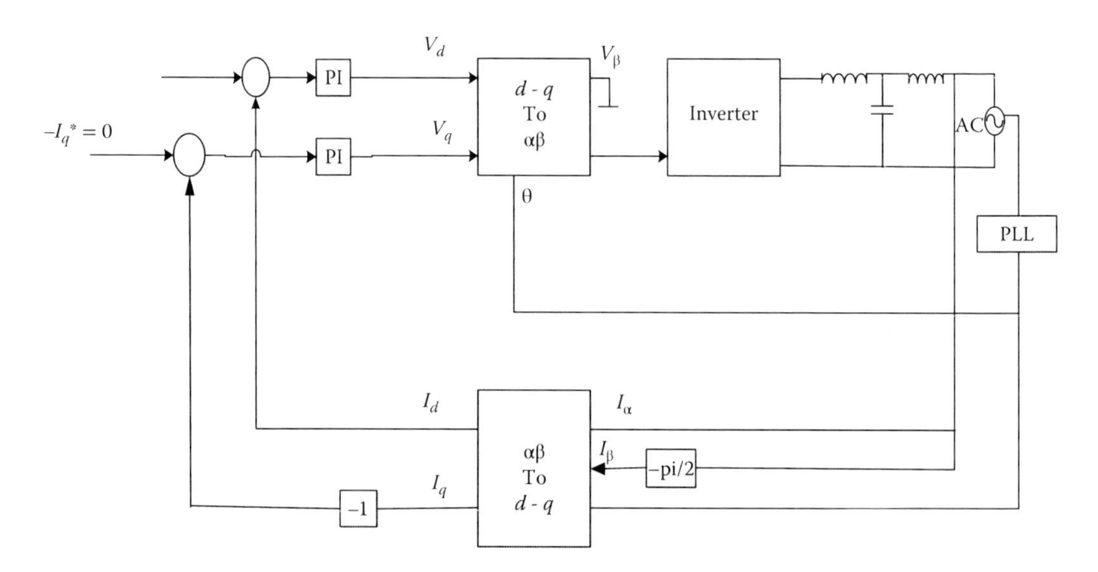

FIGURE 11.2
Reference current extraction using synchronous reference frame (SRF) theory.

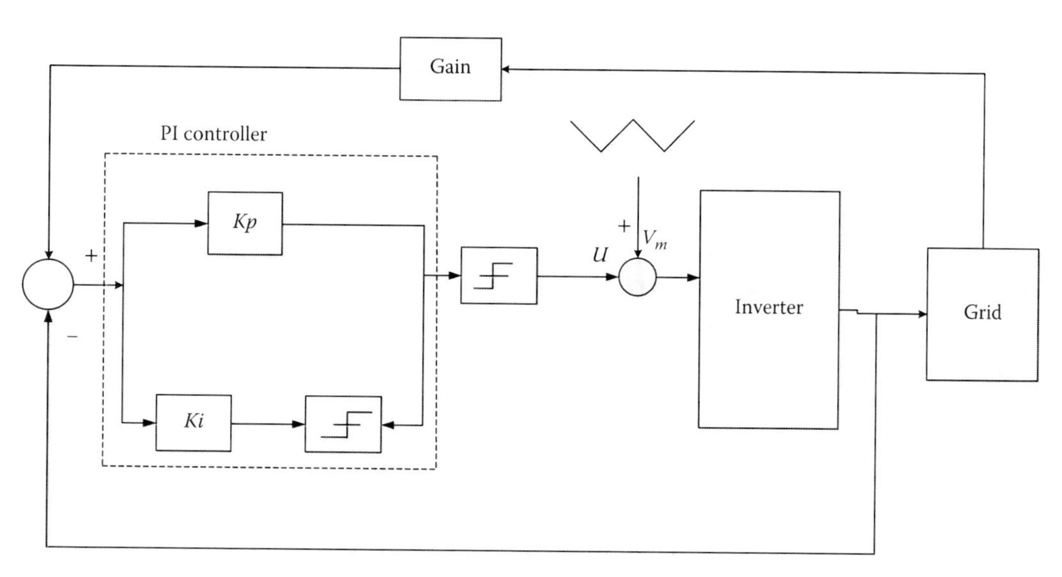

FIGURE 11.3
Inverter control using PI controller.

$$\begin{bmatrix} i_d = I\cos(\Phi - \theta_n) \\ i_q = -\,I\sin(\Phi - \theta_n) \end{bmatrix} \tag{11.4}$$

These reference currents are compared with the actual current, and errors between both will be the required amount of reactive current supplied by the inverter. The error of the current is used to generate pulses for the inverter.

11.2.2 Digital PI-Based Current Controller

The PI control method can be used digitally to control the various parameters of inverters such as current, voltage, etc. The digital PI control can be implemented into the system hardware by using microprocessors and programming the controlling algorithms.

The PI algorithm is utilized by the controller to update the response or to minimize the error between the reference current and inverter current. The reference current is the function of grid voltage and is produced by sensing the grid voltage and converting it into current. This is done to maintain the invert output current at Unity Power Factor and in phase with grid voltage. The PI controller monitors the instantaneous error between the inverter output current and the reference current. Finally, this error signal is compared with a carrier signal to obtain the PWM signal of a single-phase inverter. The PI controller algorithm can be written as following equation:

$$u(t) = Kpe(t) + Ki\int e(\tau)\mathrm{d}\tau \tag{11.5}$$

where,
$u(t) =$ control signal
$e(t) =$ error signal

Kp = proportional gain
Ki = integral gain

A discrete time domain equation for proportion gain can be directly applied as it was in the continuous form, but integral gain requires approximations. Hence, a method called the trapezoidal sum approximation can be utilized to obtain the time domain form.

$$Kpe(t) = Kpe(k) \tag{11.6}$$

$$Ki\int_{\tau=0}^{t} e\,(\tau)\mathrm{d}\tau = Ki\sum_{i=0}^{k}\frac{h}{2}\left[e(i)+e(i-1)\right] \tag{11.7}$$

$$t = k * h \tag{11.8}$$

where h denotes the sampling period and k is the discrete time index, $k = 0, 1.....$

$$u(k) = Kpe(k) + Ki\sum_{i=0}^{k}\frac{h}{2}\left[e(i)+e(i-1)\right] \tag{11.9}$$

The heart of control lies in these two equations. To ignore the need to calculate the full summation every time, the summation is expressed as a running sum:

$$\mathrm{sum}(k) = \mathrm{sum}(k-1) + \left[e(k)+e(k-1)\right] \tag{11.10}$$

$$u(k) = Kp\ e(k) + Ki\ \mathrm{sum}(k) \tag{11.11}$$

11.2.3 Adaptive Notch Filter–Based Grid Synchronization Approach

The adaptive notch filter–based technique offers the compensation of the harmonics current and reactive power with both linear and nonlinear types of loads. To maintain a constant DC link voltage, a PI controller is implemented.

The adaptive notch filter (ANF) approach is used to synchronize the interfaced PV system with the grid to maintain the amplitude, phase, and frequency parameters in power quality improvement. The signal is in the periodic form and is defined as:

$$u(t) = \sum_{i=1}^{n} A_i \sin\varphi_i \qquad \text{where} \quad \varphi_i = \omega_i + \varphi_i \tag{11.12}$$

where A_i is non-zero amplitudes, ω_i is non-zero frequencies, and phases φ_i are unknown parameters. These parameters are measured using the ANF signal to extract the required signal to be injected at the point of common coupling. The basic block of adaptive notch filter is presented in Figure 11.4. The dynamics of the signal at the time of interface is given by second order differential equations as:

$$\ddot{x} + \theta^2 x = 2\zeta\theta e(t) \tag{11.13}$$

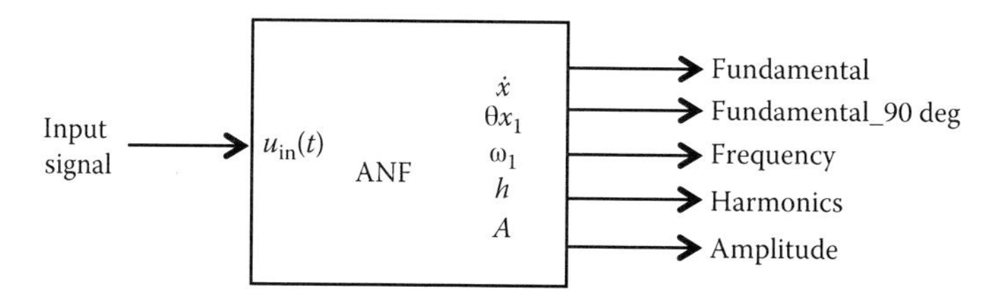

FIGURE 11.4
Basic block of adaptive notch filter.

$$e(t) = u(t) - \dot{x} \tag{11.14}$$

$$\dot{\theta} = -\gamma\theta x e(t) \tag{11.15}$$

Here, $u(t)$ is the input signal, θ represents the estimated frequency, and ζ and γ are real and positive parameters that determine the performance of the ANF in terms of proper synchronization in phase and frequency for the injection of a fundamental signal to the grid.

For extracting the fundamental component:

$$u(t) = A_1 \sin(\omega_1 t + \delta) \tag{11.16}$$

Dynamics of the signal in Equation 11.2 has a periodic function in a state equation in the form of

$$v(t) = \begin{pmatrix} x \\ \dot{x} \\ \theta \end{pmatrix} = \begin{pmatrix} -\dfrac{A_1}{\omega_1}\cos(\omega_1 + \delta) \\ A_1 \sin(\omega_1 + \delta) \\ \omega_1 \end{pmatrix} \tag{11.17}$$

The third entry of $v(t)$ is the estimated frequency, and it is identical to its correct value ω_1.

11.3 Modeling, Simulation, and Hardware Implementation of Controllers

Modeling and simulation of grid-connected PV systems have been carried out in the MATLAB® environment and are discussed in this section. An adaptive notch filter–based control technique is also proposed for the voltage source converter (VSC) for grid-connected PV systems. The proposed control technique provides better compensation of harmonics and reactive power with linear and nonlinear loads. A constant DC link voltage is maintained by using a PI controller. Modeling and simulation of a complete system have been done using the SimPower System and Simulink toolbox of MATLAB. A hardware setup and controller for the integration of PV systems into distribution network has been developed to ensure the reliable power supply during large disturbances and large swings. The results of the proposed controller are validated on the developed experimental setup. The MATLAB Simulink model for a three-phase grid-connected system has been developed and is presented in Figure 11.5. The performance of the system is analyzed by observing the various parameters such as the DC power supplied by the PV system, DC

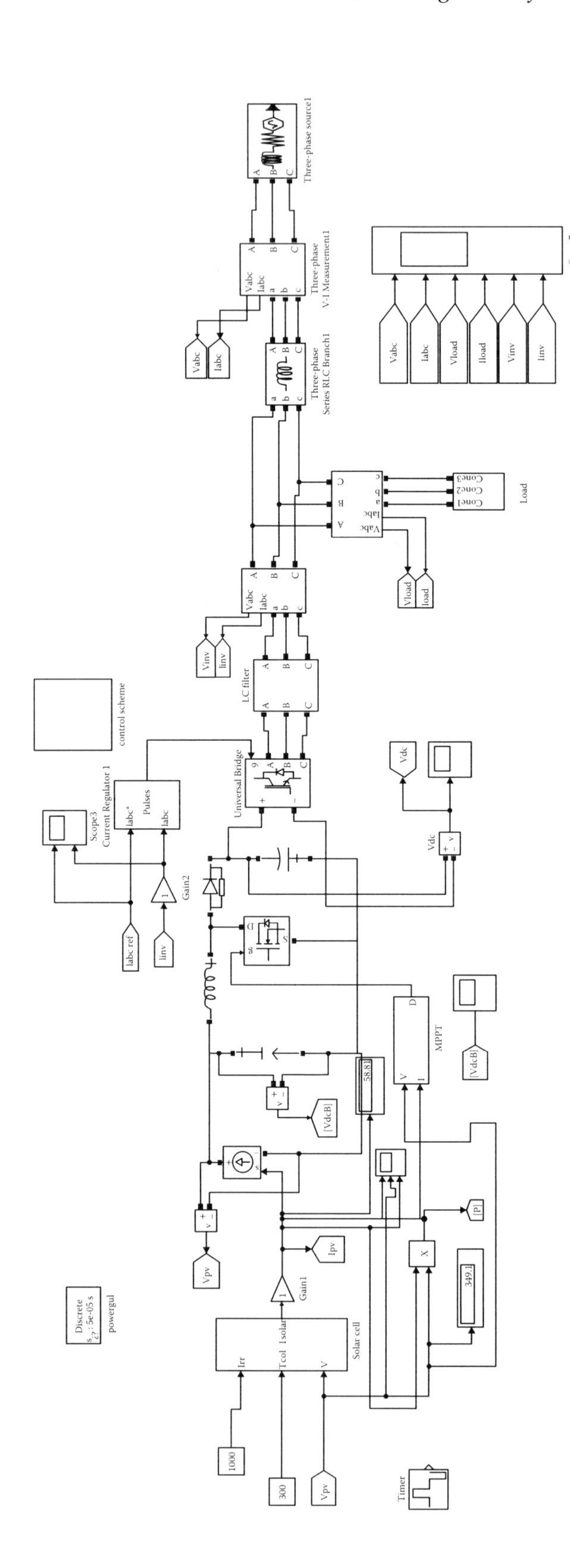

FIGURE 11.5
MATLAB® simulation of three-phase grid-connected PV system.

link voltage, three-phase voltage, and active and reactive power supplied by the system at varying meteorological parameters; the results are presented in Figures 11.6 and 11.7.

PV voltage and current are very important parameters for the performance of a PV system. These parameters are used as an input to MPPT control, which in turn generate pulses for the DC–DC boost converter. A DC–DC converter is used to maintain the DC link voltage of a desired value. A high-value capacitor is used to maintain the DC link voltage at constant rate. The DC link voltage is kept constant at a near 700 V from the boost converter (see Figure 11.8). Pulses for IGBT switches are produced by the inverter control block. The VSI uses solid-state switches, which produce current

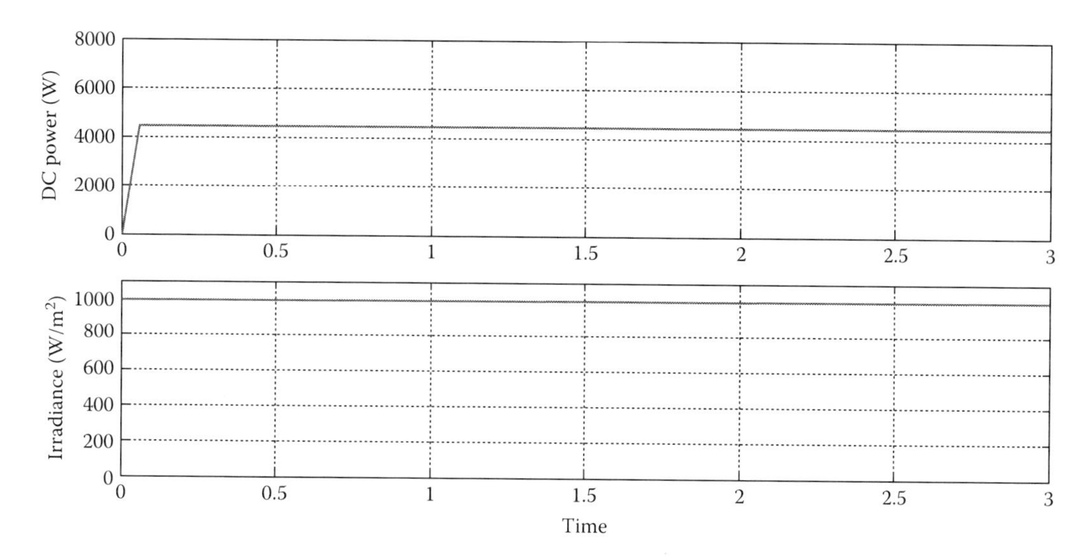

FIGURE 11.6
DC power of PV system at constant irradiance.

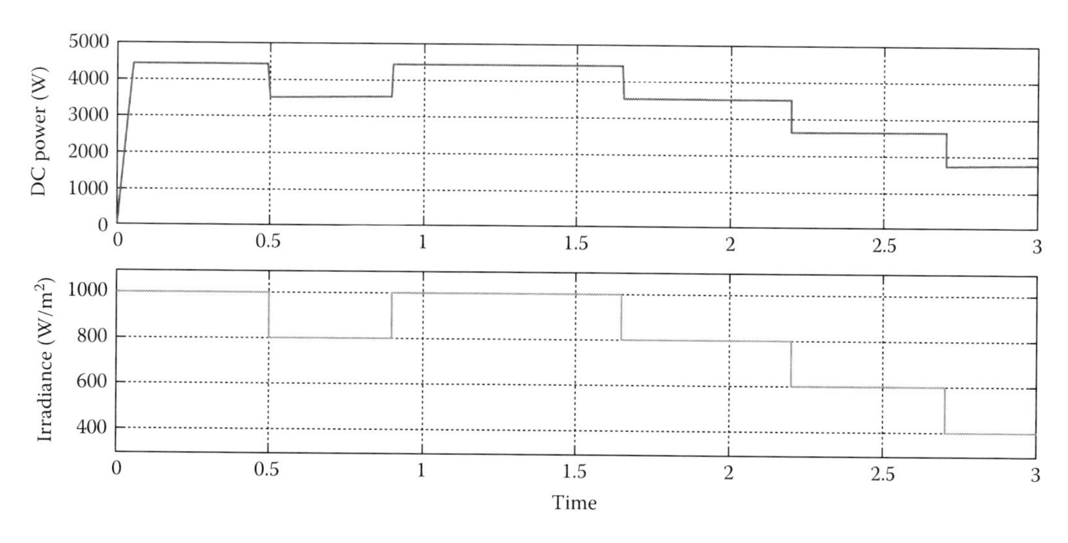

FIGURE 11.7
DC power of PV system at varying irradiance.

harmonics that can be clearly seen in Figure 11.9. As shown in Figures 11.10 and 11.11, the active power is dependent on the irradiance parameter, and the almost constant reactive power is supplied irrespective of the meteorological parameter variation. Only the partial requirement of reactive power is supplied by the PV system. The THD of the load current and its impacts at different levels of penetration are analyzed and presented in Figures 11.12 and 11.13, respectively. Further, the impacts of increasing penetration of PV on the existing grid are also analyzed through THD and voltage at the PCC point and are presented in Figure 11.14. From the results, we can conclude that the performance of the proposed controller is satisfactory and all the results are under the desired limits.

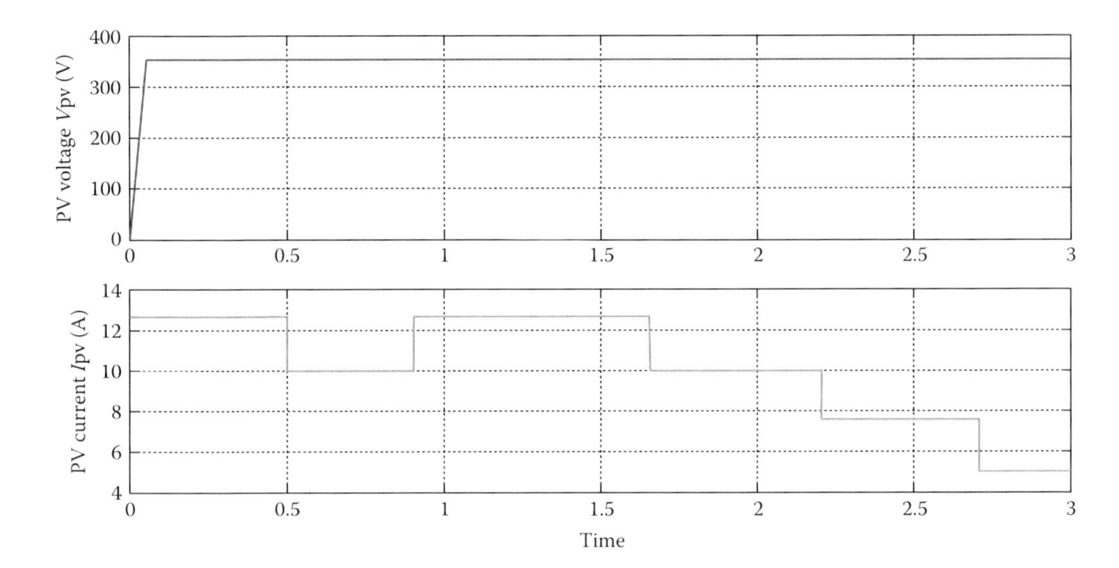

FIGURE 11.8
PV voltage and current.

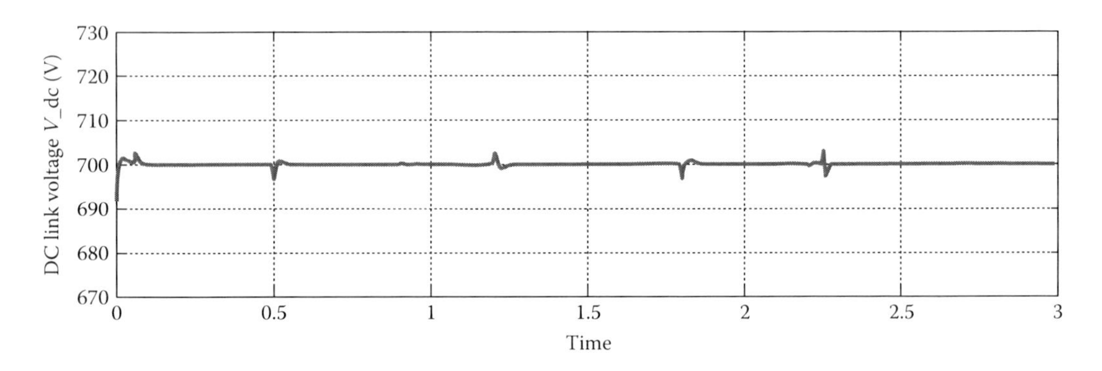

FIGURE 11.9
DC link voltage.

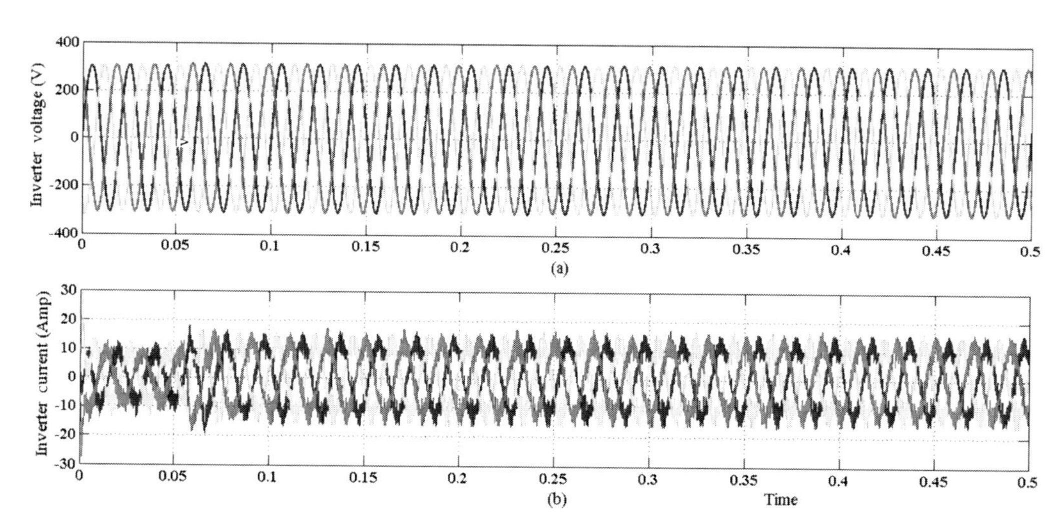

FIGURE 11.10
Inverter (a) voltage and (b) current.

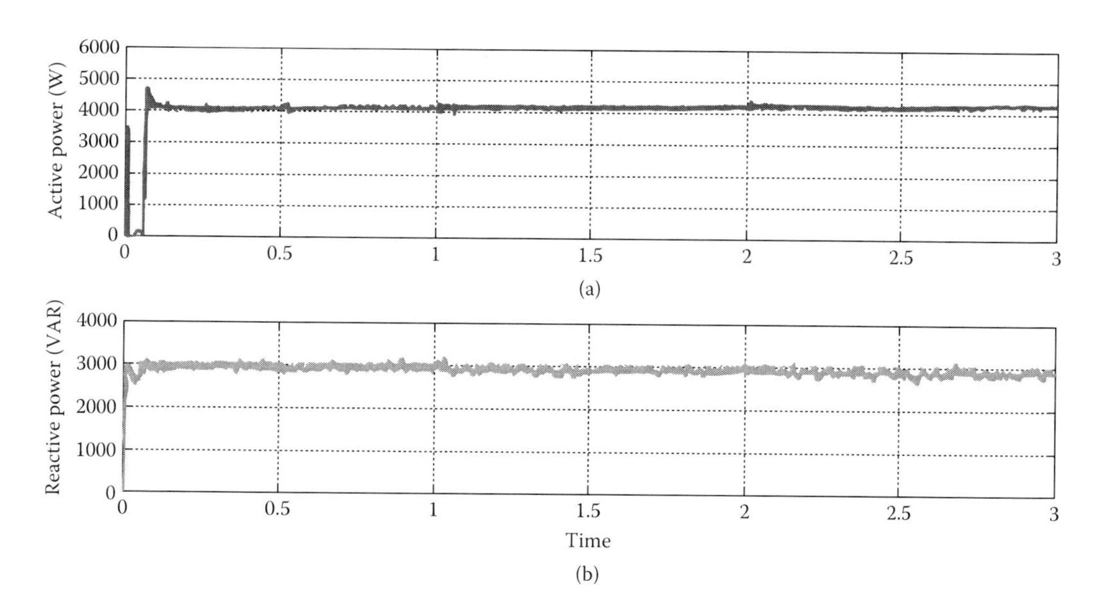

FIGURE 11.11
Power supplied to grid at constant irradiance: (a) active power and (b) reactive power.

A grid synchronization technique based on an adaptive notch filter for a PV system, along with MPPT techniques, has been developed. In this approach, a reference phase signal synchronized with the grid voltage is provided to standardize the system with grid codes and power quality standards. Therefore, this approach plays an important role in the grid-connected PV system. Because the output of the PV array fluctuates based on meteorological parameters such as solar irradiance, temperature, and wind speed, in order to maintain a constant DC voltage at VSC input, MPPT control is required to track the maximum power point from the PV array. In this work, a variable step size P&O (perturb

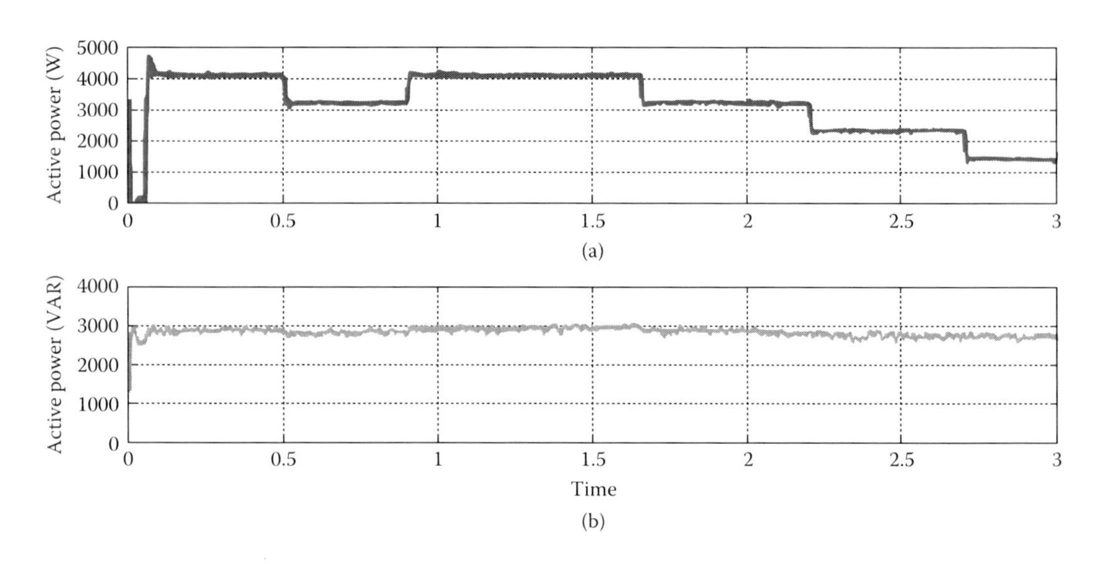

FIGURE 11.12
Power supplied to grid at varying irradiance (a) active power (b) reactive power.

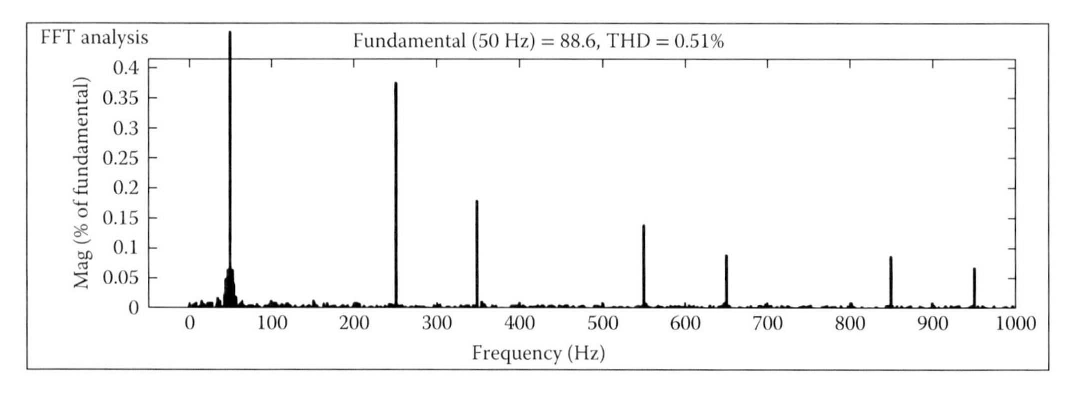

FIGURE 11.13
THD of load current.

and observe) MPPT technique with DC–DC boost converter has been used at the first stage of the system. The proposed algorithm divides the d*P*pv/d*V*pv curve of the PV panel into three separate zones (i.e., zone 0, zone 1, and zone 2). A small tracking step size is used in zone 0, whereas zone 1 and zone 2 require a large step size in order to obtain a high tracking speed. Further, the ANF-based control technique is proposed for the VSC in a PV generation system. This approach is used to synchronize the interfaced PV system with the grid to maintain the amplitude, phase, and frequency parameters, as well as power quality improvement. This technique offers the compensation of harmonics current and reactive power with both linear and nonlinear loads. To maintain a constant DC link voltage, a PI controller is also implemented and presented in this work. The complete system has been designed, developed, and simulated using the SimPower System and Simulink toolbox of MATLAB.

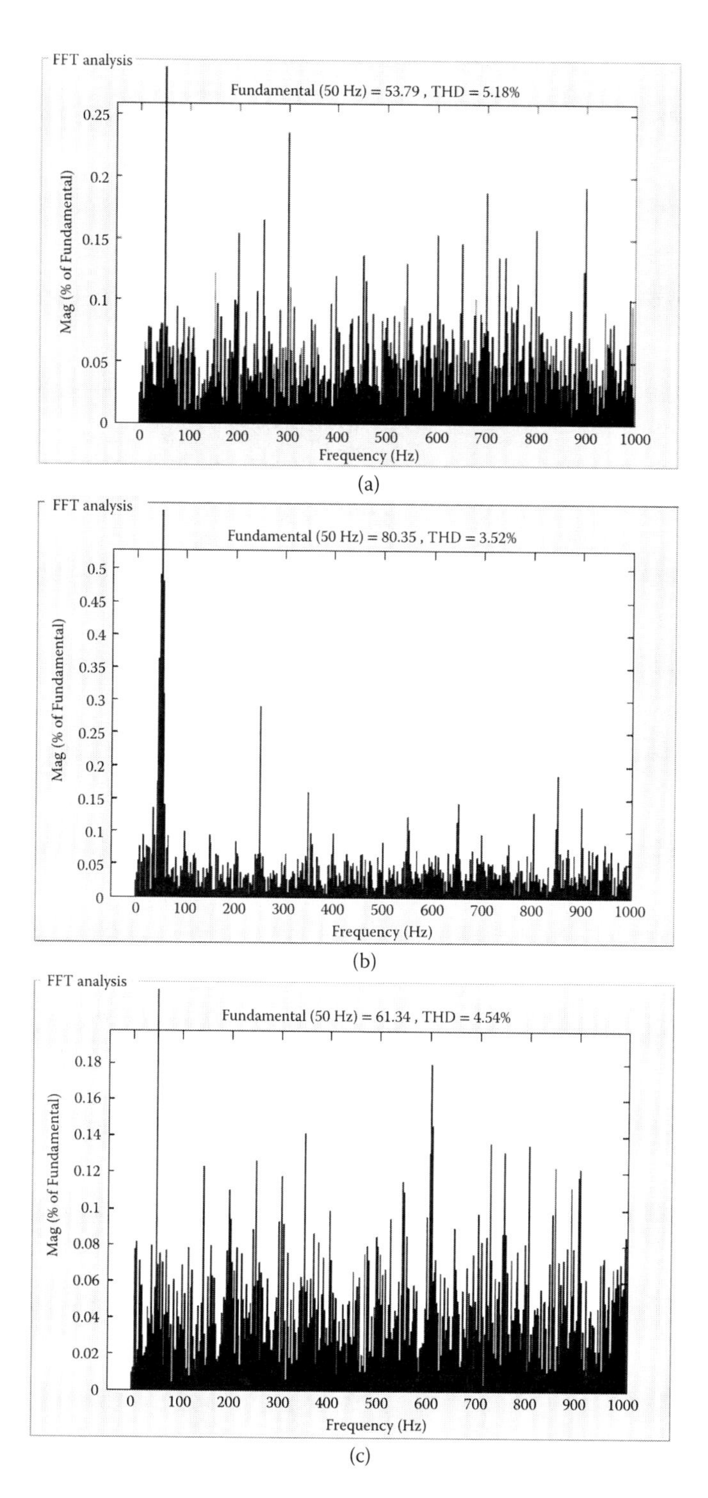

FIGURE 11.14
THD of grid at (a) 10%, (b) 15%, and (c) 20% PV penetration. *(Continued)*

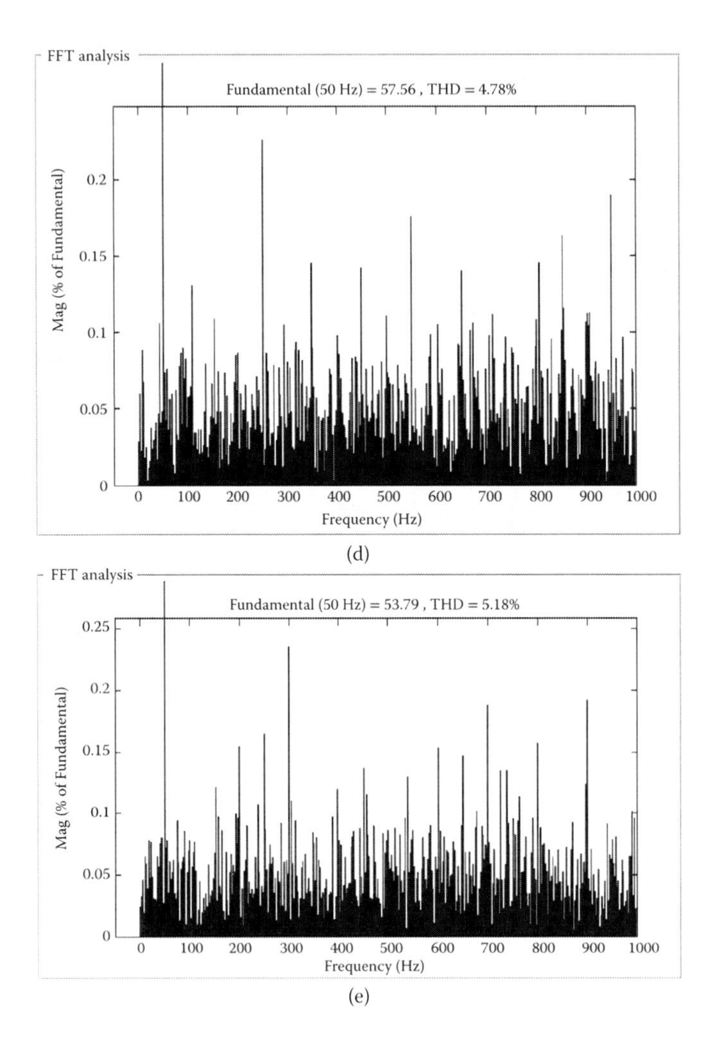

FIGURE 11.14 ***(Continued)***
THD of grid at (d) 25% and (e) 30% PV penetration.

The performance of the proposed system with linear and nonlinear load is carried out and presented (Figures 11.15 and 11.16). A linear load of 50 kW with lagging power factor is used for simulation studies. The maximum power output of the designed PV array is 20 kW. MPPT control is applied to the system to feed maximum power to the load. The reference value of the DC link voltage is kept at 700 V. A PI controller tunes the DC link voltage at its reference value. It can be clearly observed from the waveform that initially the complete load current is fed by the source only, but after the switching the inverter, the load current is shared between the grid and PV system. The performance of the system is presented in Figures 11.17 and 11.18 for linear load. A change in the irradiance of the PV array is also applied to check the accuracy and efficiency of the proposed MPPT algorithm. At 0.1 s, irradiance is raised to 1200 W/m^2 and at 0.2 s, it is reduced to 800 W/m^2. Up to 0.35 s, a part of the active power of the load is fed by the PV array. Complete reactive power compensation is provided by the inverter—that means the source is not supplying reactive power to the load. The voltages at the PCC and grid currents are in phase, thus achieving the

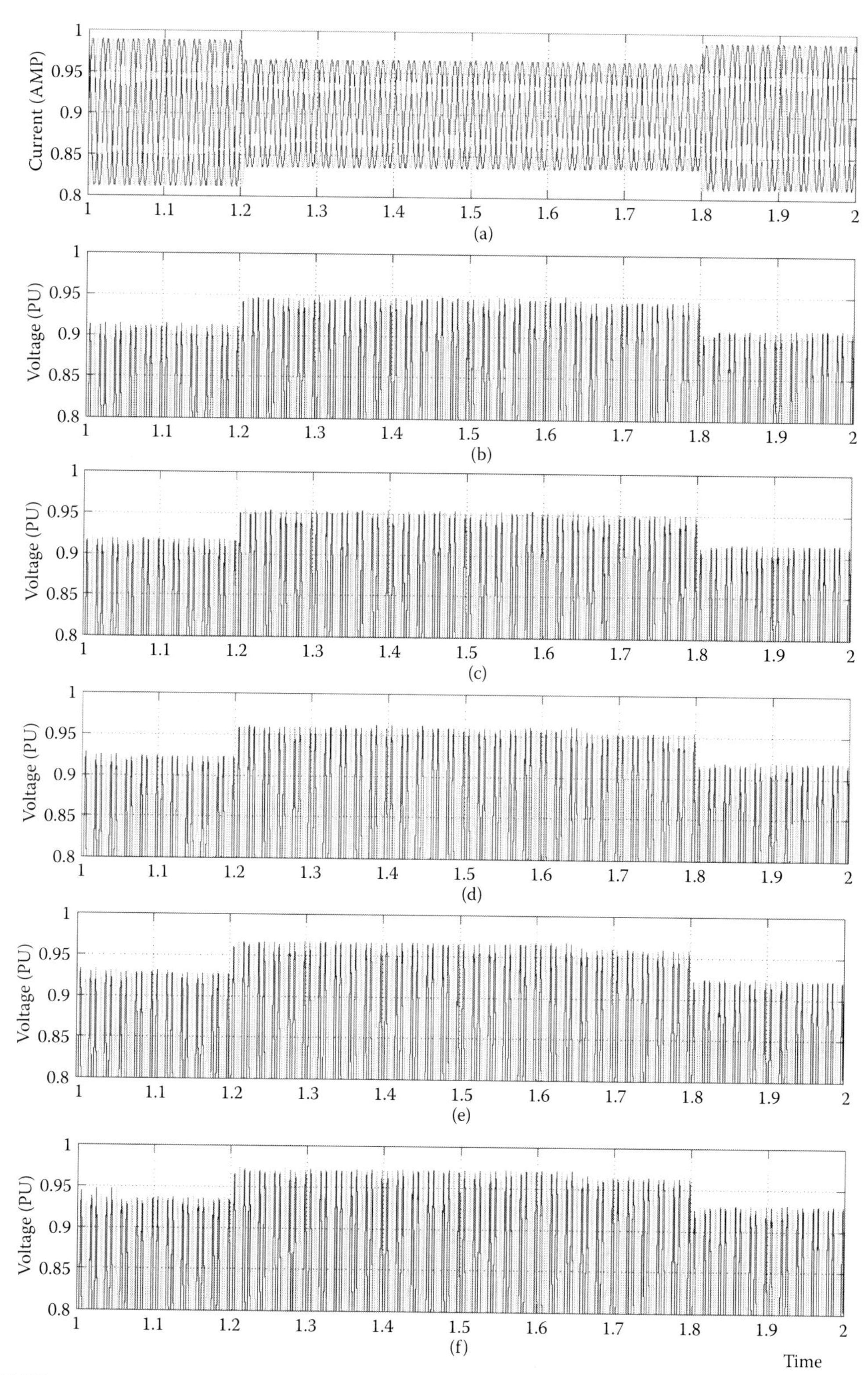

FIGURE 11.15
(a) Discontinuous load current and voltage fluctuation levels at (b) 10% (c) 15% (d) 20% (e) 25% (f) 30%.

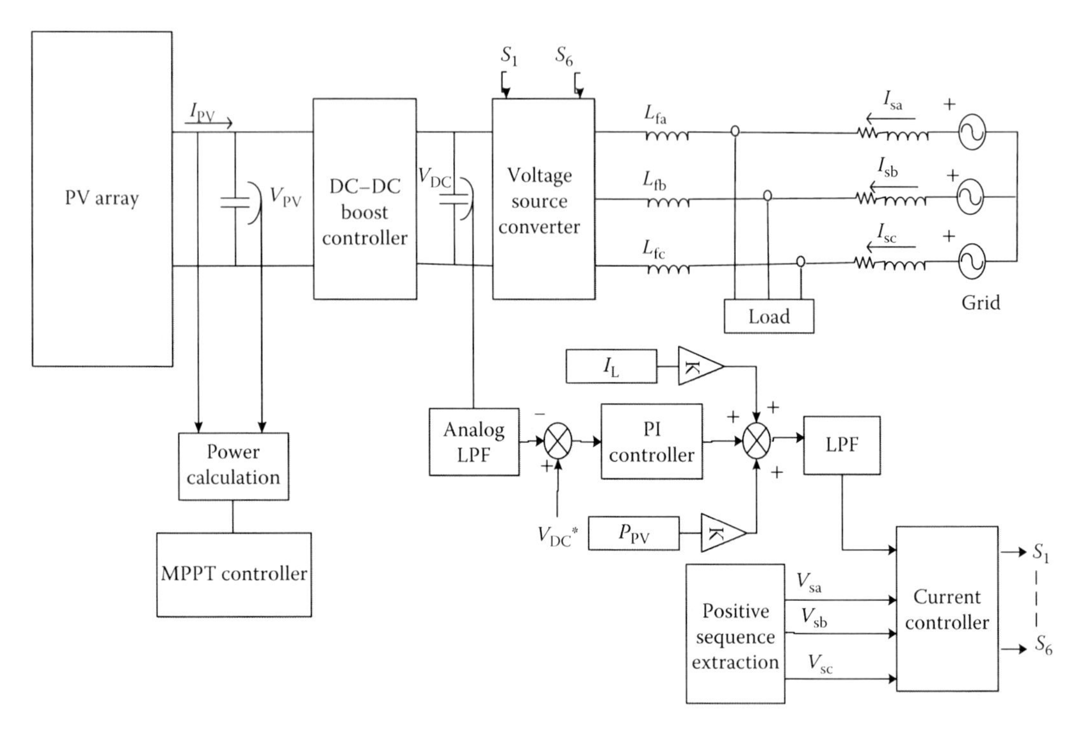

FIGURE 11.16
Block diagram of system under study.

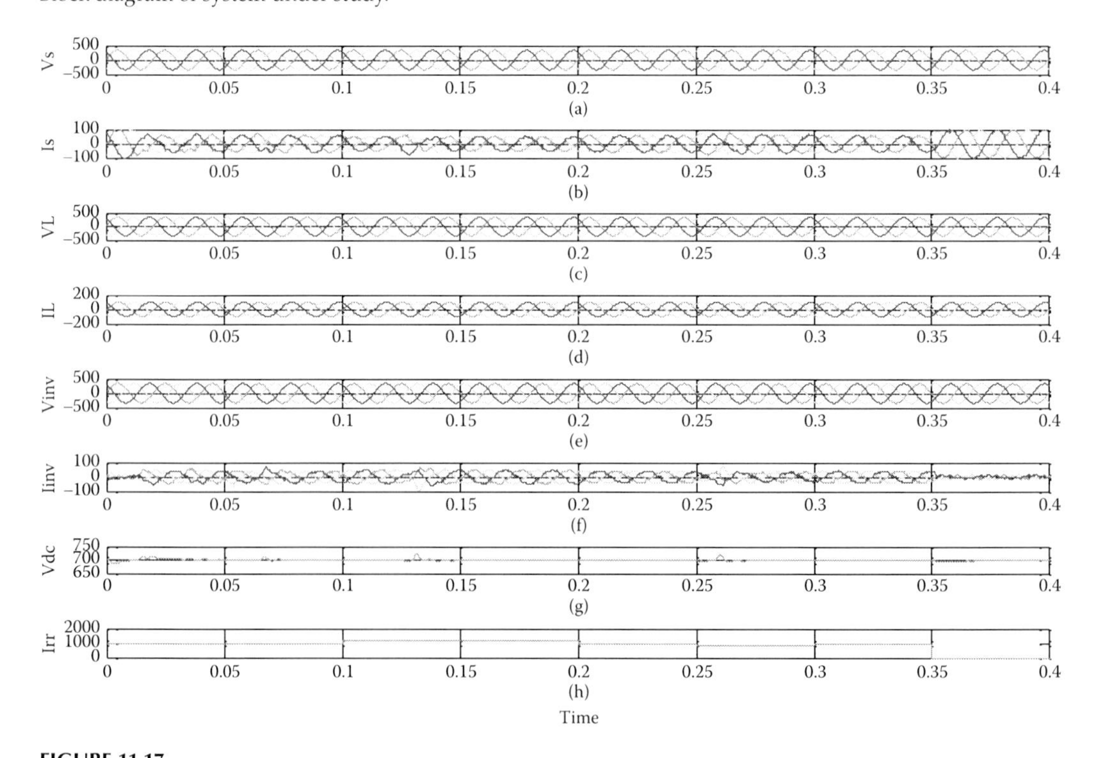

FIGURE 11.17
(a) Source side voltage (V), (b) source side current (A), (c) load voltage (V), (d) load current (A), (e) inverter side voltage (V), (f) inverter side current (A), (g) DC link voltage, and (h) change in solar irradiance.

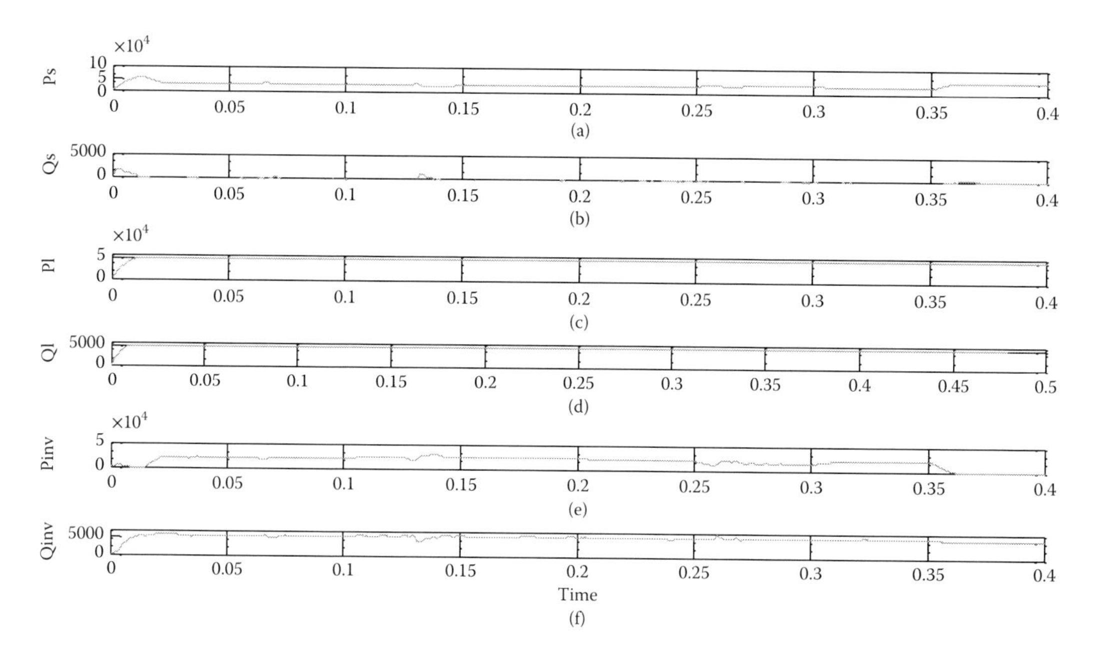

FIGURE 11.18
(a) Active power delivered by source (kW), (b) reactive power delivered by source (kVAR), (c) active power required by load (kW), (d) reactive power required by source (kVAR), (e) active power delivered by PV system (kW), and (f) reactive power delivered by VSC (kVAR).

unity power factor operation of the proposed system. The solar irradiance is reduced to zero at 0.35 s as a result of it, and the PV current becomes zero. For this mode of operation, the required compensation is provided by the voltage source converter. With the help of simulated results, it can be observed that the system performance is satisfactory under linear load conditions. Figures 11.19 and 11.20 show the various parameters of the proposed system under a nonlinear load condition. It can be observed that the grid currents are in phase with PCC voltages, achieving unity power factor operation of grid. The grid currents are balanced and sinusoidal, irrespective of the nonlinear load, due to the compensation provided by the VSC. At 0.5 s, the solar intensity becomes zero, and now the compensation of the loads is provided by the VSC. Again in this case a change in irradiance has been applied to the PV array. From the output waveforms, it can be clearly seen that the system performs well for nonlinear loads also with the proposed control algorithm.

In hardware, a power conditioning unit can be developed and may be used to ensure the reliable power supply during large disturbances and large swings. The results of the controller are validated for its robustness with respect to the intermittency and enhance its integration level in the distribution system. The experimental setup, presented in Figures 11.21 to 11.24, includes the following:

- 5 kWp PV array;
- dSPACE DS1104 R&D controller board;
- signal conditioning boards based on Hall effect voltage and current sensors;

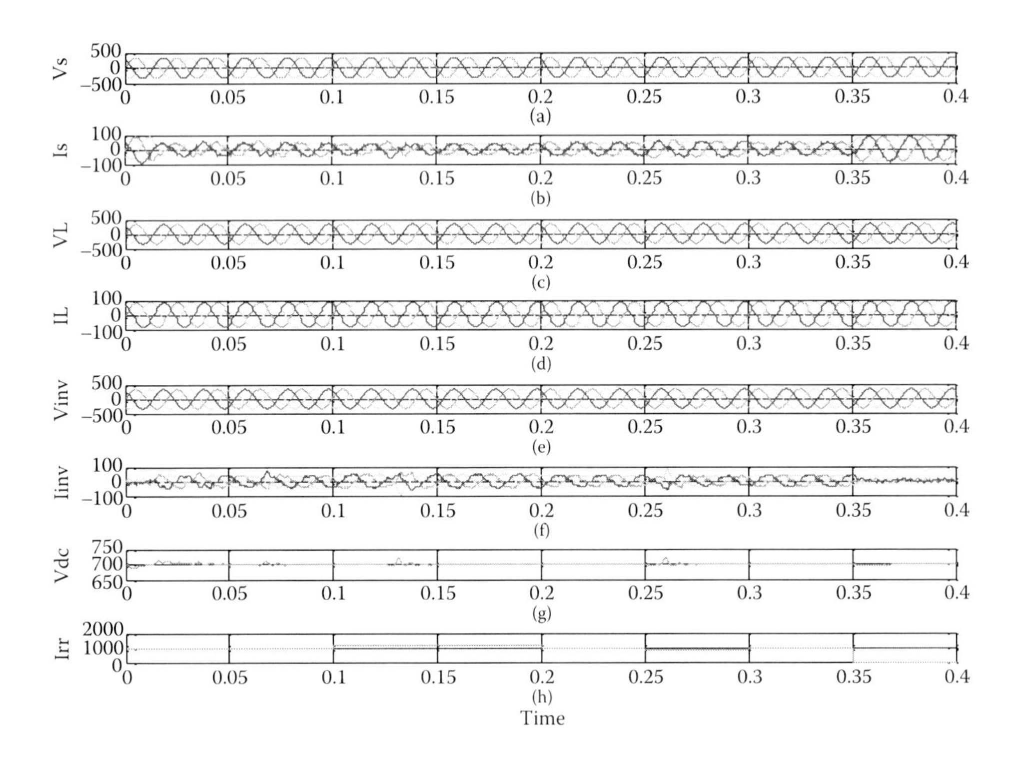

FIGURE 11.19
(a) Source side voltage (V), (b) source side current (A), (c) load voltage (V), (d) load current (A), (e) inverter side voltage (V), (f) inverter side current (A), (g) DC link voltage, and (h) change in solar irradiance.

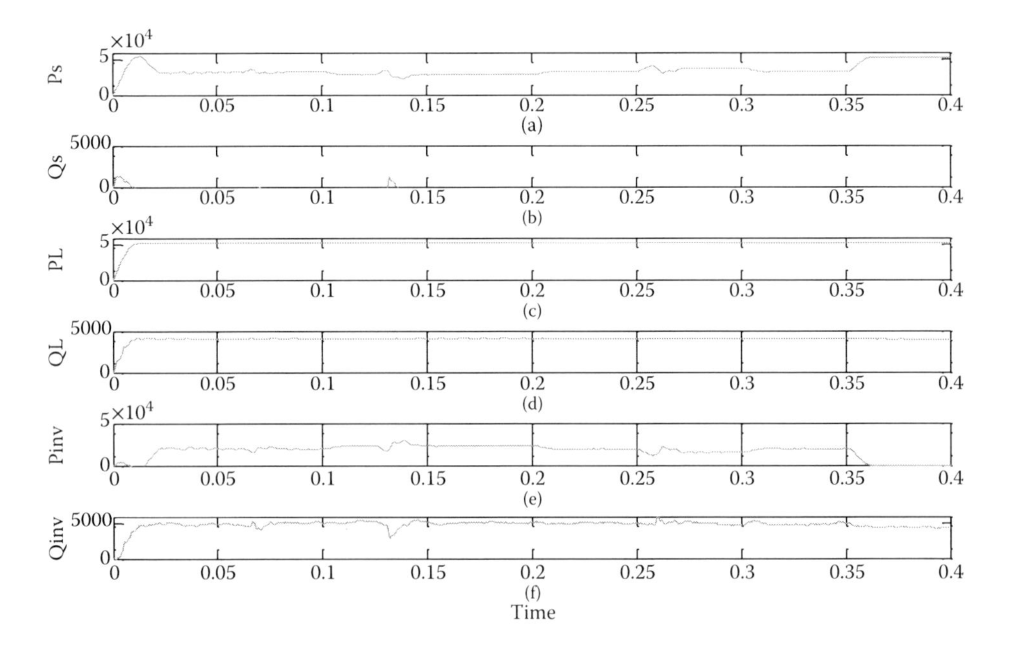

FIGURE 11.20
(a) Active power delivered by source (kW), (b) reactive power delivered by source (kVAR), (c) active power required by load (kW), (d) reactive power required by load (kVAR), (e) active power delivered by PV system (kW), and (f) reactive power delivered by VSC (kVAR).

FIGURE 11.21
5 kW_p PV system installed on the rooftop of the EEU laboratory, Delhi Technological University, Delhi, India.

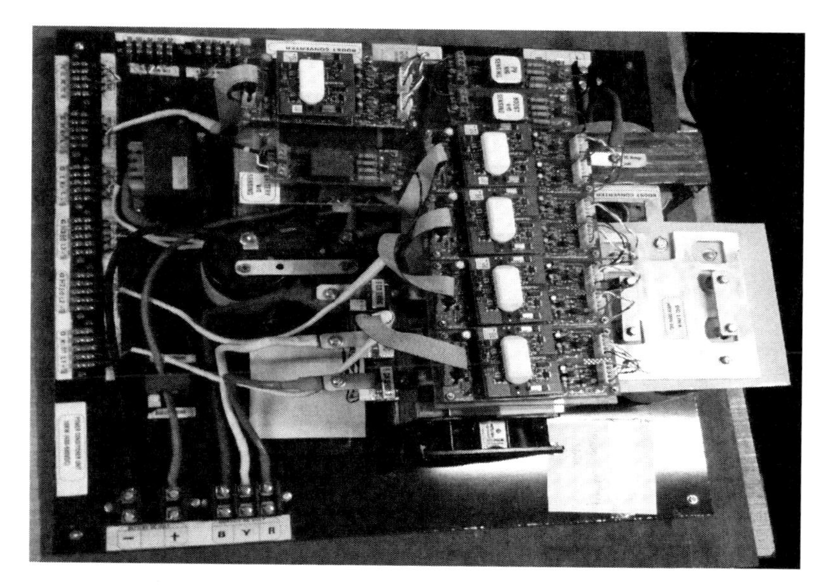

FIGURE 11.22
Power conditioning unit for the proposed system.

- power conditioning unit consisting of:
  - IGBT-based converter;
  - inverter/MPPT, gate drivers, and
  - optocoupler integrated circuits (ICs);
- isolation transformer; and
- power quality analyzer.

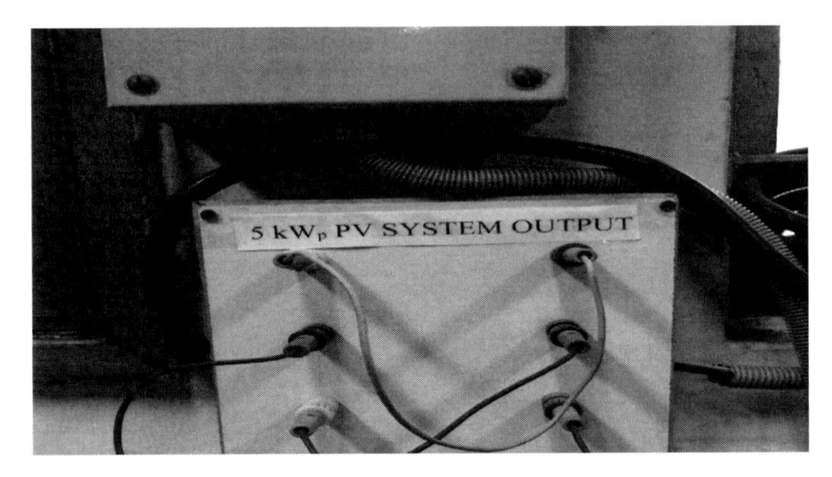

FIGURE 11.23
5 kW_p PV system output and grid connection.

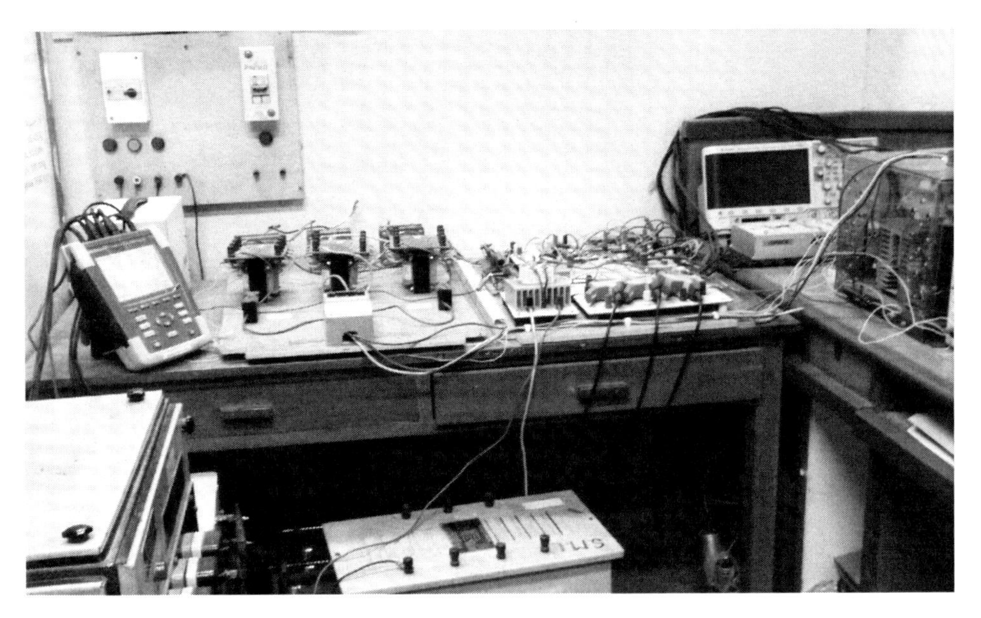

FIGURE 11.24
Experimental setup for the proposed work.

11.4 Conclusion

In this chapter, the issues and challenges in the grid integration of PV systems have been described and a case study for the development of the controller is elaborated in detail. The results of the developed control system were explained. However, it is concluded that the power generated from the PV is highly weather dependent, and the maximization of the output power from the PV generator can only be achieved by including an adequate controlling system. Also, the integration of energy storage to the PV system via a bidirectional

DC–DC converter is invaluable in mitigating problems of intermittency and variability with a PV power generator.

The following is the summary and future scope of the work at this stage:

- Modeling and simulation of a three-phase, grid-connected PV system are carried out in the MATLAB environment.
- In this work, the performance of a grid-connected PV system in terms of PV output power, PV voltage, PV current, DC link voltage, PCC voltage, grid voltage, grid current, voltage source converter current, and power supplied by the voltage source converter is found to be satisfactory for linear and nonlinear loads. The power quality analysis of the proposed system was performed, and it was noted that the THD of the source current was 3.75%, which is in the desired limit.
- An adaptive notch filter–based controller is also proposed for a VSC for grid-connected PV systems. The proposed controller provides better compensation of harmonics and reactive power with linear and nonlinear loads. The capability of the proposed controller is found to be better compared to the other controllers.
- An experimental setup was also developed, and the obtained simulation results are validated and found to be satisfactory.
- With respect to technical merit, the proposed research will be helpful in providing a solution for the power shortage via power generation and integration of solar energy resources without compromising the grid stability and reliability.

12

PV System Protection

12.1 Faults in PV System

In spite of the fact that PV systems do not have any moving parts like other types of power plants, it cannot be assumed that such a system will be fault-free. There are many challenges in the distribution networks due to increased penetration of PV systems. These challenges include power quality, protection, voltage regulation, and stability. A fault in any system can never be ruled out. PV systems may experience faults in three different parts: the AC side, the DC side, and the inverter. Faults may take place due to many technical and nontechnical reasons. A fault may be so severe that it damages the entire system, whereas some faults may be temporary and automatically die out without any damage. Faults in PV arrays may damage the PV modules and cables, as well as lead to electrical shock hazards and fire risk. Furthermore, faults in PV arrays may cause a large amount of energy loss. The most commonly occurring faults in a PV system are an open-circuit fault, earth fault, bridge fault, and mismatch fault. Internal errors of the inverter are also a common fault in PV systems, and they increase with the age of the component.

The DC side usually represents the lower amount of faults, but it is important to provide protection against them. The following types of DC-side faults may take place in the panels:

a. Open-circuit fault: An open-circuit fault occurs when one of the current-carrying paths in series with the load is broken or opened. The poor connections between cells, plugging and unplugging connectors at junction boxes, or breaks in wires can cause an open-circuit fault.
b. Earth fault: An earth fault in PV arrays is an accidental electrical short circuit involving the ground and one or more normally designated current-carrying conductors. Ground faults in PV arrays often draw people's safety concerns because it may generate DC arcs at the fault point on the ground fault path. If the fault is not cleared properly, the DC arcs could cause a fire hazard. A ground fault is the most common fault in PV and may be caused by the following reasons: Insulation failure of cables, for example, a rodent chewing through cable insulation and causing a ground fault; Incidental short circuit between normal conductor and ground, for example, a cable in a PV junction box contacting a grounded conductor incidentally; Ground faults within PV modules, for example, a solar cell short-circuiting to grounded module frames due to deteriorating encapsulation, impact damage, or water corrosion in the PV module.

   An earth fault may be a system-grounding and equipment-grounding fault. In system grounding, the negative conductor is grounded via the ground fault protection device (GFPD), whereas equipment grounding involves the noncurrent-carrying metal parts of the PV module frames.

c. Bridge fault: When the low-resistance connection is established between two points of different potential in a string of modules, then a bridge fault occurs. These faults are due to insulation failure of cables such as mechanical damage, water ingress, or corrosion.
d. Mismatch fault: Mismatches in PV modules are due to a change in electrical parameters of one or more modules. The mismatch fault will be due to the interconnection of solar cells or modules. These faults give rise to irreversible damage to PV modules and extensive power loss. They are difficult to point out because they will not lead to large currents.

The other types of faults generally occur in the cables. Such faults may be on the DC or AC side.

DC-side faults may be in the MPPT circuit and PV array. In the PV array, faults may be in the PV panel or cabling. In cabling, faults there are classified as a bridging fault.

Similarly, faults in the PV panel are classified as earth fault, bridging fault, open-circuit fault, and mismatch fault. Earth faults are further classified as upper and lower earth fault, and mismatch faults are further classified as temporary mismatch and permanent mismatch.

Temporary mismatch faults are partial shading and non–uniform temperature faults. Soldering and hot spot degradation faults are permanent mismatch faults. Cracks, delamination, and discoloration are also degradation faults.

AC-side faults are classified as grid outages and total blackout. The grid outages are of two types—one is an inverter fault and the other is a lightning fault. If there is any fault on the AC side, it leads to grid instability. It will mainly occur in rural areas and may cause unintentional islanding, which is harmful for O&M technicians.

Two particular characteristics of PV panels are their DC voltage levels and the fact that they cannot be shut off as long as PV modules are exposed to the sun. The short-circuit current produced by the PV module is too low to trigger the power supply's automatic disconnect. The most frequently used protective measures do not therefore apply to PV systems. However, because PV modules are installed outdoors, they are exposed to the elements. And because they can be installed on roofs, critical attention should be paid to the risk of fire and the protection of firefighters and emergency services staff.

There are a few diagnostic approaches for faults in PV systems:

- The fault can be diagnosed with the help of real-time observations by measuring the DC power output and impact of shadows on PV panels. The PV output is primarily affected by shadows, particularly in urban areas. These shadows may be because of trees, clouds, buildings, antennas, snow, or even other PV arrays.
- Immediate failure detection defines fast degradation in solar panels, which involves analysis of current and voltage components.
- The fault can be diagnosed using a ZigBee wireless sensor network (WSN) for photovoltaic power generation systems. The Solar Pro software can be used to simulate the photovoltaic module array and record the power generation data for solar radiation, module temperature, and fault conditions. The derived data would be used to establish the weights of the extension neural network.

The fault diagnosis in the PV power generation systems requires extracting the system's power-generating data with real-time solar radiation and module temperatures.

- Other studies use data measured by sensors to estimate energy production along with soft computing techniques. An example involves the extended correlation function and matter-element model to identify faults in a PV plant.
- An automatic monitoring system may also be used to detect a fault on the basis of power loss analysis.
- Artificial intelligent techniques and statistical data analysis are used to supervise PV systems for fault detection analysis and to clearly identify the type of fault.

12.2 Protection on the DC Side

It is obvious that photovoltaic arrays are installed either on roofs or in large and open locations. If lightning takes place in those locations, the system may be at great risk of a lightning strike. Direct discharges to the PV array, nearby strikes to earth, and cloud-to-cloud discharges may have damaging effects on the PV system and its components. The most noticeable effects from discharges are disastrous damage with visible carbonization of system components. Less manifest are the effects on the electrical system caused by the long-term exposure to repeated high transient voltages. These transients may cause premature component failure, resulting in substantial repair as well as loss of power generation. Solar panels must be designed to be safe from the effects of lightning. Lightning protection systems (LPSs) are used to protect the PV arrays against direct strikes to PV systems by a low-impedance grounding electrode system. Direct strikes to nearby grounded structures—flashes of magnitudes in excess of 100 kA—cause transient currents to be fed into the PV system wire loops. These transient voltages will appear at equipment terminals and cause insulation and dielectric failure of inverters, combiner boxes, PV modules (including bypass and blocking diodes), and other electrical and electronic equipment used for controls, instrumentation, and communications.

One method to overcome the issue of extinguishing the DC fault arc is to place a parallel fused bypass circuit around the metal oxide varistor (MOV), connected to each pole and to ground. An arc will appear across its opening contacts. However, that arc current would be redirected to the parallel path containing a fuse, where the arc would be extinguished and the fuse would operate, interrupting the fault current.

12.3 Protection on the AC Side

Direct lightning strikes to the utility service entrance or the nearby ground will cause a local ground potential rise (GPR) with regard to distant ground references. Conductors spanning long distances in the system will expose equipment to significant voltages. Surge protection should be placed at the service entrance to primarily protect the utility side of the inverter from damaging transients. The transients seen at this location are of a higher magnitude and duration—10/350 μsec versus 8/20 μsec—and therefore must be managed by surge protection with appropriately high discharge current ratings. Spark gap technology is used to discharge the full magnitude of "lightning currents" by providing equipotential bonding during the occurrence of a lightning transient. A coordinated MOV has the ability to compress the residual voltage to an acceptable level for the equipment it is designed to protect.

On the AC side there are two types of faults: total blackout measured as an exterior fault in the system against lightning, and an unbalanced voltage for AC part. The PV inverters comprising the transformers have good galvanic isolation and perfect electrical protection. When the fault in the inverter occurs, the AC output power will become low and DC output power will remain the same. This confirms that there is no possibility that a wire between modules/strings and the inverter was broken, or there could be breakdown occurs in strings and/or modules. A surge protection panel for a PV inverter, AC side, CPV240, is shown in Figure 12.1.

12.3.1 Surge Protection Panel for PV Inverter—AC Side

CPV240: AC Surge Protector Panel for 1-Phase PV Inverter: The CPV240 surge protector panels are designed to protect against the overvoltages caused by lightning, with the low-voltage side of the PV inverter linked to the low-voltage network. They integrate the following functions:

- Protection against the overvoltages (surge protector CITEL DS240) compliant with the NF EN 61643-11 standard
- Line and differential breaker
- Connection to network

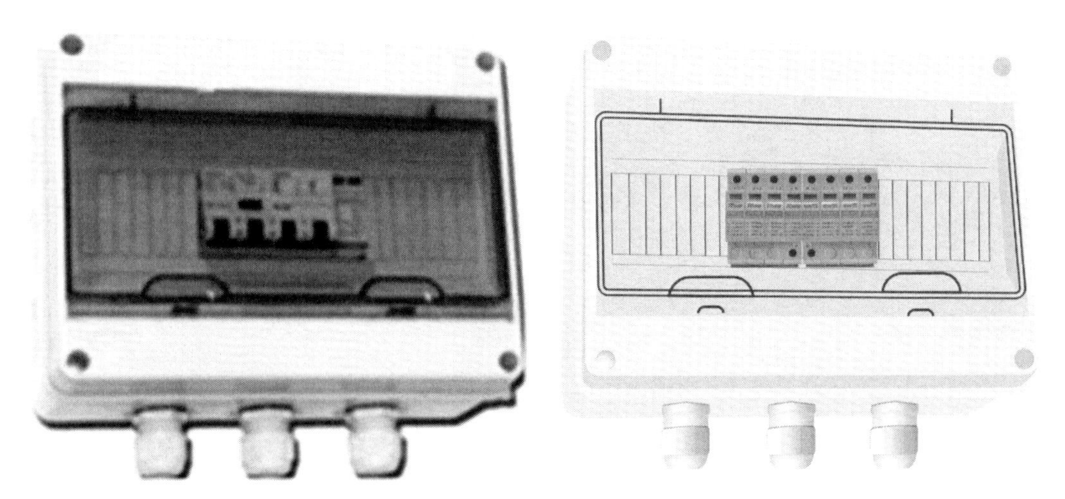

FIGURE 12.1
Surge panel for PV inverter.

The CPV240 panels are available for one-phase 230 V networks and several line currents and comply with the UTE C15-712 Guide requirement for photovoltaic installations linked to low-voltage networks. Specific versions are available by request.

Despite the high lightning risk that PV systems are exposed to, they are protected by the appropriate application of surge protection devices (SPDs) and a lighting protection system. However, care must also be given to the following:

- Proper equipotential bonding of all grounded members
- Proper grounding of the electrode system
- Strategic placement of the SPDs in the system, including integration into combiner boxes and inverters
- The connection quality from the system to the SPDs
- Adequacy of discharge rating of the SPD
- Voltage protection level of the SPD
- Suitability of the SPD for the system it is applied to (i.e., DC versus AC)
- Consideration of failure mode of SPDs and status indication
- Local and remote status indication and ease of replacing devices
- Suitability of SPDs on systems so as not to affect normal system function, specifically on nonpower systems

12.4 IEEE Standards for PV Protection

The following are the IEEE standards for PV system protections available in the literature:

- **Photovoltaic Systems—Standards:** In this category of photovoltaic system, planning and implementation are done for such systems. This includes safety regulations on PV system implementation.
- **Grid-Connected PV Systems:**
  a. IEC-60364-7-712, Electrical installations of buildings Part 7-712: Requirements for special installations or locations: Solar photovoltaic (PV) power supply systems.
  b. IEC-61727, Photovoltaic (PV) systems: Characteristics of the utility interface.
  c. IEC-61683, Photovoltaic systems: Power conditioners—Procedure for measuring efficiency.
  d. IEC-62093, Balance-of-system components for photovoltaic systems: Design qualification natural environments.
  e. IEC-62116, Test procedure of islanding prevention measures for utility-interconnected photovoltaic inverters.
  f. IEC-62446, Grid connected photovoltaic systems: Minimum requirements for system documentation, commissioning tests and inspection.
- **Monitoring**
  a. IEC-61724, Photovoltaic system performance monitoring: Guidelines for measurement, data exchange and analysis.

b. IEC-61850-7, Communication networks and systems for power utility automation Part 7-420: Basic communication structure—Distributed energy resources logical nodes.
c. IEC-60870, Tele-control equipment and systems.

12.5 General Safety Precautions

The following are safety precautions for PV systems:

1. The solar system is fully automatic with built-in safety features. Be careful while working on or repairing the system; it may cause dangerous electrical currents and may affect the human body.
2. Only a trained and certified professional must service the system.
3. Safety precautions must be taken during cleaning of the panels and avoid touching or disturbing the panels or wiring. It is important that only an authorized partner repair or touch the system.
4. Avoid stepping on the panels or allowing objects to fall on them.
5. Avoid disassembling or removing any part of the system that will void manufacturer warranties.
6. Small children and pets should be kept away from the inverter.
7. In case of any emergency such as fire, explosion, gas leakage, damage of system, or fuel spilling, contact emergency numbers at first; after that, shut down the system by turning your inverter off. Carefully check the inverter manual.

12.6 Prototype Protection Setup for Solar Power System

The power failures can be of short or long duration and indicate short-circuiting, overloading, and faults in generation, transmission, and distribution levels. Power quality is an important factor in many application areas. Generally, the voltage events are classified as undervoltage (long-term variations lasting in more than 1 minute and less than 90% of nominal value) and overvoltages (long-term variations lasting more than 1 minute and greater than 110% of nominal value). There are different types of power failures:

1. Voltage reduction in a network will damage the system's health and then leads to the malfunctioning of equipment.
2. Voltage rise due to switching and lightning phenomenon.
3. Blackouts are an extreme cause of power failures and isolate the network for a long time until the fault is cleared.

In order to provide electrical protection for standalone solar power systems, Titu Bhowmick and Dharmasa Hemadrasa (2015) proposed an Arduino microcontroller-based setup, shown in Figure 12.2. In the prototype two solid-state sensors for both sides of transformers (i.e., one on the primary side and the other on the secondary side) are used. The transformer receives 115 V as primary voltage through the contactor. The current sensors will detect the analog current reading for analog inputs of Arduino (A1 and A2) and take a 5 V supply from the Arduino, which can handle voltages between 0 and 5 V. The transformer prototype used in the circuit has a current of 1 A—for that a preset value is 0.7 A. Like most microcontroller-based protection relays (MBPRs), this has a real-time embedded system and in case of abnormal times, they operate with programmed time delays. Recently MBPRs have been developed to monitor the health of DC trip coil circuits and validate the status of input and output circuits.

In its operation, the loads are varied by a rheostat of 50 W capacity in order to decrease higher currents in the primary side. The readings taken by the sensors are sent to the

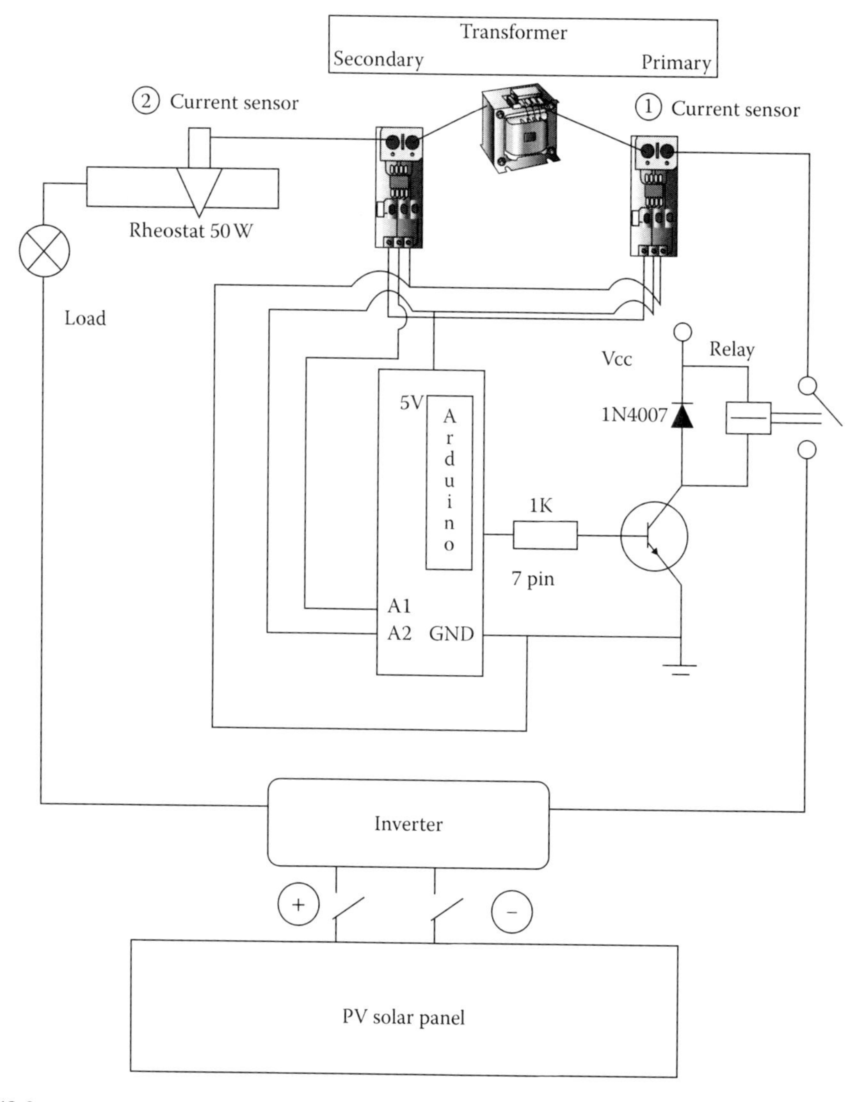

FIGURE 12.2
Protection setup for solar power system.

Arduino. A relay will act in inverse time overcurrent relay mode. The Arduino compares it with the preset value, and if the fault magnitude is more than the preset value, then Arduino sends a trip signal after a time delay. If the fault magnitude suddenly rises to a higher value than the preset value, the Arduino sends an instant trip signal. But if it reads more than the set value, the sensor sends the data to the Arduino, which compares the value with a preset one and finally the protection setup operates.

13

Economics of Grid-Connected PV Systems

13.1 Cost of Grid-Connected PV Systems

The cost to install a PV system is extremely capital-intensive. Apart from the capital costs there are various parameters that affect the cost of electricity generated by PV systems. These factors include the interest rate, installation service, etc. Moreover, in order to be valid, such cost estimates need to specify the interest rate, depreciation period, and other suppositions on which they are estimated. Life span is also an important parameter, as the payback period has to be considered to find the cost per unit of energy generated. It is safe to assume that the life span of solar modules and their wiring is anywhere from 15 to 30 years, and the life span of inverters, charge controllers, and other electronic components can be set at around 7 to 15 years. For standalone installations, battery replacement costs need to be figured in, as the life span of these components is only 2 to 10 years. The remaining upkeep cost mainly comprises servicing and repairs (particularly for inverters), insurance, and in some cases periodic solar generator cleaning.

In order to obtain reasonably accurate estimations of the cost of electricity from the PV installation, the total realization cost K_{TOT} for a complete PV installation may be broken down into the constituent costs K_G comprising the following elements: solar generator, voltage surge protection electronics components K_E such as inverters, charge controllers, etc. (7–15 years of depreciation), and battery cost K_A (2–10 years of depreciation, depending on the battery type and load).

$$\text{Total investment cost } K_{TOT} = K_G + K_E + K_A \tag{13.1}$$

where:

K_G = Costs attributed to the solar generator, site modification, wiring, voltage surge protection, and so on

K_E = Costs arising from inverters, charge controllers, and other electronic devices

K_A = Battery cost for standalone installations

Once these figures have been set, the annual cost K_J is derived from the depreciation rates that apply to the various types of costs involved, and the annual installation operating and maintenance costs K_B are determined as follows:

$$\text{Annual cost } K_{J\text{ tot}} = K_G * a_G + K_E * a_E + K_A * a_A + K_B \tag{13.2}$$

where:
a_G = Depreciation rate for net solar generator costs K_G (15–30 years of depreciation)
a_E = Electronic component depreciation rate (7–15 years of depreciation)
a_A = Battery depreciation rate (2–10 years of depreciation)
K_B = Annual operating, maintenance, and repair cost

13.2 Design of the Standalone SPV System

Figure 13.1 shows the block diagram of the household standalone PV system, where the function of the PV array is to convert the sunlight directly into DC electrical power and that of the battery is to store the excess power by using the battery charger. The inverter is used to convert the DC electrical power into AC power to match the requirements of the common household AC appliances.

The general rural area household is considered, which can fulfill the requirements of lighting and electrical appliances. The electrical load data in a residential house are given in Table 13.1.

13.2.1 PV System Design

To design a standalone PV system for this household, the following steps are required.

13.2.2 Sizing of the PV Array

The peak power of the PV generator (P_{PV}) is obtained as follows:

$$P_{PV} = \frac{E_L}{\eta_V \eta_R PSH} S_F \tag{13.3}$$

where:
E_L (daily energy consumption) = 8.44 kWh/day,
PSH (peak sun hours) = 5,
η_R (efficiency of the system component) = 0.92,
η_V (efficiency of the system component) = 0.9,
S_F (safety factor for compensation of resistive losses and PV cell temperature losses) = 1.15.

By substituting these values in Equation 13.3, we can obtain the peak power of the PV generator:

$$P_{PV} = 8.836\ kW_p \tag{13.4}$$

To install this power, a monocrystalline PV module type MBPV-CAAP (this module is from the Moser Baer Company) with a gross area of $A_{PV} = 1.65\ m^2$ and peak power of $P_{mpp} = 200\ Wp$ is selected. The number of the necessary PV modules (N_{PV}) is obtained from Equation 13.3:

$$N_{PV} = \frac{P_{PV}}{P_{mpp}} \tag{13.5}$$

$$= 44 \text{ modules are required}$$

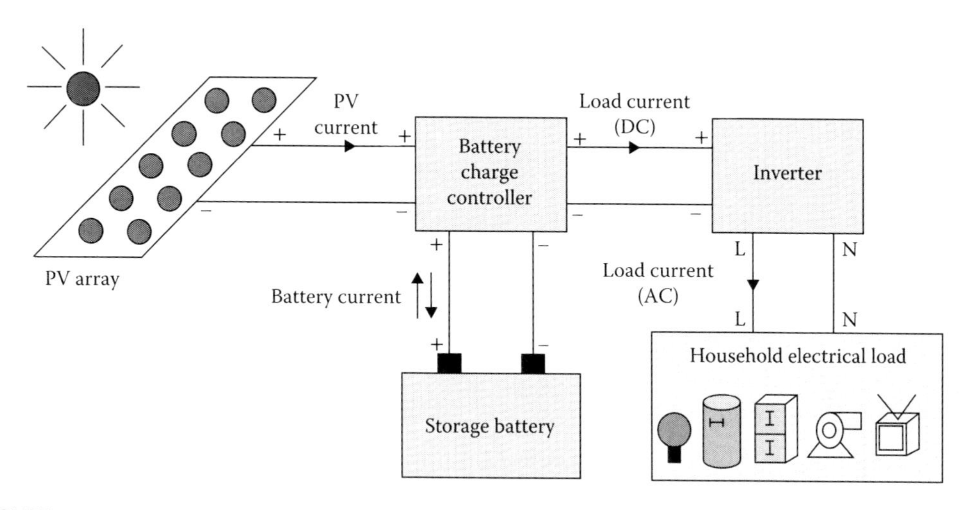

FIGURE 13.1
The standalone PV system.

TABLE 13.1
The Household Load Data

Electrical Load	No. of Units	Operating Hours Per Day	Wattage Per Unit Used	Total Units Consumed Per Day
Lighting lamp	5	4 lamps from 18 to 24 and 1 lamp from 0 to 6	60	1.8
Washing machine	1	From 11 to 13	250	0.5
Refrigerator	1	From 0 to 24	180	2.4
Water pump	1	For 3 hours daily	120	0.36
TV	1	From 17 to 24	80	0.56
Fan	3	From 0 to 24	60	1.44
Electric iron	1	For 1 hour in 2 days	1500	0.75
AC 1.5 ton	2	6 hours each daily	2000	24
Total			6530	31.81

V_{mp} for MBPV200 watt peak is 28.81 volts.

$$\text{No. of modules in series} = \frac{220 \times 1.414}{28.81} = 11$$

Hence the number of modules in parallel = 4
For the MBPV CAAP 200 Wp module:

V_{oc} (open-circuit voltage) = 35.99 volts and
I_{sc} (short-circuit current) = 7.65 amperes

Now, we obtain an open-circuit voltage and short-circuit current for this SPV array as:

$$V_{oc} = 35.99 \times 11 = 395.89 \text{ V}$$
$$I_{sc} = 7.65 \times 4 = 30.6 \text{ A.}$$

13.2.3 Sizing of the Battery Block

The storage capacity of the battery block for such systems is considerably large. Therefore, special lead-acid battery cells (block type) with a long life span, high cycling stability rate, and capability of withstanding very deep discharges should be selected. Such battery types are available, but at a much higher price than regular batteries. The ampere-hour capacity (C_{Ah}) and watt-hour capacity (C_{wh}) of the battery block necessary to cover the load demands for a period of 2 days without sun are obtained as follows from Equation 13.6 and Equation 13.7:

$$C_{Ah} = \frac{1.5 \times E_L}{V_B \times DoD \times \eta_B \times \eta_V} \tag{13.6}$$

$$C_{WH} = C_{Ah} V_B \tag{13.7}$$

where V_B and η_B are the voltage and efficiency of the battery block, and DoD is the permissible depth of discharge rate of a cell. Assuming realistic values of 6 $\eta_B = 0.85$, DoD = 0.75, and $V_B = 220\ V$, we obtain:

$$C_{Ah} = 500 \text{ Ah}$$
$$C_{wh} = 110 \text{ kWh}$$

13.2.4 Design of the Battery Charge Controller

The battery charge controller is required to safely charge the batteries and to maintain a longer lifetime for them. It has to be capable of carrying the short-circuit current of the PV array. Thus, in this case, it can be chosen to handle 31 A.

13.2.5 Design of the Inverter

The inverter must be able to handle the maximum expected power of AC loads. Therefore, it can be selected as 20% higher than the rated power of the total AC loads, and the rated power of the inverter becomes 7836 W. The specifications of the required inverter will be 7836 W, 220 VAC, and 50 Hz.

The cost of the equipment required in the proposed SPV system is presented in Table 13.2.

13.3 Life Cycle Cost Analysis

In this section the life cycle cost (LCC) estimation of the standalone PV system is discussed. The LCC of an item consists of the total costs of owning and operating an item over its lifetime, expressed in today's money.

TABLE 13.2

The Used Cost Data of All Items

Item	PV	Battery Cell for 510 Ah	Charger	Inverter	Installation	M&O/ Year
Cost	Rs 110/Wp	Rs 5500/cell	Rs 258/A	Rs 36/W	10% of PV cost	2% of PV cost

1 USD = INR 65.42 as on 22nd March 2017.

The costs of a standalone SPV system include acquisition costs, operating costs, maintenance costs, and replacement costs. All these costs have the following specifications:

- The initial cost of the system (the capital cost) is high.
- There are no fuel costs.
- Maintenance costs are low.
- Replacement costs are low (mainly for batteries).

The LCC of the PV system includes the sum of all the present worth (PW) of the costs of the PV modules, storage batteries, battery charger, inverter, the cost of the installation, and the maintenance and operation cost (M&O) of the system. The details of the cost data for all items are shown in Table 13.2.

The lifetime N of all the items is considered to be 30 years, except that of the battery, which is considered to be 5 years. Thus, an extra five groups of batteries (with six batteries each) have to be purchased after 5 years, 10 years, 15 years, 20 years, and 25 years, assuming an inflation rate, i, of 3% and a discount or interest rate, d, of 10%. Therefore, the PWs of all the items can be calculated as follows:

PV array cost $C_{PV} = 110 \times 44 \times 200 = \text{Rs. } 968000$
Initial cost of battery cells $C_B = 110 \times 5500 = \text{Rs. } 605000$

The PW of the first extra group of batteries (purchased after N = 5 years) C_{B1PW} can be calculated to be Rs. 435479 from Equation 13.6.

$$C_{B1PW} = C_B\left(\frac{1+i}{1+d}\right)^N \tag{13.8}$$

The PW of the second extra group of batteries (purchased after N = 10 years) C_{B2PW}, the third extra group (purchased after N = 15 years) C_{B3PW}, the fourth extra group of batteries purchased after N = 20 years C_{B4PW}, and the fifth extra group of batteries purchased after N = 25 years C_{B5PW} are calculated using Equation 13.4 to be Rs. 313,450.5, Rs. 225,665, Rs. 162,563.5, and Rs. 117,007, respectively.

Charger cost $C_C = 258 \times 31 = \text{Rs. } 8000$
Inverter cost $C_{Inv} = 36 \times 7836 = \text{Rs. } 282096$
Installation cost $C_{Inst} = 0.1 \times 968000 = \text{Rs. } 96800$
Operation and maintenance cost per year = Rs. 19360

The PW of the maintenance cost C_{MPW} can be calculated to be Rs. 245,383.68, using the maintenance cost per year (M/yr) and the lifetime of the system (N = 30 years), from Equation 13.7:

$$C_{MPW} = \left(\frac{M}{yr}\right) \times \left(\frac{1+i}{1+d}\right) \times \left(\frac{1-\left(\frac{1+i}{1+d}\right)^N}{1-\left(\frac{1+i}{1+d}\right)}\right) \tag{13.9}$$

Therefore, the LCC of the system can be calculated to be Rs. 3,459,566 from Equation 13.10:

$$LCC = C_{PV} + C_B + C_{B1PW} + C_{B2PW} + C_{B3PW} + C_C + C_{Inv} + C_{Inst} + C_{MPW} \tag{13.10}$$

It is sometimes useful to calculate the LCC of a system on an annual basis. The annualized LCC (ALCC) of the PV system in terms of present-day Rupees can be calculated to be Rs. 255,579.6/yr, from Equation 13.11:

$$ALCC = LCC\left[\frac{1-\left(\frac{1+i}{1+d}\right)}{1-\left(\frac{1+i}{1+d}\right)^N}\right] \tag{13.11}$$

Once the ALCC is known, the unit electrical cost (cost of 1 kWh) can be calculated to be Rs. 22/kWh from Equation 13.2:

$$\text{Unit Electricity Cost} = \frac{ALCC}{365E_L} \tag{13.12}$$

13.4 Case Study 1

13.4.1 Methodology

To find out the cost analysis for a grid-connected solar PV plant in India, many factors are considered. Because solar radiations differ from months to months, the average monthly outputs are established and related graphs are plotted to show the variations. A case study is presented here reported by Viney et al. The project ran from January to December. For calculating the output, the efficiency of the PV module is taken as 13.2%. The chosen area for the estimated plant capacity is 101,533 m^2.

The system cost is around Rs. 60,600,000.

13.4.2 Difference in Power Consumption Bill

The difference in the bill annually after connecting the SPV generation system was found to be Rs. 7,589,307. The typical cost of PV Panel, three phase inverter, three phase transformer, cost of system without battery, cost of system with all auxiliary and misc. cost is provided in Table 13.3 to Table 13.7.

TABLE 13.3

PV Panel Cost with Subsidy

Cost of 1 kWp rooftop solar PV	1	100,000
Cost of 350 kWp rooftop solar PV	400 kW	40,000,000
Subsidy @ 30%	0.3	12,000,000
Net cost after subsidy		28,000,000
Accelerated depreciation @ 80%	0.8	22,400,000
Tax rate @ 35%	0.35	7,840,000
Tax saved through accelerated depreciation		20,160,000

TABLE 13.4

Cost of Three-Phase Inverter

Cost of three-phase inverter in (Rs/Watt)	25
Size of inverter	500 kVA
Total cost for three-phase inverter	12,500,000

TABLE 13.5

Cost of Three-Phase Transformer

Cost of three-phase step-up transformer	20
Size of transformer	500 kVA
Total cost for three-phase inverter	10,000,000

TABLE 13.6

Cost of System Without Online Battery

Net cost after subsidy	28,000,000
Total cost of three-phase inverter	12,500,000
Total cost of three-phase transformer	10,000,000
Net cost	50,500,000

TABLE 13.7

Cost of System With all Auxiliary and Misc. Cost

Subtotal above by 20% to cover balance of system cost	0.2
Cost estimate for balance of system (net cost for whole setup * 0.20)	50,500,000
Total estimated PV system cost in RS	10,100,000

13.4.3 Payback Period Calculation

The payback period is given by X/Y:

where:
X = Total cost of PV system with all auxiliary equipment
Y = Total annual cost saving after installation of PV system
Payback period = 8 years

Profit after payback period until useful life of SPV

The useful life of the PV is considered to be 25 years, and the payback period of the solar PV is calculated to be 8 years, so by subtracting the useful life from the payback period and then multiplying the difference in the bill amount, a profit of Rs. 13 crore is reported.

Profit = (Useful life payback period) * difference amounted in bill after connection of PV

13.4.4 Operation and Maintenance Costs

Key factors that influence operation and management costs are:

1. Availability of salt-free water and manpower to clean the modules as prescribed (around 15 k to 20 k liters of water is required per MW per wash depending on the level of scale formation)

TABLE 13.8
Calculation if required Roof Area for 500 kW Solar PV Installation

Length	1690 mm
Breadth	990 mm
Area covered by one panel	1,673,100 mm^2
Applying trigonometry for effective area calculation	
Number of panels required for 500 kW	1667
Extra area added for clearance	25 mm^2
Area covered by 1667 panels (m^2)	69,712
Present build up area in SDM College	101,533 m^2

FIGURE 13.2
Maintenance of PV arrays.

The Calculation of Roof Area for 500 kW Solar PV Installation and the Maintenance of PV arrays are presented in Table 13.8 and Figure 13.2 respectively.

2. Control growth of shrubs and plants around the modules
3. Because a 100 MW solar system is spread out and involves monitoring more than 1,500 strings and roughly around 24 modules connected to each string, continuous monitoring so as to ensure desired output is required

13.4.5 PV Module

Although the PV module does not have any moving parts, the yield from the module is highly dependent on the panels being cleaned.

A rule of thumb of Rs. 5 to 6.5 lakh/MW can be taken as the guideline for maintenance and operation costs.

13.4.6 Electric Subsystems

In India, rodents are known to cause damage to underground cables, and it is important to monitor this regularly. Earthing protection also needs to be checked often.

13.4.7 Civil and Structural Subsystem

One of the tasks of the operation and maintenance personnel will be to make sure that the growth of shrubs and other vegetation are kept under control.

The other problems are related to the mounting structures; in some cases structures can bend due to improper design.

During the rainy season it has been observed that the topsoil gets washed away due to improper drainage systems. Caving in of the foundation and the structure caused by improper compaction of the ground has also been observed.

13.4.8 Communication

Most of the solar power plants are located in remote places with unreliable communication infrastructures. The remote monitoring system needs an Internet connection, and in the absence of a reliable connection, there could be a problem of a lack of data logging for long period of time.

13.4.9 Comparison of Investment in Solar PV and Wind Power Plants

It really depends on geographical locations and energy requirements, but given the substantial investment involved with either option, it is critical to select the right system for your needs from the outset.

A solar power system:

1. Has no moving parts
2. Has better reliability and a 25-year warranty
3. Requires less monitoring
4. Provides a better value for the money in sites with an average wind speed less than 5 m/s
5. Is less conspicuous than a wind turbine
6. Is totally silent in operation
7. Is less susceptible to high wind damages
8. Is less susceptible to lighting damages
9. Requires less space in most cases, as the panels can be installed on a roof

Lower cost per kWh produced

Annual output-16,500 kWh

Forty-four solar panels will produce 16,500 kWh annually. The gross system cost is approximately Rs. 23.40 lakhs. After tax credits and depreciation, the net system cost is Rs. 5.22 lakhs.

A 10 kW wind turbine in an area with decent wind exposure (12 mph) will produce 16,500 kWh. The gross installed system is around Rs. 33 lakhs. After tax credits and depreciation, the net system cost is Rs. 13.28 lakhs.

Thus we can see that solar is the obvious choice for small-scale needs. From maintenance and noise to warranties and economics, solar is much better than wind.

13.5 Case Study 2

The considered SPV system is shown in Figure 13.3 with the basic components shown in different blocks.

13.5.1 Sizing of the SPV System

The module selected for the proposed SPV system is MBPV CAAP (commercially available) of 230 Wp; the specifications of this module are given in Table 13.9. Before designing the system certain assumptions regarding efficiencies of various components of the SPV system have been taken into consideration, as mentioned in Table 13.10. There are various assumptions to be taken in the design of PV system. The assumptions for a typical system are presented in Table 13.11.

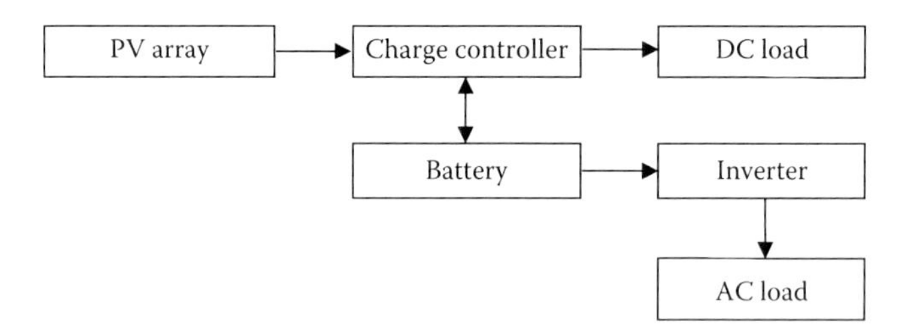

FIGURE 13.3
Basic block diagram of SPV system.

TABLE 13.9
Associated Accessories of the PV System

S. No.	Items
1.	SPV module (230 Wp ±2%)
2.	Module mounting structure
3.	Array junction box (5 In–1 Out)
4.	Main junction box (3 In–1 Out)
5.	1 core 4 sq. mm PVC Cu cable
6.	2 core 16 sq. mm PVC Cu cable
7.	1 core 70 sq. mm PVC Cu cable
8.	1 core 25 sq. mm PVC Cu cable
9.	Power conditioning unit-SCC: 45 kVA, and inverter and charger 20 kVA, 03 phase @ 230V
10.	Battery bank 150 Ah @ 12V Tubular LMLA
11.	DC distribution board
12.	AC distribution board
13.	Lightning arrestors
14.	Electrical accessories
15.	Earthing kit
16.	Circuit breakers
17.	Installation cost/civil works
18.	Maintenance and operating costs

TABLE 13.10

Specifications of the Selected Module

MBPVCAAP	230 Wp
Maximum power, P_{max} (W)	230
Voltage at P_{max}, V_{mp} (V)	30.0
Current at P_{max}, I_{mp} (A)	7.67
Open circuit voltage V_{oc} (V)	36.83
Short circuit current I_{sc} (A)	7.95
Temperature coefficient of P_{max} (%/K)	–0.43
Temperature coefficient of V_{oc} (%/K)	–0.344
Temperature coefficient of isc (%/K)	0.11
Power tolerance (%)	± 3
Maximum system voltage (IEC/UL) (V DC)	1000/600
Cells per by-pass diode (Nos)	20
System voltage	120 V DC
Maximum voltage	250 V

13.5.2 Calculation of Plant Rating for IEEE33 Bus System

Energy required per day = 2633 kW * 24 hrs = 63,192 = 64,000 kWh

Power in kW required for generating required energy in 8 hrs = 64,000/8 hrs = 8000 kW = 8 MW.

Taking into consideration losses in different stages, only 60% of power is available, so to have a total installation capacity of 8 MW/0.6 = 13.33 MW = 13.5 MW.

13.5.3 Calculation of Plant Rating for IEEE69 Bus System

Energy required per day = 783 * 24 = 18,792 kWh = 18,800 kWh

Power in kW required for generating required energy in 8 hrs = 18,800/8 hrs = 2350 kW = 2.35 MW.

Taking consideration losses in different stages, only 60% of power is available, so to have a total installation capacity of 2.35 MW/0.6 = 3.91 MW = 4 MW.

13.5.4 Battery Bank Calculations

The storage capacity of the battery should be considerably large so as to provide power during the period when solar radiations are insufficient to provide adequate supply. Another desirable feature of battery storage is that it should be able to bear the deep discharges.

Based on the calculations presented in the previous section, the specifications of the battery bank selected are given in Table 13.12. The battery bank calculations are given in Table 13.13 for IEEE 33 bus systems and in Table 13.14 for IEEE 69 bus systems.

13.5.5 Calculation of Battery Rating for IEEE 33 Bus Systems

Energy generated = 64,000 kWh

Charging voltage = 12 volts

Ampere-hour required is = 64,000/ (12 * 0.8) = 6666.66 kAh = 6,666,666.66 Ah

TABLE 13.11

Assumptions for PV System Design

Battery efficiency	0.85
Annual average sunshine hours	5.5 pk. hrs.
Inverter efficiency	0.9
Derating of module output	0.95
Other losses	0.95
Total load	854 Ah/day

TABLE 13.12

Battery Specifications

Battery type	150 Ah
Individual cell voltage	2 Volts
Number of batteries in series	6
Battery bank voltage	12 Volts

TABLE 13.13

Battery Bank Calculations for IEEE 33 Bus Systems

Load operating duration	As per load energy sizing sheet
Number of days of autonomy	1.0
Maximum DoD permissible	0.8
Correction factor for 16°C temperature	1.0
Battery capacity required	2135 Ah
Discharge current	110.42 amperes
Discharge rate	19 hour rate
Required battery capacity	2102 Ah

TABLE 13.14

Battery Bank Calculations for IEEE 69 Bus Systems

Load operating duration	As per load energy sizing sheet
Number of days of autonomy	2.0
Maximum DoD permissible	0.8
Correction factor for 16°C temperature	1.0
Battery capacity required	2135 Ah
Discharge current	110.42 amperes
Discharge rate	19 hour rate
Required battery capacity	2102 Ah

One battery rating is 150 Ah
So number of batteries required = 6,666,666.66/150 = 44,444.44 = 44,445

13.5.6 Calculation of Battery Rating for IEEE 69 Bus Systems

Energy generated = 18,800 kWh
Charging voltage = 12 volts
Ampere-hour required is = 18,800/(12 * 0.8) = 1958.33 kAh = 1,958,333.34 Ah

One battery rating is 150 Ah
So number of batteries required = 1,958,333.34/150 = 13,055.55 = 13,056

13.5.7 Charge Controller

The charge controller regulates the charge entering in the battery (i.e., it controls the charging current according to the state of the battery). If the battery is fully charged, then it disconnects the battery from the PV module; if the battery is fully discharged, then it disconnects the load from the battery.

To prevent the battery from deep discharging, load shedding or complete disconnection of the load is provided by the load management system. The SPV system charge controller can be specified by the following parameters:

1. DC system voltage
2. Array amps
3. Maximum DC load amps

13.5.8 Inverter

Inverters are used to convert DC voltage into the equivalent AC voltage. Inverters can be classified according to the wave shape of the output voltage viz. rectangle, trapezoid, or sine shaped. The best-quality inverters are the ones where the output voltage is matched perfectly to the sine wave, but at the same time they are the most expensive. The input voltage of the inverter depends on its power rating. Inverters with large power ratings need large voltage at the input terminals so that the inverter losses can be reduced.

13.6 Result and Discussion

An economic analysis is carried out for the feasibility of the proposed system. The installation and running costs of the proposed systems are compared with the cost of power received by the conventional grid. Table 13.17 and Table 13.18 show the cost of electricity for one year, including fixed charges, variable energy charges, and taxes. The cost of total expenditure is calculated for 25 years, assuming the inflation rate of 10%.

13.6.1 Cost Evaluation of the IEEE 33 Bus Distribution System

The connection MW rating is taken as the amount of power injected from the distributed generation (DG) and the saving in losses when the DG is operating, so the total MW rating of 2.735 MW (i.e., power injected from DG as well as saved losses) is considered for IEEE 33 bus systems for DG at a single location. For a load of 3.72 MW, the power supplied by the grid is the total power 3.93 MW (i.e., load consumption 3.72 MW plus the losses 0.210 MW). When the power to the load is supplied through DG in addition to the grid, the total power drawn from the grid reduces. If the DG is connected at one location with a capacity of 2.633 MW, not only does the power drawn from the grid decrease, but also the losses reduce to a greater extent (i.e., 0.11031 MW). In this case the only power drawn from the grid is calculated as 1.197 MW. If the DG supplies the same power of 2.633 MW at two locations, now the total power drawn from the grid reduces to 1.18 MW. It should be noted that not only does the net power drawn from the grid decrease, but also the losses are further reduced to 0.09349 MW. When the DG with the same power capacity of 2.633 MW supplies power at three locations, the power supplied by the grid further reduces to 1.177 MW in addition to the reduction in losses up to 0.0924 MW. It appears that the reduction in losses does not make an appreciable difference for the DG at two and three locations, respectively. However, when the calculations are made in terms of economic analysis, the financial savings increased from $484.34 million to $486.95 million for the DG at one and two locations. When the DG is considered at three locations, the savings are further increased to $487.53 million, and it also proved that it improves the system voltage profile.

13.6.2 Cost Evaluation of IEEE 69 Bus Distribution System

The economic analysis for the IEEE 69 bus system has been presented in Table 13.18. For a load of 3.80 MW, the power supplied by the grid is the total power 4.036 MW (i.e., load consumption 3.8 MW plus the losses 0.23632 MW). When the power to the load is supplied through the DG in addition to the grid, the total power drawn from the grid reduces. If the DG is connected at one location with a capacity of 0.783 MW, not only does the power drawn from the grid decrease, but also the losses reduce to a greater extent (i.e., 0.19166 MW). In this case, the only power drawn from the grid is calculated as 3.2083 MW. Now if the DG supplies the same power of 0.783 MW at two locations, the total power drawn from the grid reduces to 3.1883 MW. It should be noted that not only does the net power not drawn from the grid decrease, but also the losses are further reduced to 0.17159 MW.

TABLE 13.15

Cost of Energy Calculation When Power is Fed From Grid for IEEE 33 Bus Test System

	DG Location		
Annual Electricity Bill Calculation	**DG at 1 Location**	**DG at 2 Location**	**DG at 3 Location**
Total power not drawn from grid in kW	2732.94	2749.76	2753
If we take the connection of 2735 kW @ 0.9 pf (kVA)	3038.89	3055.29	3058.89
Fixed charge of per kVA @ 125 **(A)**	4,558,333.33	4,582,933.33	4,588,333.33
Energy per year at pf 0.9 = kWh/0.9 = kVAh	26,620,666.67	26,764,330.67	26,795,866.67
Energy charge per kVAh @ 8.40 = **(B)**	223,613,600	224,820,377.6	225,085,280
TOD surcharge on energy charges @ 20% = **(C)**	44,722,720	44,964,075.52	45,017,056
Surcharge on fixed charge @ 8% = **(D)** = (A*0.08)	364,666.66	366,634.66	367,066.66
Surcharge on energy charge charge @ 8% = **(E)** = ((B + C) * 0.08)	21,466,905.6	21,582,756.25	21,608,186.88
Electricity tax @ 5% on energy charge = **(F)** = ((B + C + E) * 0.05)	14,490,161.28	14,568,360.47	14,585,526.14
Total bill for one year = **(T)** = (A + B + C + D + E + F)	309,216,386.9	310,885,137.8	311,251,449
Annual electricity bill calculation for 25 years assuming yearly inflation of 10%	3041.05 Crores	3057.46 Crores	3061.06 Crores
Figures in million dollars	484.337	486.951	487.525

TABLE 13.16

Cost of Energy Calculation When Power is Fed From Grid for IEEE 69 Bus Test System

	DG Location		
Annual Electricity Bill Calculation	**DG at 1 Location**	**DG at 2 Location**	**DG at 3 Location**
Total power not drawn from grid is in kW	827.66	847.73	852.64
If we take the connection of 828 kW @ 0.9 pf (kVA)	919.62	941.92	947.37
Fixed charge of per kVA is @ 125 **(A)**	1,379,433.33	1,412,883.33	1,421,066.66
Energy per year at pf 0.9 = kWh/0.9 = kVAh	8,055,890.66	8,251,238.66	8,299,029.33
Energy charge per kVAh is @ 8.40 = **(B)**	67,669,481.6	69,310,404.8	69,711,846.4
TOD surcharge on energy charges @ 20% = **(C)**	13,533,896.32	13,862,080.96	13,942,369.28
Surcharge on fixed charge @ 8% = **(D)** = (A*0.08)	110,354.66	113,030.66	113,685.33
Surcharge on energy charge charge @ 8% = **(E)** = ((B + C) * 0.08)	6,496,270.23	6,653,798.86	6,692,337.25
Electricity tax @ 5% on energy charge = **(F)** = ((B + C + E) * 0.05)	4,384,982.40	4,491,314.23	4,517,327.64
Total bill for one year = **(T)** = (A + B + C + D + E + F)	93,574,418.56	95,843,512.85	96,398,632.58
Annual electricity bill calculation for 25 years assuming yearly inflation of 10%	920.27 Crores	942.59 Crores	948.05 Crores
Figures in million dollars	146.569	150.123	150.992

When the DG with the same power capacity of 0.783 MW is supplying power at three locations, it is observed that the power supplied by the grid further reduces to 3.1834 MW in addition to the reduction in losses up to 0.16668 MW. It appears that the reduction in losses is not having an appreciable effect on the DG at two and three locations, respectively. But when the calculations are made in terms of economics, the financial savings increased from $146.57 million to $150.12 million for the DG at one and two locations. When the DG is considered at three locations, the savings are further increased to $150.99 million, and it also proved that it improves the system voltage profile. Cost of Energy Calculation When Power is Fed From Grid for IEEE 33 and 69 Bus Test System is presented in Table 13.15 and Table 13.16 respectively.

13.6.3 Cost Evaluation of the SPV System

In Table 13.17 the installation and maintenance costs of a solar photovoltaic–based DG has been calculated. To provide 2.633 MW of power from the SPV-based DG continuously, 13.5 MW solar plants are required for 33-bus system and to provide 0.783 MW of power from an SPV-based DG continuously 4 MW solar plants are required for a 69-bus system. Because solar power can be generated only during the daytime and the DG has to supply power for the entire day, it must store the power in the battery storage system. During the daytime solar-based DG power will be fed to the system as well as storing the power in battery; when no power is generated, the stored power will be supplied. The conversion efficiency of the solar power plant is considered to be 60%. In second column, detailed costs of a 1 MW plant is given, and in the third and fourth columns, the interpolated cost of 13.5 MW and 4 MW, including storage costs, is calculated for the lifetime of 25 years for

TABLE 13.17

Installation and Maintenance Costs of SPV-Based DG for IEEE 33 and IEEE 69 Bus Test System

Fixed Cost for 1 MW SPV Plant		13.5 MW for 33-Bus System	4 MW for 69-Bus System
1. Solar panels German tech. For 1 MW	0.958		
2. Central inverters (4)	0.162		
3. Combiner + junction boxes	0.048		
4. Protective gears arrangement	0.016	1.225*13.5 MW = 16.540	1.225*4 MW = 4.9
5. SCADA and data logger system	0.016		
6. Land bank	0.008		
7. Erection of project	0.016		
Total cost for 1 MW SPV plant	1.225		
Maintenance cost for 1 MW SPV plant			
1. Human resource per year	0.0323	Maintenance cost for 25 years, including 10% inflation = 47.212	Maintenance cost for 25 years, including 10% inflation = 13.892
2. PV maintenance per year	0.0016		
3. Site maintenance per year	0.0016		
Total maintenance cost	0.0355		
Consumable material		44,445*12 cycle	13,056*12 cycle
1. Batter per piece (150 Ah) @ **0.000145* 44,445**		77.584	22,634
Total cost of the project for 25 years		141.337	41.39

*All figs in million dollar ($)

TABLE 13.18

Result of Economic Comparison for IEEE 33 and IEEE 69 Bus System at 11 kV

	33-Bus System		69-Bus System	
Cases	**Cost of Grid Energy**	**Cost of SPV Based DG**	**Cost of Grid Energy**	**Cost of SPV Based DG**
DG at 1 location	484.37	141.337	146.57	41.39
DG at 2 location	486.95		150.12	
DG at 3 location	487.53		151.00	

SPV. Installation and maintenance costs of the SPV-based DG consist of the panel, storage system, land, protective equipment, inverter, etc., and 10% inflation is estimated over the period of 25 years.

13.7 Comparison of PV and Conventional Electricity Costs

In Table 13.18, the cost of the proposed SPV-based MLDG is compared with the cost to purchase energy when power is drawn from the conventional grid. The proposed SPV-based MLDG is very cost-effective and gives a total minimum savings of around $343.033 million for the IEEE 33-bus primary test distribution system for a single location, and for the MLDG at two and three locations, the savings are $345.613 million and $346.193 million, respectively. For the IEEE 69-bus test system, the total minimum savings is $105.18 million for a single location, $108.73 million at two locations, and $109.61 million at three locations.

14

System Yield and Performance

14.1 Determination of the Energy Yield of a Grid-Connected PV System

The use of appropriate performance parameters facilitates the comparison of grid-connected PV systems that may differ with respect to design, technology, or geographic location. Four performance parameters that define the overall system performance are final PV system yield, reference yield, performance ratio, and Photovoltaics for Utility Scale Applications (PVUSA) rating. All four parameters are described here.

The final PV system yield (Y_f) is the net energy output, *E*, divided by the nameplate DC power P_0 of the installed PV array. It represents the number of hours that the PV array would need to operate at its rated power to provide the same energy. The Y_f normalizes the energy produced with respect to the system size; consequently, it is a convenient way to compare the energy produced by PV systems of differing size:

$$Y_f = \frac{E}{P_0} \text{ (kWh/kW) or (hours)} \tag{14.1}$$

The reference yield (Y_r) is the total in-plane irradiance *H* divided by the PV's reference irradiance G. It represents an equivalent number of hours at the reference irradiance. If G equals 1000 W/m², then Y_r is the number of peak sunshine hours or the solar radiation in units of kWh/m². The Y_r defines the solar irradiance resource for the PV system. It is a function of the location, orientation of the PV array and weather variability (like month-month or year to year):

$$Y_r = \frac{H}{G} \tag{14.2}$$

The performance ratio is the ratio between actual yield Y_f (i.e., annual production of electricity delivered at AC) and the target yield Y_r:

$$\text{PR} = \frac{\text{Actual Yield}_{AC}}{\text{Target Yield}_{DC}} = \frac{E}{hA\eta_{nom}} = \eta_{pre}\eta_{rel}\eta_{sys} \tag{14.3}$$

The performance ratio, often called the "quality factor," is independent from the irradiation and therefore useful to compare systems. It takes into account all pre conversion losses, inverter losses, thermal losses, and conduction losses. It is useful to measure the performance ratio throughout the operation of the system, as a deterioration could help pinpoint causes of yield losses.

The energy, E, delivered by a system with area, A, can be estimated from:

$$\left(\frac{E}{A}\right) = h\eta_{pre}\eta_{sys}\eta_{rel}\eta_{nom} = PRh\eta_{nom} \begin{cases} \eta_{pre} & \text{Pre-conversion efficiency} \\ \eta_{sys} & \text{System efficiency} \\ \eta_{rel} & \text{Relative module efficiency} \\ \eta_{nom} & \text{Nominal module efficiency} \\ h & \text{Yearly sum of global irradiance [kWh/m}^2\text{]} \end{cases} \tag{14.4}$$

The pre-conversion efficiency reflects the losses incurred before the beam hits the actual semiconductor material, caused by shading, dirt, snow, and reflection off the glass. The system efficiency reflects electrical losses caused by wiring, inverters, and transformers. The module itself is defined by a nominal efficiency and a relative efficiency.

14.1.1 Energy Per Rated Power

Sometimes, the energy yield is expressed in terms of the peak power of the module, which is independent from the area of the module. It is (with H_0= 1,000 W/m^2):

$$\left(\frac{E}{P_{peak}}\right) = \frac{h}{H_0}\eta_{pre}\eta_{sys}\eta_{rel} = \left(\frac{E}{A}\right)\frac{1}{H_0\eta_{nom}} = PR\frac{h}{H_0} \tag{14.5}$$

This is a very useful ratio, because the energy yield E is a measure of the earnings potential, and the peak power reflects the cost of the system.

PR values are typically reported on a monthly or yearly basis. Values calculated for smaller intervals, such as weekly or daily, may be useful to identify occurrences of component failures. Because of losses due to PV module temperatures, PR values are greater in the winter than in the summer and normally fall within the range of 0.6 to 0.8. If PV module soiling is seasonal, it may also affect differences in PR from summer to winter. Decreasing yearly values may indicate a permanent loss in performance.

The PVUSA rating method uses a regression model and system performance and meteorological data to calculate power at PVUSA test conditions (PTC), where PTC are defined as 1000 W/m^2 plane-of-array irradiance, 20°C ambient temperature, and 1 m/s wind speed. PTC differ from standard test conditions (STC) in that its test conditions of ambient temperature and wind speed will result in a cell temperature of about 50°C, instead of the 25°C for STC. This is for a rack-mounted PV module with relatively good cooling on both sides. For PV modules mounted close to the roof or integrated into the building with the airflow restricted, PTC will yield greater cell temperatures. The difference between the nameplate DC power rating and the system PVUSA rating is an indication of the total system losses associated with converting DC module energy into AC energy. As with decreasing PR values, decreasing PVUSA ratings over time may indicate a permanent loss in performance.

TABLE 14.1

Various types of energy losses

Losses	Description
Pre-module/panel losses: The losses may be often up to 5%	
Shadow losses	The Shadows may be caused by trees, chimneys etc. Depending on the stringing of the cells, partial shading may have a significant effect.
Dirt losses	It varies depending on wather condition like due to dirt up to 4% in temperate regions with some frequent rain and it may go up to 25% in arid regions with only seasonal rain and dust.
Snow losses	It depends on location and maintenance of PV panels.
Reflection losses	Reflection losses depends on angle of incidence. It increases with angle of incidence. It is less in cloudy conditions or more diffuse irradiance.
Module/panel losses	
Conversion losses	The nominal efficiency is given by the manufacturer for standard conditions.
Thermal losses	These losses depend on temperature. If the temperature is more the conversion losses will be more increase. Also these losses depends on irradiance (i.e. location), mounting method (glass, thermal properties of materials), and wind speeds. Roughly it is around 8%
System Losses	
Losses in wires	Any cables have some resistance and therefore it causes losses.
MPP losses	Ability of the MPP tracker to consistently find the maximum power point.
Inverter losses	Losses in the inverters.
Inverter sizing	The proper size of inverter is utmost important, otherwise it may lead to losses. For example, power is clipped for high intensity light for undersized inverter. If it is oversized, the inverter's efficiency will be too low for low intensity of light.
Transformer losses	Transformer losses in case electricity has to be connected to a high-voltage grid.
Operation & Maintenance losses	
Downtime losses	Downtime for maintenance is usually very low for photovoltaic systems.

14.2 Energy Losses

The energy losses in solar PV systems are discussed below:

1. Pre-photovoltaic losses: The incoming light is attenuated through shading, dirt, snow and reflection before it hits the photovoltaic panel. The losses from concentration devices is also included in concentrated PV systems.
2. Module and thermal losses: Reflecting the efficiency and temperature dependance of the solar module
3. System losses: Reflecting losses in the electrical components including wiring, inverters and transformers. The various types of energy losses are presented in Table 14.1.

14.3 Yield Calculation of Grid-Connected PV System Using PVSYST

For any project, it is required to test the system viability before implementing it practically. In the case of solar power systems, this task is undertaken by many commercial software products that have been utilized by the industries or in academics, such as PVSYST,

PV*SOL, and so on. With the help of this software, design engineers can calculate the expected yield by the rated solar system at a particular geographical location. A study shows that if losses have been considered correctly, a variation of only 3% to 5% exists between the expected generation predicted from this software and the actual generation.

To understand the concept and how to calculate yield using this software, an example has been considered using the PVSYST software, which is one of the most commonly used. A test system has been considered with the following inputs:

1. A 50 kW grid-connected rooftop PV system is to be installed at New Delhi, India.
2. All solar modules should be of equal size and of rating 250 Wp.
3. A single central inverter of rating 50 kW is to be considered.
4. The plane angle and azimuth angle should be at 0°.
5. The following losses have been considered:
   a. Soil loss @ 1%
   b. Module mismatch losses @ 2%
   c. Module quality loss @ 1%

To start working on designing the system using PVSYST, the following steps have to be taken to determine the yield calculations:

Step 1:
Open the PVSYST software. The window has the following three options:

1. Preliminary Design
2. Project Design
3. Tools

The preliminary design is for those who want to get a rough idea of the system. To design the system or simulate the solar PV system, the Project Design options should be selected. The Tools option is for adding any new entry.

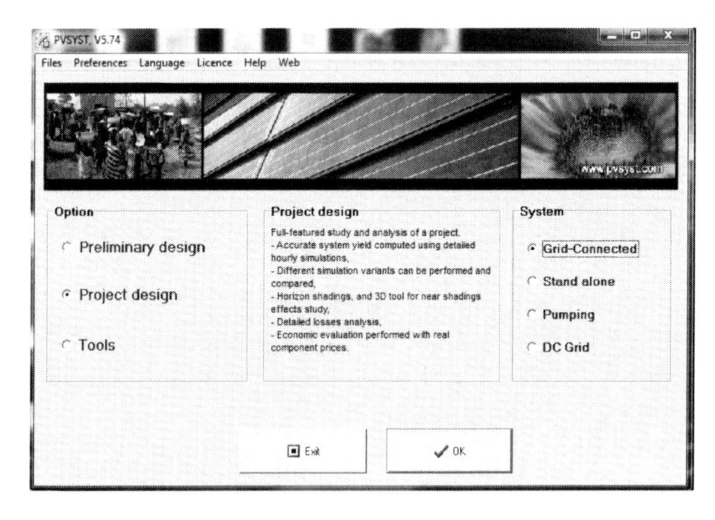

FIGURE 14.1
Design options in PVSYST software.

As shown in Figure 14.1, Project Design is selected to design the 50 kW system at New Delhi, India. Because it is a grid-connected system, the Grid-Connected option is selected.

Step 2:
After option and system type, the next stage has to do with the project location (Figure 14.2). The system has to be installed at New Delhi, India. For any location, the exact latitude and longitude should be known so that the solar irradiance (already prefilled) for the location can be considered.

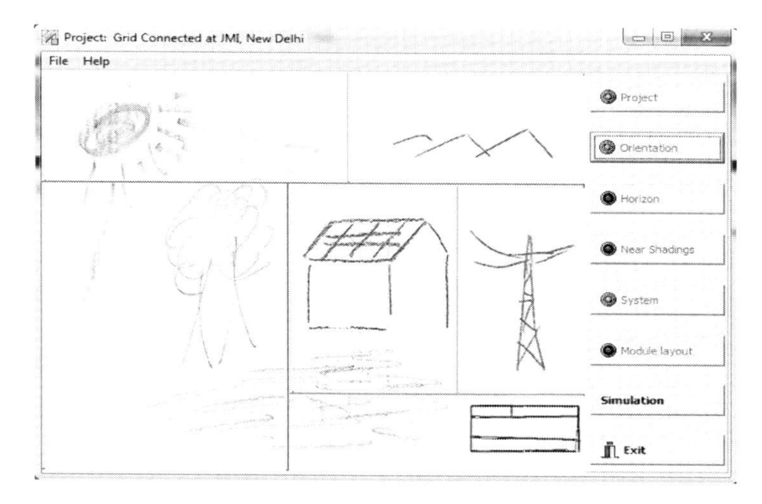

FIGURE 14.2
Option for location in PVSYST software.

Step 3:
After clicking Site and Meteo, the geographical location can be modified, if required and if it varies from the pre-field data (Figure 14.3).

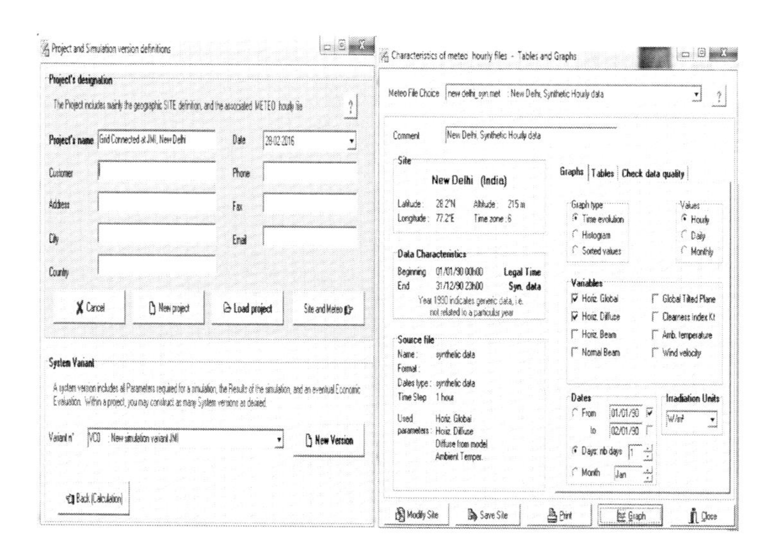

FIGURE 14.3
Options for different locations in PVSYST software.

Step 4:
Next is orientation (Figure 14.4). In this section, there are two parts: plane tilt and azimuth angle. In the Field type section, there are 12 types of system such as fixed tilt, seasonal tilt, or different types of tracking system. Based on the type of system, field parameters have to be filled in. We know that the plane tilt angle and azimuth angle both are 0°.

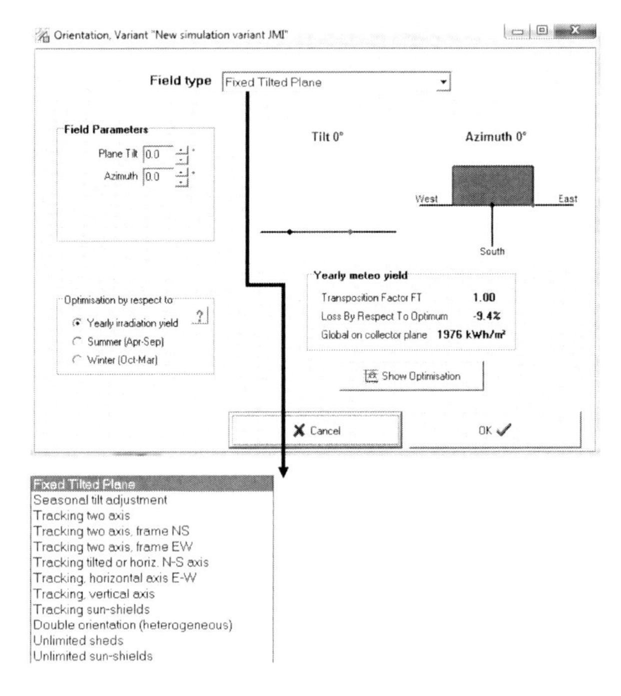

FIGURE 14.4
Design options for field parameters in PVSYST software.

Step 5:
The next step is to see the horizon of the project (Figure 14.5). It is basically a graph between sun height and azimuth angle.

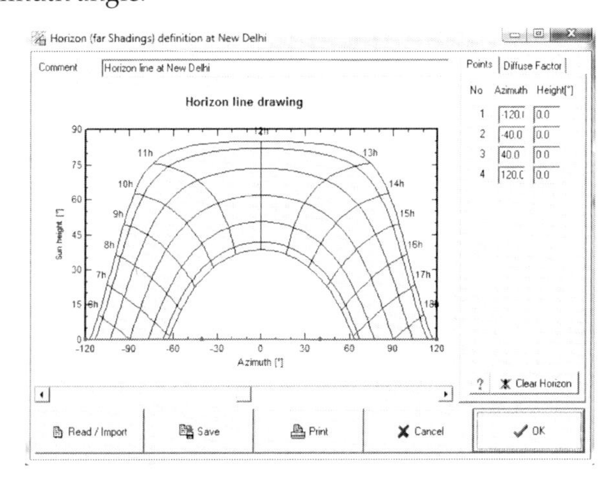

FIGURE 14.5
Graph between sun height and azimuth angle in PVSYST software.

Step 6:
Next we need to provide the near shading details. The shading can be constructed according to the location details, for example, if there is any building or pole or any physical construction that can produce shadows at the location or site of installation. We select No Shadings as no parameters have been specified regarding shading in the problem statement (Figure 14.6).

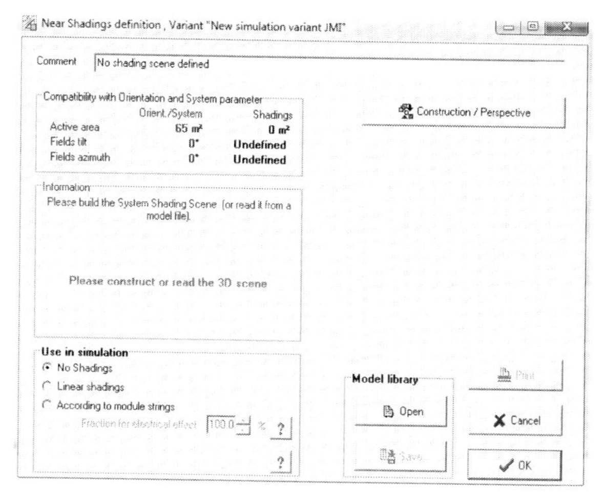

FIGURE 14.6
Shading options in PVSYST software.

Step 7:
Figure 14.7 shows where the whole system is simulated. In this section, the sizing of the project (i.e., either by rated capacity or area available), PV module type and rating, inverter type and rating, and detailed losses have been considered. Accordingly, the numbers of

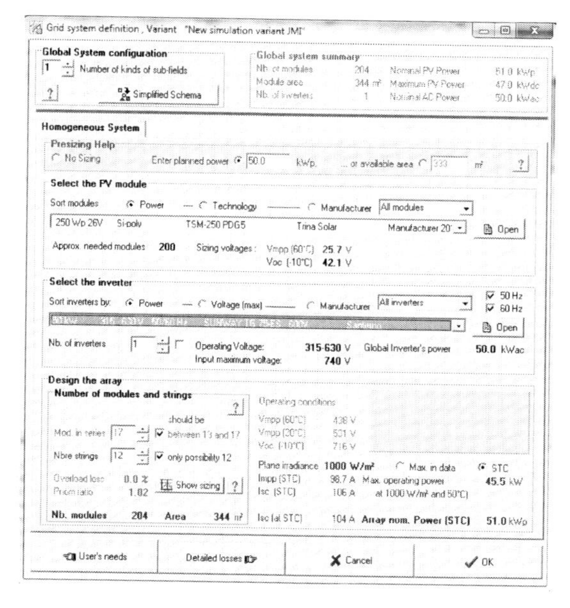

FIGURE 14.7
Design options in PVSYST software.

modules in series and in string are formed. This combination can vary, considering overvoltage and current in one string. Here, all modules have the same rating (i.e., 250 Wp) and the central inverter for the 50 kW test system is to be simulated at New Delhi. The losses are also considered according to the location of the project.

After filling in all the inputs of the project, the PV system is simulated to produce the PVSYST report, which contains details such as system configuration with losses, produced energy, performance ratio, horizontal global irradiance, global incidence, and total energy injected into grid.

The following sample report (Figures 14.8 through 14.10) is generated from the PVSYST software as per the inputs of the problem statement:

PVSYST V5.74		20/03/17	Page 1/3

Grid-Connected System: Simulation parameters

Project : **Grid Connected at JMI, New Delhi**

Geographical Site		**New Delhi**	**Country**	**India**
Situation	Latitude	28.2°N	Longitude	77.2°E
Time defined as	Legal Time	Time zone UT+6	Altitude	215 m
	Albedo	0.20		

Meteo data : New Delhi, Synthetic Hourly data

Simulation variant : **New simulation variant JMI**

Simulation date 20/03/17 18h34

Simulation parameters

Collector Plane Orientation	Tilt	0°	Azimuth	0°
Horizon	Free Horizon			
Near Shadings	No Shadings			

PV Array Characteristics

PV module Si-poly	Model	**TSM-250 PDG5**		
	Manufacturer	Trina Solar		
Number of PV modules	In series	17 modules	In parallel	12 strings
Total number of PV modules	Nb. modules	204	Unit Nom. Power	250 Wp
Array global power	Nominal (STC)	**51.0 kWp**	At operating cond.	45.5 kWp (50°C)
Array operating characteristics (50°C)	U mpp	461 V	I mpp	99 A
Total area	Module area	**344 m²**		
Inverter	Model	**SUNWAY TG 75-ES - 600V**		
	Manufacturer	Santerno		
Characteristics	Operating Voltage	315-630 V	Unit Nom. Power	50 kW AC

PV Array loss factors

Thermal Loss factor	Uc (const)	20.0 W/m²K	Uv (wind)	0.0 W/m²K / m/s
=> Nominal Oper. Coll. Temp. (G=800 W/m², Tamb=20°C, Wind=1 m/s.)			NOCT	56 °C
Wiring Ohmic Loss	Global array res.	364 mOhm	Loss Fraction	6.9 % at STC
Array Soiling Losses			Loss Fraction	1.0 %
Module Quality Loss			Loss Fraction	1.0 %
Module Mismatch Losses			Loss Fraction	2.0 % at MPP
Incidence effect, ASHRAE parametrization	IAM =	1 - bo (1/cos i - 1)	bo Parameter	0.05

System loss factors

Wiring Ohmic Loss	Wires	0 m 3x16 mm²	Loss Fraction	0.0 % at STC

User's needs : Unlimited load (grid)

FIGURE 14.8
PVSYST report.

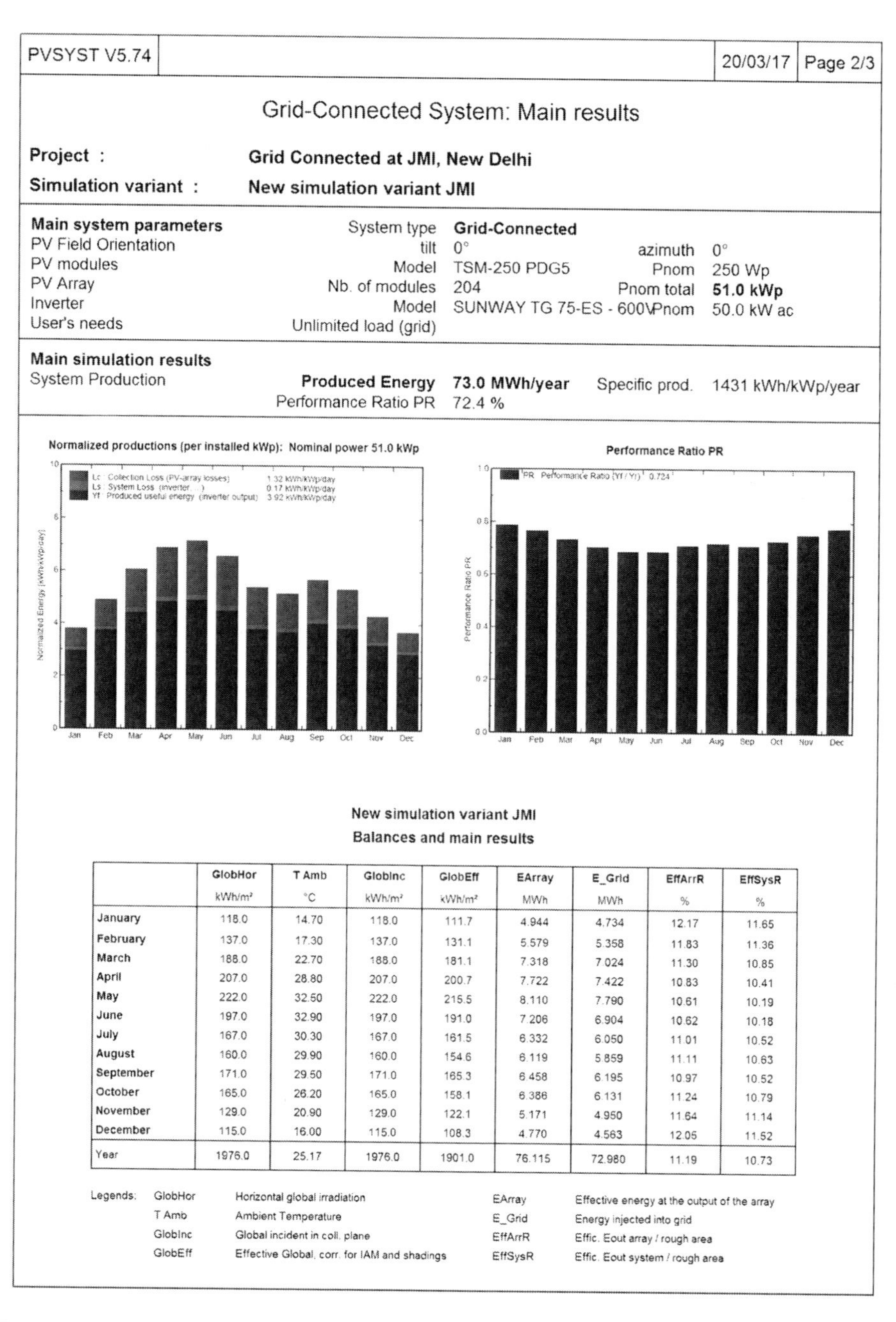

PVSYST V5.74		20/03/17	Page 2/3

Grid-Connected System: Main results

Project : **Grid Connected at JMI, New Delhi**
Simulation variant : **New simulation variant JMI**

Main system parameters — System type **Grid-Connected**
PV Field Orientation — tilt 0° — azimuth 0°
PV modules — Model TSM-250 PDG5 — Pnom 250 Wp
PV Array — Nb. of modules 204 — Pnom total **51.0 kWp**
Inverter — Model SUNWAY TG 75-ES - 600V Pnom 50.0 kW ac
User's needs — Unlimited load (grid)

Main simulation results
System Production — **Produced Energy** **73.0 MWh/year** — Specific prod. 1431 kWh/kWp/year
Performance Ratio PR 72.4 %

New simulation variant JMI
Balances and main results

	GlobHor	T Amb	GlobInc	GlobEff	EArray	E_Grid	EffArrR	EffSysR
	kWh/m²	°C	kWh/m²	kWh/m²	MWh	MWh	%	%
January	118.0	14.70	118.0	111.7	4.944	4.734	12.17	11.65
February	137.0	17.30	137.0	131.1	5.579	5.358	11.83	11.36
March	188.0	22.70	188.0	181.1	7.318	7.024	11.30	10.85
April	207.0	28.80	207.0	200.7	7.722	7.422	10.83	10.41
May	222.0	32.50	222.0	215.5	8.110	7.790	10.61	10.19
June	197.0	32.90	197.0	191.0	7.206	6.904	10.62	10.18
July	167.0	30.30	167.0	161.5	6.332	6.050	11.01	10.52
August	160.0	29.90	160.0	154.6	6.119	5.859	11.11	10.63
September	171.0	29.50	171.0	165.3	6.458	6.195	10.97	10.52
October	165.0	26.20	165.0	158.1	6.386	6.131	11.24	10.79
November	129.0	20.90	129.0	122.1	5.171	4.950	11.64	11.14
December	115.0	16.00	115.0	108.3	4.770	4.563	12.05	11.52
Year	1976.0	25.17	1976.0	1901.0	76.115	72.980	11.19	10.73

Legends:
GlobHor — Horizontal global irradiation
T Amb — Ambient Temperature
GlobInc — Global incident in coll. plane
GlobEff — Effective Global, corr. for IAM and shadings
EArray — Effective energy at the output of the array
E_Grid — Energy injected into grid
EffArrR — Effic. Eout array / rough area
EffSysR — Effic. Eout system / rough area

FIGURE 14.9
Simulated results obtained from PVSYST software.

Here the total energy generated is 97,100 units from the 50 kW system at New Delhi. But the total energy injected into the grid is 73,000 units. The rest units are dropped as a loss in conversion and are due to environmental factors.

This is how the solar PV system has been simulated in the software to calculate the yield at a particular geographical location.

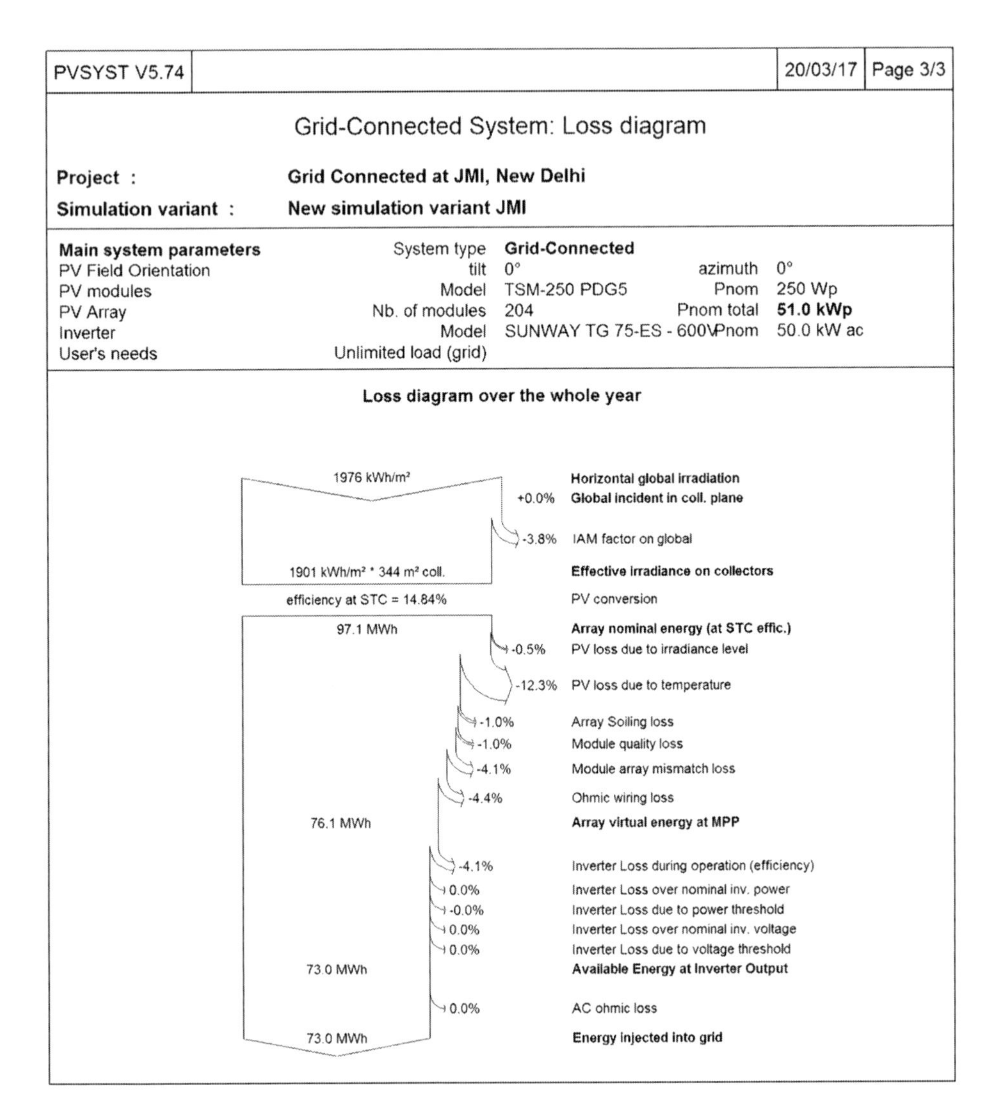

FIGURE 14.10
Energy loss diagram obtained from PVSYST software.

14.4 Tutorials

Based on the software described earlier, let us try the following problem statements:

Problem Statement #1:
Construct a test system with the following inputs:

1. A 100 kW grid-connected PV system is to be installed at Mumbai, India.
2. All solar modules should be of equal size and have a rating of 200 Wp.
3. A single central inverter with a rating of 100 kW is to be considered.
4. The plane angle and azimuth angle should be at 0°.

5. The following losses have been considered:
   a. Soil loss @ 2%
   b. Module mismatch losses @ 2%
   c. Module quality loss @ 1%

Problem Statement #2:
Construct a test system with the following inputs:

1. A 1 MW grid-connected PV system is to be installed at Aligarh, Uttar Pradesh, India, with latitude and longitude 27.8974° N, 78.0880° E, respectively.
2. All solar modules should be of equal size and have a rating of 300 Wp.
3. Two central inverters with a rating of 500 kW each are to be considered.
4. The plane angle and azimuth angle should be at 0°.
5. The following losses have been considered:
   a. Soil loss @ 1%
   b. Module mismatch losses @ 2%
   c. Module quality loss @ 2%

Problem Statement #3:
Construct a test system with the following inputs:

1. A 2 MW grid-connected PV system is to be installed at Melbourne, Australia, with a seasonal tilt system.
2. All solar modules should be of equal size and have a rating of 250 Wp.
3. Four central inverters with a rating of 500 kW each are to be considered.
4. The plane angle and azimuth angle should be at 20°.
5. The following losses have been considered:
   a. Soil loss @ 0.5%
   b. Module mismatch losses @ 1%
   c. Module quality loss @ 1%

15

Design of Transformers, Inductors, and Filters for PV Applications

15.1 Inductor Design

Even though inductors and transformers are both magnetic components, there is a very important difference in their functioning and design aspects. In a transformer, the core flux (or the flux density) is determined by the magnetizing current. The load current virtually has no say in determining the core flux (the flux due to the load current is nullified by the counterflux produced by the primary component of the load current), whereas in an inductor, the core flux is decided only by the load current. Thus if the load current increases, there is a possibility that the core may saturate and inductance will come down. So the primary consideration in an inductor is that one has to know the maximum load current and have a core that does not saturate at this current. This can lead to a huge core size if the current to be handled is large. The core size can be reduced considerably by introducing an appropriate air gap in the magnetic circuit.

There are several approaches to inductor design, two of which are mentioned here:

1. Trial and error approach, often guided by the "Hanna curves"
2. Area product approach

Here the area product approach is discussed as it is a sound design technique and is easy to follow.

Figures 15.1 and 15.2 show the inductor geometry using an E–I core. Figure 15.3 shows the model of magnetic circuit for the inductor. Using Faraday's law, we get:

$$e = N\frac{\mathrm{d}\phi}{\mathrm{d}t} \tag{15.1}$$

Note that R_c and R_g are the core and air gap reluctances, respectively.

The flux in the circuit is given by:

$$\phi = \frac{mmf}{R_c + R_g} = \frac{Ni}{\dfrac{l_c}{\mu_0\mu_r A_c} + \dfrac{l_g}{\mu_0 A_c}} \tag{15.2}$$

$$e = N\frac{\mathrm{d}\phi}{\mathrm{d}t} = \frac{N^2}{\dfrac{l_c}{\mu_0\mu_r A_c} + \dfrac{l_g}{\mu_0 A_c}}\frac{\mathrm{d}i}{\mathrm{d}t} = L\frac{\mathrm{d}i}{\mathrm{d}t} \tag{15.3}$$

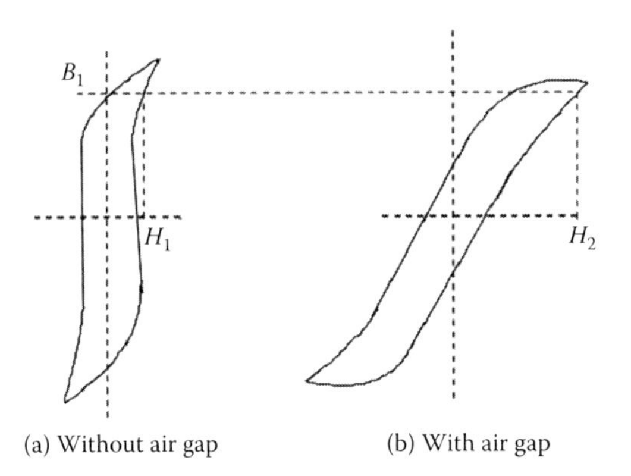

FIGURE 15.1
B–H characteristics: (a) without air gap and (b) with air gap.

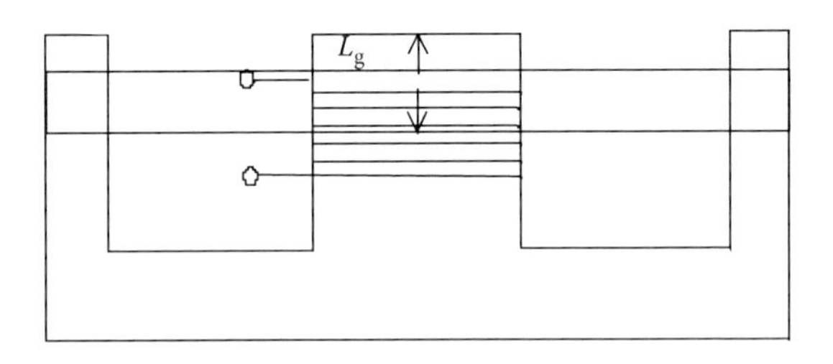

FIGURE 15.2
Inductor geometry with E–I core.

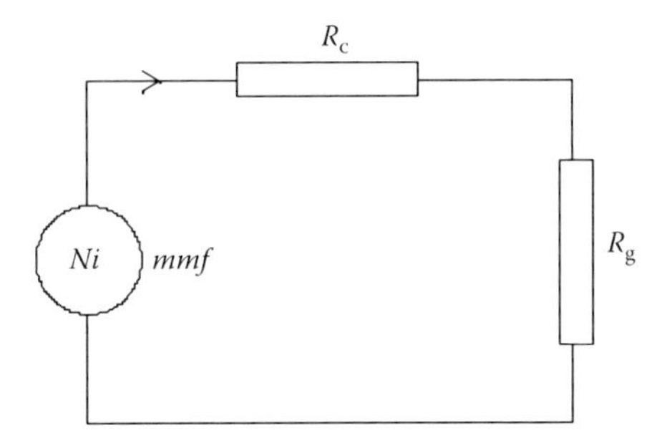

FIGURE 15.3
Model of magnetic circuit for the inductor.

$$L = \frac{N^2}{\dfrac{l_c}{\mu_0 \mu_r A_c} + \dfrac{l_g}{\mu_0 A_c}} \tag{15.4}$$

For zero, air gap l_g is zero, then

$$L = \frac{\mu_0 \mu_r A_c N^2}{l_m} \tag{15.5}$$

Which is a familiar expression of the inductance.

15.1.1 Designing

Inductor value: The first step in inductor design is to find out the value of L for the particular application. Faraday's equation $e = N\frac{d\phi}{dt}$ is used to find the value of L for any circuit. This equation is best suited for switched mode applications. For circuits based on the resonant principle, the L value is determined from the resonant frequency and the Q of the circuit, etc. So depending on the type of application and the configuration of the circuit, the value of L has to be arrived at.
Area product: The energy to be handled by the inductor core is given by:

$$E = \frac{1}{2}(LI^2) \tag{15.6}$$

E is the energy in joules, L is the inductance in henrys, and I is the current in amperes. The window area of the core should accommodate N turns of wire cross-section area a

$$K_w A_w = Na \tag{15.7}$$

Here, $a = I/J$, where I is the rms current through the inductor in amps and J is the current density in A/mm². Thus,

$$K_w A_w = N(I/J) \tag{15.8}$$

Defining crest factor K_c

$$K_c = I_m / I = \text{peak/rms} \tag{15.9}$$

$$K_w K_c A_w J = NI_m \tag{15.10}$$

Therefore, from Faraday's equation

$$e = N\frac{d\phi}{dt} = NA_c\frac{dB}{dt} \tag{15.11}$$

$$LI_m = NA_c B_m \tag{15.12}$$

Therefore, the equation is

$$E = \frac{1}{2}NI_m A_c B_m \tag{15.13}$$

Substituting for Im and rearranging, the area product for the core is given by:

$$A_p = A_w A_c = \frac{2E}{K_w K_c J B_m} \tag{15.14}$$

The number of turns is given by:

$$N = \frac{LI_m}{A_c B_m} \tag{15.15}$$

Gauge of wire: The cross-sectional area of the wire can be calculated from the formula

$$a = I/J \tag{15.16}$$

The gauge of the wire can be determined by comparing the calculated wire cross-section from the earlier equation with wire gauges given in Table 15.1.

The following remarks regarding inductor design are to be noted carefully:

- Because the permeability of the core is not a constant parameter and the energy stored in the core material cannot be totally neglected, this design, when implemented, may not give the exact value of the required *L*. However, this design procedure is satisfactory for most cases. The actual value of the desired *L* may have to be trimmed by adjusting the air gap.
- The assembly of the core is done in a slightly different manner than that of the transformer, when laminations are used to make the inductor. In the case of the inductor, all *E* laminations are put together, and all *I* laminations are held together, and the air gap is put between the *E* laminations and the *L* laminations. Thin Hylam/leatheroid/fiberglass/Melinex can be used as the air gap. The assembly is then tightened by means of clamps.
- In the case of pot cores, *EE* cores, and *EI* cores, a gap of $l_g/2$ only need be introduced in the center to achieve an effective air gap of l_g, due to the geometry of the core.
- The inductor fabricated is not a pure *L* but has a resistance as well. Its value can be computed using the same procedure used for the transformer. If there is a particular constraint on *R* (e.g., low copper loss requirement), then the wire size has to be chosen properly. One should not forget to cross-check the space requirement in case the wire size is altered. Due to the copper losses, the actual energy handled by the core is slightly higher than that given by Equation 15.6. An efficiency factor can be incorporated in Equation 15.14 so that the required *Ap* is slightly higher.
- When using ferrite cores, once the core is chosen, calculating the number of turns is often simplified by using the so-called *AL* value of the core. The *AL* value represents the inductance factor. The formula for the number of turns is given by:

$$L = ALN^2$$

When an air gap is to be placed, a preadjusted ferrite core can be used, and its *AL* value can be used in the design.

15.2 Design of Isolation Transformers

Transformers with a turns ratio of unity (i.e., number of turns of primary windings is equal to the number of turns of secondary windings) are primarily used as isolation transformers. There might be a slight difference in the number of turns to compensate for voltage

TABLE 15.1

Wire Size and Corresponding Gauge

Area of Bare Conductor, mm^2	0.003973	0.005189	0.006567	0.008107	0.009810	0.011675	0.013700	0.018240	0.023430	0.029270	0.035750	0.042890	0.050670	0.059100
SWG	45	44	43	42	41	40	39	38	37	36	35	34	33	32
Diameter with Enamel, mm	0.086	0.097	0.109	0.119	0.132	0.142	0.152	0.175	0.198	0.218	0.241	0.264	0.287	0.307
R/Km @ 20 °C ohms	4340	3323	2626	2127	1758	1477	1258	945.2	735.9	589.1	482.2	402	340.3	291.7
Weight Kg/m	36.9	48.1	61	75	90.8	107.9	126.22	167.9	220.2	268.6	328.1	393.2	465	540.8

Area of Bare Conductor, mm^2	0.06818	0.07791	0.09372	0.11100	0.13630	0.16420	0.20270	0.24520	0.29190	0.39730	0.51890	0.65670	0.81070	1.16700
SWG	31	30	29	28	27	26	25	24	23	22	21	20	19	18
Diameter with Enamel, mm	0.33	0.351	0.384	0.417	0.462	0.505	0.561	0.612	0.665	0.77	0.874	0.978	1.082	1.293
R/Km @ 20 °C ohms	252.9	221.3	184	155.3	126.5	105	851	70.3	59.1	43.4	33.2	26.3	21.3	14.8
Weight Kg/m	624.5	712.1	855.9	1014	1245	1499	1851	2233	2655	3607	4702	5939	7324	10537

drops; otherwise, the secondary voltage would be slightly less than the primary voltage. These transformers employ insulation between the primary and the secondary windings. This insulation provides a degree of protection from electric faults, equivalent to double insulation.

The separation between the windings will limit the risk of accidental simultaneous contact with the exposed conductive part and the live parts (or metal parts that can become live in the event of an insulation fault). Two types of insulation are commonly introduced between the windings:

- By placing a magnetic safety shield (usually copper) between the primary and secondary windings and connecting it to ground. This could also be another winding or a metal strip surrounding a winding. If the isolation breaks, the current flows to the ground, providing the required safety. The voltage spikes are attenuated and filtered by the shield, and hence the coupling of common mode noise is reduced by decreasing capacitive coupling.
- By using reinforced insulation. This type of insulation is made up of layers. All layers must pass tests required by established standards. If one layer breaks, the next layer provides the required safety. Normally, leakage currents of these transformers are lower than those that require a shield.

15.2.1 Transformer Design

Isolation transformers (Figure 15.4) are found in many different electronic devices that have circuits that are too sensitive to handle an alternating current directly. They can be used as power adapters for laptop computers, cell phones, and most other electronic devices. This is because the voltage levels that are supplied to homes and other establishments are much higher than the voltage that electronic devices require due to low electrical loss at higher voltages and higher transfer rates over long distances. As a result, most electronic devices must include an isolation transformer to decrease the voltage before it reaches the device.

In a power transformer design two main goals need to be addressed:

- Keeping the core out of saturation
- Minimizing core losses

Materials for transformers typically have high permeability, usually ferrites or tape-wound cores. The following characteristics are kept in consideration when choosing between tape-wound cores and ferrites:

- Frequency of operation
- Temperature of operation
- Unit cost, size, and shape

Ferrite cores offer several benefits, such as low losses, low cost, and a wide variety of available shapes and sizes. Pot cores have the advantage of protective shielding, which is quite important in EMI/RFI-sensitive designs. Planar *E* cores offer ease of assembly, consistent results, and a low profile. Ferrites are typically considered for use at frequencies of 10 kHz and above. Above 20 kHz, the ferrite design is typically loss-limited,

FIGURE 15.4
Isolation transformer.

whereas below 20 kHz, the design is typically limited by the flux capacity of the unit. **Tape-wound cores** have higher B max and saturation flux density, so the overall design can be smaller. The actual picture of isolation transformer is shown in Figure 15.4.

15.2.2 Designing with Magnetic Powder Cores

- **Powder core calculations** include winding factor, mean length of turn (MLT), DC resistance (DCR), wound coil dimensions, and temperature rise calculations.
- The **powder core loss calculation** provides a step-by-step method to calculate losses generated by powder cores under certain conditions.

Rapid expansion in the solar power conversion, wind power conversion, hybrid vehicle, uninterruptible power supplies, and electric drive transportation markets has increased the demand for high-current (100–300 amps) inductors. For many high-current applications, the limiting factor is not necessarily the ability of the material to provide enough inductance at DC bias; it is the ability to fit enough turns of the heavy wire or foil to provide the necessary inductance. To address that need, the company Magnetics came up with a family of larger *E* cores and *U* cores that are able to accommodate all sizes. Isolation transformers increase and decrease alternating currents [4]. They can be built to scale and generally handle any amount of voltage necessary, depending on how many times the coils are wound. Isolation transformers reduce electrical and electromagnetic interruptions from other power sources and devices and improve signal strength in radio and wireless communication systems.

15.3 Designing of Filters

The recent advances in semiconductors and magnetic components also have led to a wide application of voltage source converters (VSCs), such as for PV panels, wind turbines, electrical vehicles (EVs), and high-speed trains. Although harmonic distortion of the AC-side current is usually below 5%, the current contains high-frequency components near switching frequency that are commonly caused by the pulse width modulation (PWM) switching process (frequencies: 2–15 kHz). Therefore, a low-pass passive filter (i.e., *LCL* and *LC* filters) is suggested to be installed between a PV inverter bridge and the power utility. The capacitor of the filters offers a low-impedance path for the high-frequency current ripple and, thus, the content of current ripple injecting into the utility is attenuated. In general, *LCL* filters are superior to *L* filters and *LC* filters, as their switching harmonic attenuation, with smaller reactive elements, is more effective. Thus *LCL* filters are more frequently adopted in low- and medium-voltage systems. The basic guideline for the selection of the *LCL* filter parameters is presented in various texts in the literature. Generally, the converter-side inductor ($L1$) is sized based on the current ripple at the switching frequency, whereas the capacitor rating C is limited by the fundamental reactive power, and the grid-side inductor ($L2$) is characterized based on the high-frequency current attenuation from the inverter to the power grid. Damping resistance (R_d) is optional to attenuate resonance peaks. In fact, it can be said that only *LCL* and *LC* types of PV filters can resonate with an inductive system. Thus, ignoring the effects of other loads, the equivalent impedance at the point of common coupling (PCC) is determined by both filter impedance and system impedance versus frequency. And therefore to assess the characteristics of the PCC impedance, a sensitivity study on the installation number and damping resistance has to be made.

The damping resistor plays an important role in damping the resonance peak. However, the filter is also designed for an additional function that is mitigating an amount of high-order harmonics, because the capacitor branch shows a low-impedance path for high-frequency harmonics. The increase of the damping resistor eventually degrades the filtering performance. In order to evaluate the effect of the damping resistor on both resonance damping and filter effectiveness, the amplification ratio of resonance impedance to the corresponding system is evaluated, and impedance at resonance frequency (Amp) and total demand distortion (TDDi) of the VSC output current are employed.

Filter design is a process of designing a signal-processing filter that satisfies a set of requirements, of which some are contradictory. The purpose of this is to find a filter that meets each of the requirements to a sufficient degree to make it useful.

Some of the requirements that are considered during the design process are as follows:

1. It should have a specific frequency response.
2. It should have a specific phase shift or group delay.
3. It should have a specific impulse response.
4. It should be causal.
5. It should be stable.
6. The computational complexity of the filter should be low.

In this consideration, the filter to be designed tells the engineer what type of filter is needed. For example:

1. Low-pass filter, which passes the signal at a frequency below the cut-off frequency
2. High-pass filter, which passes the signal at a frequency above the cut-off frequency
3. Band-pass filter, which passes the signal between a frequency range
4. Band-stop filter, which passes the signal above and below a frequency range

Filters, like all pass filters, pass all the frequency, but changes the phase of the signal can be used to equalize the group delay of recursive filters. These filters are also used in phaser effects.

Other considerations in the filter design process, broadly speaking, are:

1. Analog filter
2. Digital filter
   a. Finite impulse response filters
   b. Infinite

Although a resonance point exists at the impedance curve, all peak impedances should be below 150% of the grid impedance at the resonance frequency. Therefore, it can be said that the resonance issues associated with PV inverters are determined by the following factors:

- The value of damping resistance R_d has a significant damping impact on the resonance. But at the same time, it produces additional power loss. The resonance amplification decreases with the increasing of R_d. Thus, the resonance can be mitigated when a suitable value of R_d is designed in series with the capacitor of the filters.
- The number of PVs is another cause for the harmonic resonance conditions. Even though the number of PVs does not determine whether the resonance will occur, it does affect the resonance frequency and the magnitude of the system. A large number of PV installations will not aggravate the resonance problem, whereas the filters will provide a low-impedance path at nonresonance frequency bands. Therefore, it can be concluded that the large installation of the PVs in a distribution system is not of concern.
- For resonance amplification, the impedance of the supply system is not the dominant factor, contributing only insignificantly.
- An individual PV inverter will excite harmful resonance if the passive damping impedance is below 2Ω. Consequently, only a small proportion of power loss is required. However, the high-frequency components of the power loss have to be effectively mitigated by optimizing the shunt branch of the filters.
- The large increment in resistor and the filter gain in the high-frequency region reduce the effectiveness of the filter. Therefore, the selection of the damping resistor is limited.

15.4 Tutorials

Q.1 Design an inductor for a maximum power point tracker with the following specifications:
Frequency of operation = 20 kHz
DC current carrying capacity = 5 amp
Inductance = 500 μH
AC current = 0.8 amp (peak to peak)

Solution:
The current can be determined such that the core does not saturate:

$$I = I_o + I_{AC}/2$$
$$= 5 + 0.8/2 = 5.4 \text{ amp}$$
$$\text{Energy} = LI^2/2 \text{ Ws}$$
$$= (500 \times 10^{-6} \times 5.42)/2$$
$$\text{Energy} = 0.007290 \text{ Ws}$$

Calculating the area product with $B = 0.3$ T, $K = 0.3$ and $J = 186$ A/cm^2

$$\text{AP} = [(2 \times \text{Energy} \times 10^4)/BKJ] \text{ cm}^4$$
$$= (2 \times 0.007290 \times 10^4)/(0.3 \times 0.3 \times 186)$$
$$= 8.709 \text{ cm}^4$$

The ferrite core EE 5521 is selected.
This core has turns = (3.74 × 0.3)/0.02627 = 42.7 ≈ 43 (approx.)
Now the inductance factor can be calculated from the relation

$$AL = L/\text{N}^2$$
$$= (500 \times 10^{-6})/(43 \times 43)$$
$$= 270 \text{ nH/turn}$$

A standard value for the core is AL = 250 nH/turn.
Core EE 4220 will also work, but it will heat up.

Q.2 Discuss EMI filter design. Design an EMI filter for a power factor controller MC34262 operating at a switching frequency of 2 kHz. The PFC input voltage varies from 120 to 290 V and has an output of 500 w.

Solution: Line filters prevent excessive noise from being conducted through any electronic equipment, particularly those involving power switching, to an AC line so that the latter is not contaminated. The power of an AC line is designed to carry a pure 50 Hz, 220 V power to all the equipment working in a certain area. If some of these generate large quantities of electromagnetic interference (EMI) or radio frequency interference (RFI), the other instruments working on the same line will be affected. The direction of common mode noise, which occurs simultaneously on both the lines, referred to as earth common, is from the load and into the filter, where the noise common to both lines becomes sufficiently attenuated. The resulting common mode output of the filter onto the AC line through the impedance-matching circuitry is then low. There are two paths for noise to travel: common mode or differential mode.

Q.3 For a switching frequency of 25 kHz, find the suitable value of a capacitor to provide an attenuation of 2 dB considering two cut-off frequencies.

Solution: Assuming that the conversion efficiency of the PFC is 95% of its input power = 500/0.95 = 526.31. Therefore at 120 V it draws a current of 526.31/120 = 4.38 amp. This means that all of the inductors should be able to carry at least this current. If the switching frequency is 25 kHz, then the filter cut-off frequencies, f_{RDM} and f_{RCM}, should be at least half this value to provide an attenuation of 12 dB. Thus:
f_{RDM} = 12,500 Hz
f_{RCM} = 12,500 Hz
Assume C_y= 4700 pF
Therefore,

$$L_c = [(1/(2\pi \times 12{,}500))^2(1/(2 \times 4700 \times 10^{-12})]$$
$$= 17 \times 10^{-3} \text{ H}$$

Because this is quite a large inductance, its leakage inductance can be used as L_D. Leakage inductance can be 1% of this value of L_{DM}. Therefore:
$L_{DM} = 170 \times 10^{-6}$ H
Now:

$$C_x = C_{x1} = C_{x2} = (1/(2\pi \times f_{\mathrm{RDM}})^2 \times 1/L_{DM}$$
$$= (1/(2\pi \times 12{,}500))^2 \times 1/(170 \times 10^{-6}) = 9.52 \times 10^{-7}$$
$$= 1\ \mu\mathrm{F}$$

C_{x1} and C_{x2} reduce the noise when the rectifiers are On and Off, respectively. Thus, if two cut-off frequencies are required, then the two capacitors can be of different values. In this case C_{x2} will be retained at 1 μF for an f_{RDM} of 12,500 Hz for the condition when the rectifiers are Off. If, however, f_{RDM} is chosen to be 40 kHz when the rectifiers are On, then a new value of C × 1 may be calculated as:
$C = [(1/(2\pi \times 40{,}000))^2(1/170 \times 10^{-6})] = 0.09312 \times 10^{-6}\ \mathrm{F} = 0.1\ \mu\mathrm{F}$

Bibliography

Abdelsalam, A.K., Massoud, A.M., Ahmed, S., and Enjeti, P. Highperformance adaptive perturb and observe MPPT technique for photovoltaic-based microgrids. *IEEE Transactions on Power Electronics*, 26(4): 1010–1021, April 2011.

Abdulkadir, M., Samosir, A.S., and Yatim, A.H.M. Modeling and simulation based approach of photovoltaic system in Simulink model. *ARPN Journal of Engineering and Applied Sciences*, 7(5): 616–623, 2012.

Abdulkadir, M., Samosir, A.S., Member and Yatim, A.H.M. Modeling and simulation of maximum power point tracking of photovoltaic system in Simulink model. *IEEE International Conference on Power and Energy*, Kota Kinabalu, Malaysia, 325–330, 2012.

Adamidis, G., Bakas, P., and Balouktsis, A. Photovoltaic System MPP Tracker Implementation using DSP engine and buck-boost DC-C converter. Power Electronics and Motion Control Conference, 2008.

AKPABIO, L. E., and ETUK, S. E. Relationship between global solar radiation and sunhine duration for Onne, Nigeria. *TÜBİTAK*, 2003.

Alam, S. Assessment of diffuse solar energy under general sky condition using artificial neural network. *Applied Energy*, 86(4): 554–564, April 2009.

Alam, S. Computation of beam solar radiation at normal incidence using artificial neural network. *Renewable Energy*, 31(10): 1483–1491, August 2006.

Alam, S. Prediction of direct and global solar irradiance using broadband models: Validation of REST model. *Renewable Energy*, 31(8): 1253–1263, July 2006.

Aletr, D.M. Thermoelectric cooler control using TMS320F2812DSP and a DRV592 power amplifier. Texas Instruments Application Report.

Ali, A.N.A., Saied, M.H., Mostafa, M.Z., and Abdel Moneim, T.M. *A Survey of Maximum PPT Techniques of PV Systems*. Cleveland, OH: IEEE Conference Publications, 2011.

Al-moudi, A., and Zhang, L. Application of radial basis function networks for solar-array modeling and maximum power point prediction *IEE Proceedings*, 147(5): 310–316, 2000.

Altas, I.H., and Sharaf, A.M. A photovoltaic array simulation model for MATLAB-Simulink GUI environment IEEE, clean electrical power. International Conference on Clean Electrical Power (ICCEP '07), Ischia, Italy, 14–16 June 2007.

Amorndechaphon, D., Premrudeepreechacharn, S., and Higuchi, K. *Small Grid Connected PV System with Lossless Passive Soft Switching Technique*. Fukuoka, Japan: Fukuoka International Congress Center, 2009.

Amudhavalli, D., Meyyappan, M., Imaya, S., and Preetha Kumari, V. Interleaved soft switching boost converter with MPPT for photovoltaic power generation system. 2013 International Conference on Information Communication and Embedded Systems (ICICES 2013). Chennai, Tamilnadu, India 21–22 February 2013, pp. 1214-1219.

Anees, A.S. *Integration of Renewable Energy Sources and Grid Management*, PhD thesis, Jamia Millia Islamia, New Delhi, India, 2016.

Anees, A.S., and Jamil, M. Economical analysis of solar photovoltaic based distributed generation for 33 bus primary distribution system. In National Conference on Emerging Trends in Electrical And Electronics Engineering (ETEEE-2015), at Jamia Millia Islamia, New Delhi, 2–3 February 2015.

Arizona Solar Solutions. General Safety Precautions, 2004. Retrieved from www.az-solarsolutions.com/clients/solar-safety.html.

Azmi, S.A., Adam, G.P., Ahmed, K.H., Finney, S.J., Williams, B.W. Grid interfacing of multimegawatts photovoltaic inverters. *IEEE Transaction on Power Electronics*, 28(6): 2770–2784, June 2013.

Bahgat, A.B., Helwab, N.H., Ahmadb, G.E., and Shenawyb, E.T. Maximum power point tracking controller for PV systems using neural networks. *Renewable Energy*, 30: 1257–1268, 2005.

Balfour, J.R., Shaw, M., and Bremer Nash, N. *Advanced Photovoltaic System Design*. Burlington, MA: Jones & Bartlett Learning, 2013.

Battarsha, I. *Power Electronics Circuits*. Orlando, FL: John Wiley and Sons, 2004.

Battery parameter in a pv system. Wikipedia. https://en.wikipedia.org/wiki/Rechargeable_battery.

Berger, J.J. *Charging Ahead: The Business of Renewable Energy and What It Means for America*. New York, NY: Henry Holt and Company, 1997.

Bird J. *Electrical Circuit Theory and Technology*. Burlington, MA: Elsevier, 2010.

Bhowmick, T., and Hemadrasa, D. Development of prototype protection setup for standalone solar power system. *American Journal of Electrical Power and Energy Systems*, 4(6): 100–105, 2015.

Blasko, V., and Kaura, V. A novel control to actively damp resonance in input LC filter of a three-phase voltage source converter. *IEEE Transaction on Industry Electronics*, 33(2): 542–550, March/April 1997.

Boyle, G. (ed.) *Renewable Energy: Power for a Sustainable Future*. London, UK: Oxford University Press, 1996.

Bureau of Indian Standards. IS 12834: Solar photovoltaic energy systems—Terminology [ETD 28: Solar Photovoltaic Energy Systems]. New Delhi, India: Bureau of Indian Standards, 1989.

Bureau of Indian Standards. IS 875-3: Code of Practice for Design Loads (Other than Earthquake) for Buildings and structures, Part 3: Wind Loads [CED 37: Structural Safety]. New Delhi, India: Bureau of Indian Standards, 1987.

Capogna, V. *Development of Monitoring and Automatic Fault Detection Solutions for Grid-connected Photovoltaic systems*, MSc SELECT thesis, Universitat Politècnica de Catalunya, 2012.

Chao, K.H., Chen, C.T., Wang, M.H., and Wu, C.F. A novel fault diagnosis method based-on modified neural networks for photovoltaic systems. In Y. Tan, Y. Shi, and K.C. Tan (Eds.), *Advances in Swarm Intelligence: First International Conference, ICSI 2010, Beijing, China, June 12–15 2010, Proceedings, Part II* (pp. 531–539). Berlin/Heidelberg: Springer-Verlag.

Chao, K.H., Chen, P.Y., Wang, M.H., and Chen, C.T. An intelligent fault detection method of a photovoltaic module array using wireless sensor networks. *International Journal of Distributed Sensor Networks*, 2014, 1–12, May 2014.

Chao, P.C.P., Chen, W.-D., and Chang, C.-K. Maximum power tracking of a generic photovoltaic system via a fuzzy controller and a two stage DC/DC converter. *Microsystem Technologies*, 18: 1267–1281, 2012.

Chaturvedi D.K. *Soft Computing: Techniques and Its applications in Electrical Engineering*, 2008, New York, NY: Springer Publications.

Chaturvedi, D.K. Improved generalized neuron model for short-term load forecasting. *Soft Computing - A Fusion of Foundations Methodologies and Applications*, 8(5), 01 April 2004.

Chaturvedi, D.K. Improved generalized neuron model for short-term load forecasting, *Soft Computing - A Fusion of 16 Foundations Methodologies and Applications*, 8(1): 10–18, 01 October 2003.

Chine, W., Mellit, A., Pavan, A.M., and Kalogirou, S.A. Fault detection method for grid connected photovoltaic plants. *Renewable Energy*, 66: 99–110, 2014.

Chouder, A. Modeling and simulation of a grid connected PV systems based on the evaluation of main PV module parameters. *Simulation Modelling Practice and Theory* 20(2012): 46–58.

Chouder, A., and Silvestre, S. Automatic supervision and fault detection of PV systems based on power losses analysis. *Energy Conversion and Management*, 51: 1929–1937, 2010.

Cocina, V.C. *Economy of Grid-Connected Photovoltaic Systems and Comparison of Irradiance/Electric Power Predictions vs. Experimental Results*. PhD thesis, Politecnico di Torino, Torino, Italy, March 2014.

Consul Neowatt Power Solutions. "Solar—Hybrid Bi-Directional, Grid tied Inverters, PCU, Power Packs Power Back Up—UPS, STS, Industrial Inverters, Industrial UPS Power Conditioning—Stabilizers, Isolation Transformers, Active Harmonic filters Custom Power Electronic Solutions." Retrieved from http://slideplayer.com/slide/9132578.

Dannehl, J., Liserre, M., and Fuchs, F.W. Filter-based active damping of voltage source converters with LCL filter. *IEEE Transaction on Industrial Electronics*, 58(8): 3623–3633, August 2011.

Davarifar, M., Rabhi, A., and El Hajjaji, A. Comprehensive modulation and classification of faults and analysis their effect in DC side of photovoltaic system. *Energy and Power Engineering*, 5: 230–236, July 2013.

Desai, H., Patel, P., and Patel, H.K. Maximum power point algorithm in PV generation an overview. *IEEE Proceedings of Power Electronics and Drive Systems*, pp. 624–630, 2007.

Diagram of a photovoltaic system, 2014. Retrieved from http://pv.vadium.sk.

Dorvlo, A. S. S. Solar radiation estimation using artificial neural networks. *Applied Energy*, 71(4): 307–319, April 2002.

Drews, A., de Keizer, A.C., Beyer H.G., Lorenz, E., Betcke, J., van Sark, W.G.J.H.M., Heydenreich, W., Wiemken, E., Stettler, S., Toggweiler, P., Bofinger, S., Schneider, M., Heilscher, G., and Heinemann, D. Monitoring and remote failure detection of grid-connected PV systems based on satellite observations. *Solar Energy*, 81(4): 548–564, 2007.

Du, Y., Lu, D.D.C., James, G., and Cornforth, D.J. Modelling and analysis of current harmonic distortion from grid connected PV inverters under different operating conditions. *Solar Energy*, 94: 182–194, August 2013.

Ducange, P., Fazzolari, M., Lazzerini, B., and. Marcelloni. F. An intelligent system for detecting faults in photovoltaic fields. Proceeding of the 11th International Conference on Intelligent Systems Design and Applications (ISDA), Cordoba, Spain, 1341–1346, 2011.

El Khateb, A.H., Abd Rahim, N., and Selvaraj, J. Cuk—buck converter for standalone photovoltaic systems. *Journal of Clean Energy Technologies*, 1(1): 69–74, January 2013.

Endecon Engineering. *A Guide to Photovoltaic (PV) System Design and Installation*. Report for California Energy Commission. San Ramon, CA: Endecon Engineering, 2001.

Energy Information Administration. Renewable Energy Annual 1997: Volume I. Washington, DC: US Department of Energy, 1998. Retrieved from www.osti.gov/scitech/servlets/purl/573149.

Energy Saving Trust. "Solar Panels: Generate Cheap, Green Electricity from Sunlight," 2017. Retrieved from www.energysavingtrust.org.uk/domestic/content/solar-panels.

Enslin, J.H.R., and Snyman, D.B. Combined low-cost, high-efficient inverter, peak power tracker and regulator for PV applications. *IEEE Transactions on Power Electronics*, 6(1): 73–82, 1991.

Esram, T., and Chapman, P.L. Comparison of photovoltaic array maximum power point tracking techniques. *IEEE Transactions on Energy Conversion*, 22(2): 439–449, June 2007.

Fan, L.L., Yuvarajan, S., and Kavasseri, R. Harmonic analysis of a DFIG for a wind energy conversion system. *IEEE Transaction on Energy Conversation*, 25(1): 181–190, March 2010.

Faranda, R., Leva, S., and Maugeri, V. MPPT techniques for PV systems: Energetic and cost comparison. Proceedings of IEEE PES GM 2008, July 2008.

Farhoodnea, M., Mohamed, A., Shareef, H., and Zayandehroodi, H, Power quality analysis of grid connected photovoltaic system in distribution networks. *Przegla Elektrotechniczny*, 2013.

Francis, B.A., and Wonham, W.M. The internal model principle for linear multivariable regulators. *Applied Mathematics Optimization*, 2: 170–194, 1975.

Gadhavi Akash, G., and Kundaliya, D.D. Design and analysis of solar panel support structure—A review. *International Journal of Advance Research in Engineering, Science & Technology*, 2(5), 1–6 May 2015.

Giga Solar. Brand Module Data Table, June 2015. Retrieved from www.gigasolarpv.com/wp-content/plugins/download-attachments/includes/download.php?id=1604.

Garg, H. P. Fundamentals and characteristics of solar radiation. *Renewable Energy*, 3(4–5): 305–319, June/July 1993.

Giraud, F., and Salameh, Z.M. Analysis of the effects of a passing cloud on a grid-interactive photovoltaic system with battery storage using neural networks. *IEEE Transactions on Energy Conversion*, 14(4): 1572–1577, 1999.

Global Solar Installations Forecast to Reach Approximately 64.7 GW in 2016, Reports Mercom Capital Group, Clean Energy News report data dated 14 December 2015.

Go, S.-I., Ahn, S.-J., Choi, J.-H., Jung, W.-W., Yun, S.-Y., and Song, I.-K. Simulation and analysis of existing MPPT control methods in a PV generation system. *Journal of International Council on Electrical Engineering*, 1(4): 446–451, 2011.

Goetzberger, A., Knobloch, J., and Voss, B. *Crystalline Silicon Solar Cells*. Chichester, UK: John Wiley & Sons, 1998.
Gokmen, N., Karatepe, E., Celik, B., and Silvestre, S. Simple diagnostic approach for determining of faulted PV modules in string based PV arrays. *Solar Energy*. 86: 3364–3377, 2012.
Gow, J.A., and Manning, C.D. Development of a photovoltaic array model for use in power-electronics simulation studies. *IEE Proceedings on Electric Power Applications*, 146: 193–200, 1999.
Gueymard, C. A two-band model for the calculation of clear sky solar irradiance, illuminance, and photosynthetically active radiation at the earth's surface. *Solar Energy*, 43(5): 253–265, 1989.
Gueymard, C.A. REST2: High-performance solar radiation model for cloudless-sky irradiance, illuminance, and photosynthetically active radiation - Validation with a benchmark dataset. *Solar Energy*, 82(3): 272–285, 2008.
Hamidon, F.Z., Abd Aziz, P.D., and MohdYunus, N.H. Photovoltaic array modeling with P&O MPPT algorithms in MATLAB. International Conference on Statistics in Science, Business, and Engineering, Langkawi, Malaysia, 1–5, 2012.
Hamidon, F.Z., Abd Aziz, P.D., and Mohd Yunus, N.H. *Photovoltaic Array Modeling with P&O MPPT Algorithms in MATLAB*. Langkawi, Malaysia: IEEE Conference Publications, 2010.
Hanut India. Solar PV Array Junction Boxes, October 2010. Retrieved from www.hanut-india.com/solar PV junction boxes India.asp
Hassaine, L., Olías, E., Quintero, J., and Barrado, A. Power control for grid connected applications based on the phase shifting of the inverter output voltage with respect to the grid voltage. *Electrical Power and Energy Systems*, 57: 250–260, 2014.
Hiyama, T., and Kitabayashi, K. Neural network based estimation of maximum power generation from PV module using environmental information. *IEEE Transactions on Energy Conversion*, 12: 241–247, 1997.
Hiyama, T., Kouzuma, S., and Imakubo, T. Identification of optimal operating point of PV modules using neural network for real time maximum power tracking control *IEEE Transactions on Energy Conversion*, 10(2): 360–367, 1995.
Hohm, D., and Ropp, M. Comparative study of maximum power point tracking algorithms. *Progress in Photovoltaics*, 11: 47–62, 2003.
Honsberg, C., and Bowden, S. Module Materials. PVEducation.org. Retrieved from http://pveducation.org/pvcdrom/modules/module-materials.
Honsberg, C., and Bowden, S. Welcome to PVCDROM | PVEducation, 2017. Retrieved from www.pveducation.org/pvcdrom.
Howes, R., and Fainberg, A. *The Energy Sourcebook: A Guide to Technology, Resources, and Policy*. New York, NY: American Institute of Physics, 1991.
http://mercomcapital.com/news-analysis.
Hussein, K.H., Muta, I., Hoshino, T., and Osakada, M. Maximum photovoltaic power tracking: An algorithm for rapidly changing atmospheric conditions. *IEEE proceedings Generation, Transmission and Distribution*, 142: 59–64, 1995.
IEEE Standards by IEEE Standards Association.
Ishaque, K., Salam, Z., and Tahri, H. Accurate MATLAB Simulink PV systems simulator based on a two-diode model. *Journal of Power Electronics*, 85(9): 2217–2227, 2011.
Ishaque, K., Salam. Z., and Tahri, H. Accurate MATLAB Simulink PV systems simulator based on a two-diode model. *Journal of Power Electronics*, 11: 179–187, 2011.
Islam, M.R., Guo, Y., and Zhu, J. Power Converters for Small- to Large-Scale Photovoltaic Power Plants. In *Power Converters for Medium Voltage Networks* (pp. 17–49). Berlin, Heidelberg: Springer-Verlag.
Jain, S., and Agarwal, V. Comparison of the performance of maximum power point tracking schemes applied to single-stage grid connected photovoltaic systems. IET Electric Power Application, 1(5): 753–762, September 2007.
Jamil, M., and Anees, A.S. Optimal sizing and location of SPV based MLDG in distribution system for loss reduction, voltage profile improvement with economical benefits. *Elsevier International Journal Energy*, 103: 231–239, 2016.

Jamil, M., Khwaja, M.R., and Ahuja, N. Customized biogas based distributed power generation. *International Journal for Renewable Energy*, 2(4): 182–190, 2012.

Jamil, M., Kirmani, S., and Chatterjee, H. Techno-economic viability of three different energy-supplying options for remote area electrification in India. *International Journal of Sustainable Energy*, 33(2): 470–482, 2014.

Jamil, M., Rizwan, M., Kalam, A., and Ansari, A.Q. Generalized neural network and wavelet transform based approach for fault location estimation of a transmission line. *Applied Soft Computing (Elsevier)*, UK, 19:322–332, 2014. ISSN: 1568–4946.

Jamil, M., Rizwan, M., Kirmani, S., and Kothari D.P., Fuzzy logic based modeling and estimation of global solar energy using meteorological parameters. *Energy: The International Journal (Elsevier)*, 68:685–691, 2014. ISSN: 0360–5442.

Jang, Y., Senior Member, and Jovanovic, M.M. New two inductor boost converter with auxiliary transformer. *IEEE Transactions on Power Electronics*, 19(1): 169–175, January 2004.

Jawaharlal Nehru National Solar Mission (JNNSM) Report, Government of India, 2009, www.mnre.gov.in.

Jiang, S., Cao, D., and Peng, F.Z. Grid-connected boost-half-bridge photovoltaic microinverter system using repetitive current control and maximum power point tracking. *In IEEE Transactions on Power Electronics*, 27(11): 4711–4722, November 2012.

Jiang, S., Cao, D., Peng, F.Z. and Li, Y. Grid connected boost half bridge photovoltaic micro inverter system using repetitive current control and maximum power point tracking. *IEEE Applied Power Electronics Conference and Exposition*, 27: 590–597, 2012.

Jiang, S., Cao, D., Peng, F.Z., and Li, Y. *Grid Connected Boost Half Bridge Photovoltaic Micro Inverter System Using Repetitive Current Control and Maximum Power Point Tracking*. Orlando, FL: IEEE, 2012.

Johansson, T.B., Kelly, H., Reddy, A.K.N., and Williams, R.H. (eds.) *Renewable Energy: Sources for Fuels and Electricity*. Washington, DC: Island Press, 1993.

Jung, H.-Y. Ji, Y.-H. Won, C.-Y. Song, D.-Y. and Kim, J.-W. *Improved Grid-synchronization Technique based on Adaptive Notch Filter*. Sapporo, Japan: International Power Electronics Conference, 1494–1498, 2010.

Junior, L.G., de Brito, M.A.G., Sampaio, L.P., and Canesin, C.A. Evaluation of integrated inverter topologies for low power PV systems. Ischia, IT: IEEE, 2011.

Kandpal, T.C., and Garg, H.P. *Financial Evaluation of Renewable Energy Technologies*. New Delhi, India: Macmillan, 2003.

Kasa, N., Iida, T., and Chen, L. Flyback inverter controlled by sensorless current MPPT for photovoltaic power system. *IEEE Transaction on Industrial Electronics*, 52(4): 1145–1152, August 2005.

Katiraei, F., Iravani, R., Hatziargyriou, N., and Dimeas, A. Microgrid management Control and operation aspect of microgrids. *IEEE Power & Energy Magazine*, 6(3): 54–65, May/June 2008.

Kazimierczuk, M.K. *Pulse-width Modulated DC–DC Power Converters*. Dayton, OH: John Wiley and Sons, Publication, 2009.

Khadem, S.K., Basu, M., and Conlon, M.F. Power quality in grid connected renewable energy systems: Role of custom power devices. *International Conference on Renewable Energies and Power Quality (ICREPQ '10)*, Granada, Spain, March 2010.

Kirmani, S., Jamil, M., and Rizwan, M. Empirical correlation of estimating global solar radiation using meteorological parameters. *International Journal of Sustainable Energy*, 34(5): 327–339, 2015.

Kirmani, S. *Techno Economic Feasibility Analysis of Solar Energy Based Distributed Generation System*, PhD thesis, Jamia Millia Islamia, New Delhi, India, 2014.

Kjaer, S.B., and Blaabjerg, F. Design optimization of a single phase inverter for photovoltaic applications. In Proceeding of IEEE PESC, Acapulco, Mexico, 1183–1190, 2003.

Kjaer, S.B., Pedersen, J.K., and Blaabjerg, F. A review of single phase grid-connected inverters for photovoltaic modules. *IEEE Transactions on Industry Applications*, 41(5): 1292–1306, September/October 2005.

Kobayashi, K., Takano, I., and Sawada, Y. A study on a two stage maximum power point tracking control of a photovoltaic system under partially shaded insolation conditions. *IEEE Power Engineering Society General Meeting*, 4: 2612–2617, 2003.

Kolhe, M. Techno-Economic Optimum Sizing of a Stand-Alone Solar Photovoltaic System. *IEEE Transaction on Energy Conversion*, 24: 511–519, 2009.

Koutroulis, E., and Kalaitzakis, K. Novel battery charging regulation system for photovoltaic applications. *Proceedings of Institute of Electrical Engineering, Electrical Power Applications*, 151(2): 191–197, 2004.

Koutroulis, E., Kalaitzakis, K., and Voulgaris, N.C. Development of a microcontroller-based, photovoltaic maximum power point tracking control system. IEEE Transactions on Power Electronics, 16(21): 46–54, January 2001.

Kuo, Y.-C., Liang, T.-J., and Chen, J.-F. Novel maximum-power point tracking controller for photovoltaic energy conversion system. *IEEE Transactions on Industrial Electronics*, 48(3): 594–601, June 2001.

Lasseter, R.H. Microgrids and distributed generation. *Journal of Energy Engineering, American Society of Civil Engineers*, 133(3): 144–149, September 2007.

Lee, K.-J., Lee, J.-P., Shin, D., Yoo, D.-W. and Kim, H.-J. A novel grid synchronization PLL method based on adaptive low-pass notch filter for grid-connected PCS. *IEEE Transactions on Industrial Electronics*, 61(1): 292–301, 2014.

Lettl, J., Bauer, J., and Linhart, L. Comparison of different filters types for grid connected inverters. PIERS Proceedings, Marrakesh, MOROCCO, 1426–1429, 2011.

Li, H., Peng, F.Z., and Lawler, J.S. Modeling, simulation, and experimental verification of soft-switched bi-directional dc-dc converters. In Proceedings of IEEE APEC, Anaheim, CA, 736–742, 2001.

Li, H., Xu, Y., and Adhikari, S. *Real and Reactive Power Control of a Three Phase Single Stage PV System and PV Voltage Stability*. San Diego, CA: IEEE, 2012.

Li, Q., and Wolfs, P., A review of the single phase photovoltaic module integrated converter topologies with three different DC link configurations. *IEEE Transactions on Power Electronics*, 23(3): 1320–1333, May 2008.

Li, Y.W. Control and resonance damping of voltage source and current source converters with LC Filters. *IEEE Transaction on Industry Electronics*, 56(5): 1511–1521, May 2009.

Lim, Y.H., and Hamill, D.C. Simple maximum power point tracker for photovoltaic arrays. *Electronics Letters*, 36: 997–999, May 2000.

Liserre, M., Blaabjerg, F., and Hansen, S. Design and control of an LCL-filter-based three-phase cctive rectifier. *IEEE Transactions on Industry Applications*, 41(5): 1281–1291, September/October 2005.

Liu, D., and Li, H. A ZVS bi-directional DC–DC converter for multiple energy storage elements. *IEEE Transactions on Power Electronics*, 21(5): 1513–1517, September 2006.

Lo, Y.K., Lin, J.Y., and Wu, T.Y. Grid connection technique for a photovoltaic system with power factor correction. *IEEE Power Electronics and Drives System*, 1: 522–525, 2005.

Luigi, G., de Brito, M.A.G., Sampaio, L.P., and Canesin, C.A. Evaluation of integrated inverter topologies for low power PV systems. *IEEE International Conference on Clean Electrical Power (ICCEP)*, Ischia, Italy, 2011.

Marion, B., Adelstein, J., Boyle, K., Hayden, H., Hammond, B., Fletcher, T., Canada, B., Narang, D., Shugar, D., Wenger, H., Kimber, A., Mitchell, L., Rich, G., and Townsend, T. Performance parameters for grid-connected PV systems. *31st IEEE Photovoltaics Specialists Conference and Exhibition Lake Buena Vista, Florida*, 3–7 January 2005.

Martins, D.C., and Demonti, R. Grid connected PV system using two energy processing stages. In Proceedings of IEEE Photovoltaic Specialists Conference, New Orleans, LA, 1649–1652, 2002.

Masoum, M.A.S., Dehbonei, H., and Fuchs, E.F. Theoretical and experimental analysis of photovoltaic systems with voltage and current based maximum power-point tracking. *IEEE Transactions on Energy Converters*, 7(4): 514–522, 2002.

Mellit, A. Development of an expert configuration of standalone power PV system based on adaptive neuro-fuzzy inference system (ANFIS). *IEEE Mediterranean Electro Technical Conference, Malaga*, Spain, 893–896, 2006.

Mellit, A., and Kalogirou, S.A. ANFIS-based modelling for photovoltaic power supply system: A case study. *Renewable Energy*, 36: 250–258, 2011.

Messaia, A., Mellitb, A., Pavanc, A.M., Guessoumd, A., and Mekkia, H. FPGA-based implementation of a fuzzy controller (MPPT) for photovoltaic module. *Energy Conversion and Management*, 52(7): 2695–2704, 2011.

Meydbray, J., and Dross, F. PV Module Reliability Scorecard Report 2016. Bærum, Norway: DNV GL, 2016. Retrieved from www.gutami-solar.com/nl/merken/file/download/57.

Mohan, N., Undeland, T.M., and Robbins, W.P. *Power electronics converters, applications, and design*. Third Edition. The University of Michigan: Wiley India Press, Reprint 2009.

Moore, L.M., and Post, H.N. Five years of operating experience at a large, utility-scale photovoltaic generating plant. *Progress in Photovoltaics: Research and Applications*, 16: 249–259, 2008.

Mosal, M., Abu Rubl, H., Ahmed, M.E., and Rodriguez, J. *Modified MPPT with Using Model Predictive Control for Multilevel Boost Converter*. Montreal, QC: IEEE, pp. 5080–5085, 2012.

Mosal, M., Abu Rubl, H., Ahmed, M.E., and Rodriguez, J. *Modified MPPT with using model predictive control for multilevel boost converter*. IEEE Conference on Industrial Electronics Society, 38: 5080–5085, 2012.

Muñoz, M., Correcher, A., Ariza, E., Garcia, E., and Ibanez, F. Fault detection and isolation in a photovoltaic system. International Conference on Renewable Energies and Power Quality (ICREPQ'15) La Coruna, Spain, 25–27 March, 2015.

Muoks, P.I., Hawel, M.E., Gargoomed, A., and Negnevitsky, M. *Modeling and Simulation and Hardware Implementation of a PV Power Plant in a Distributed Energy Generation System*. Washington, DC: IEEE, 2013.

Murtaza, A.F., Ahmed Sher, H., Chiaberge, M., Boero, D., De Giuseppe, M., and Addoweesh, K.E., *A Novel Hybrid MPPT Technique for Solar PV Applications Using Perturb and Observe and Fractional Open Circuit Voltage Techniques*. Prague, Czech Republic: IEEE Conference Publications, 2010.

Muselli, M., Notton, G., Canaletti, J.L., and Louche, A. Utilization of meteosat satellite-derived radiation data for integration of autonomous photovoltaic solar energy systems in remote areas. *Energy Conversion and Management*, 39(1–2): 1–19, 1998.

Nansen, R. *Sun Power: The Global Solution for the Coming Energy Crisis*. Washington, DC: Ocean Press, 1995.

National Renewable Energy Laboratory, www.nrel.gov.

Nilsson, D. Fault detection in photovoltaic systems. Masters thesis, KTH Royal Institute of Technology, School of Computer Science and Communications. 2014.

Notte, Francine V. "The Micro-Inverter." Retrieved from www.esf.edu/outreach/spare/documents/TheMicro-Inverter.ppt.

Pandiarajan, N., and Ranganath, M. Mathematical modeling of photovoltaic module with Simulink. *IEEE Xplore Conference: Electrical Energy Systems (ICEES)*, Newport Beach, CA, 258–263, 2011.

Pandiarajan, N., Ramaprabha, R., and Ranganath, M. Application of circuit model for photovoltaic energy conversion system. *International Journal of Advanced Engineering Technology*, 2012: 118–127, 2011.

Papavasilou, A., Papathanassiou, S.A., Manias, S.N., and Demetriadis, G. *Current Control of a Voltage Source Inverter Connected to Grid via LCL Filter*. Orlando, FL: IEEE Conference Publications, 2004.

Parker, S.G., McGrath, B.P., and Holmes, D.G. Regions of active damping control for LCL filters. *IEEE Transactions on Industry Applications*, 50(1): 424–443, January/February 2014.

Partain, L.D. (ed.) *Solar Cells and Their Applications*. New York, NY: John Wiley & Sons, 1995.

Patel, H., and Agarwal, V. Maximum power point tracking scheme for PV systems operating under partially shaded conditions. *IEEE Transactions Industrial Electronics*, 55(4), 1689–1698, 2008.

Paul, M.M.R., Mahalakshmi, R., Karuppasamypandiyan, M., Bhuvanesh, A., and Ganesh, R.J. Classification and detection of faults in grid connected photovoltaic system. *International Journal of Scientific & Engineering Research*, 7(4): 149–154, April, 2016.

Peng, F.Z., Li, H., Su, G., and Lawler, J.S. A new ZVS bidirectional DC–DC converter for fuel cell and battery application. *IEEE Transactions on Power Electronics*, 19(1): 54–65, January 2004.

Prayas Energy Group. Grid Integration of Distributed Solar Photovoltaics (PV) in India: A review of technical aspects, best practices and the way forward, Pune, India, July 2014.

Pukhrem, S., Conlon, M., and Basu, M. The challenges faced while integrating large PV plants into the electrical grid. PhD presentation, University College Dublin, School of Electrical and Electronic Engineering, Dublin, Ireland, February 24, 2015.

Purcell, E.M., and Morin, D.J. *Electricity and Magnetism*. Massachusetts: Cambridge University Press, pp. 364. ISBN 1107014026, 2013.

PVsyst SA. *PVsyst: Photovoltaic software*. Software program. Satigny, Switzerland: PVsyst SA, ©2012–2017.

Rafi, K.M., Jamil M., Mubeen, A. Energy potential & generation through IRES in City of Faridabad-India. IEEE Conference Publications, Global Humanitarian Technology Conference-USA, pp. 238–241, Digital Object 10.1109/GHTC.2011.87, Meeting, July 2009.

Ragasudha, C.P., Algazar, H., Monier, A.L., El-halim, H.A., and Salem, M.E.K. Maximum power point tracking using fuzzy logic control. *Electrical Power and Energy Systems*, 392: 1–8, 2012.

Rashid, M.H. *Introduction to Buck-Boost Converter Design*, Third Edition. Boston: Pearson Education.

Reisi, A.R., Moradi, M.H., and Jamsab, S. Classification and comparison of maximum power point tracking techniques for photovoltaic system: A review. *Renewable and Sustainable Energy Reviews*, 19: 433–443, 2013.

Reisi, A.R., Moradi, M.H., and Jamsab, S. Classification and comparison of maximum power point tracking techniques for photovoltaic system: A review. *Renewable and Sustainable Energy Reviews*, 19: 433–443, 2013.

Rivera. O. *Maximum Power Point Tracking using the Optimal Duty ratio for DC-DC Convertors and Load Matching in Photovoltaic Applications*. Austin, TX: IEEE, pp. 987–991, 2008.

Rizwan, M., Jamil, M., and Kothari, D. P. Assessment of SPV system using ANN and VHDL. *2010 Joint International Conference on Power Electronics Drives and Energy Systems & 2010 Power India*, 2010.

Rizwan, M., Jamil, M., and Kothari, D. P. Generalized neural network approach for global solar energy estimation in India. *IEEE Transactions on Sustainable Energy*, 3(3): 576–584, 2012.

Said, S., Massoud, A., Benammar, M., and Ahmed, S.A. MATLAB/Simulink based photovoltaic array model employing Simpower system toolbox. *Journal of Energy and Power Engineering*, 6: 1965–1975, 2012.

Salah, C.B., and Ouali, M. Comparison of fuzzy logic and neural network in maximum power point tracker for PV systems. *Electric Power Systems Research*, 81: 43–50, 2011.

Salameh, Z., Dagher, F., and Lynch, W. Step-down maximum power point tracker for photovoltaic systems. *Solar Energy*, 46(5): 270–282, 1991.

Salas, V., Olias, E., Barrado, A., and Zaro, A.L. Review of the maximum power point tracking algorithms for stand-alone photovoltaic system. *Solar Energy Materials & Solar cells*, 90: 1555–1578, 2006.

Samlex Solar. Solar Panel Characteristics, 2004. Retrieved from www.samlexsolar.com/learning-center/solar-panels-characteristics.aspx

Schimpf, F., and Norum, L.E. Grid connected converters for photovoltaic, state of art, ideas for improvement of transformer-less inverters. *Nordic Workshop on Power and Industrial Electronics, (NORPIE)*, Espoo, Finland, June 9–11, 2008.

Selvaraj, J., Rahim, N.A., and Krismadinata, C. *Digital PI Control for Grid Connected PV Inverters*. Singapore: IEEE Conference Publications, 2008.

Senjyu, T., and Uezato, K. Maximum power point tracker using fuzzy control for photovoltaic arrays. Proceedings of IEEE International Conference Industrial Technology, Guangzhou, China, 143–147, 1994.

Shademan, M., and Hangan, H. *Wind Loading on Solar Panels at Different Inclination Angles*. Paper presented at 11th Americas Conference on Wind Engineering, San Juan, Puerto Rico, June 22-26, 2009. Retrieved from http://www.iawe.org/Proceedings/11ACWE/11ACWE-Shademan.pdf.

Shanthi, T., and Ammasai Gounden N. Power electronics interface for grid connected PV array using boost converter and line-commutated inverter with MPPT. *International Conference on Intelligent and Advanced Systems*, Kuala Lumpur, Malaysia, 882–886, 2007.

Shanthi, T., and Gounden, N.A. Power electronics interface for grid connected PV array using boost converter and line-commutated inverter with MPPT. International Conference on Intelligent and Advanced Systems, pp. 882-886 2007.

Sharma, A. A comprehensive study of solar power in India and World, *Renewable and Sustainable Energy*, 15(4): 1767–1776, 2011.

Sharma, O.P., and Trikha, P. *Geothermal Energy and Its Potential in India*. Indira Gandhi National Open University, New Delhi. 18 August 2013.

Shen, G., Xu, D., Cao, L., and Zhu, X. An improved control strategy for grid-connected voltage source inverters with an LCL filter. *IEEE Transaction on Power Electronics*, 23(4): 1899–1906, July 2008.

Shetty, V.J., and Kulkarni, K. Estimation of cost analysis for 500 kW grid connected solar photovoltaic plant. *International Journal of Current Engineering and Technology*, 4(3): 1859–1861, June 2014.

Silva, M.F.A. Analysis of new indicators for fault detection in grid connected PV systems for BIPV applications. Master's thesis, Universidade de Lisboa, 2014. Retrieved from http://repositorio.ul.pt/bitstream/10451/11392/1/ulfc107345_tm_Mario_Silva.pdf.

Silvestre, S., Chouder, A., and Karatepe, E. Automatic fault detection in grid connected PV systems. Solar Energy, 94: 119–127, 2013.

Silvestre, S., Chouder, A., and Karatepe, E. Automatic monitoring and fault detection of grid connected PV systems. *Solar Energy*, 94, 119–127, May 2013.

Silvestre, S., Chouder, A., and Karatepe, E. Automatic fault detection in grid connected PV systems. Solar Energy, 94: 119–127, 2013.

Simoes, M.G., Franceschetti, N.N., and Friedhofer, M. A fuzzy logic based photovoltaic peak power tracking control. In Proceedings IEEE International Symposium Industrial Electronics, Pretoria, South Africa, 300–355, 1998.

Singh, S.N. *Renewable Energy Generation in India: Present Scenario and Future Prospects*. Calgary, AB: IEEE Power & Energy Society General, 2009.

SMA Solar Technology. *Performance ratio: Quality factor for the PV plant* (Technical Information). Niestetal, Germany: SMA Solar Technology AG, 2010. Retrieved from http://files.sma.de/dl/7680/Perfratio-TI-en-11.pdf.

Solar Choice Pty. Ltd., New South Wales, Australia, ©2008-2017. Retrieved from www.solarchoice.net.au/

"Solar Inverter Overview and System Topologies," Texas Instruments. Retrieved from http://www.ti.com.

SolarPV.co.uk. Solar PV Arrays. Retrieved from www.solarpv.co.uk/solar-pv-arrays.html.

Stettler, S., Toggweiler, P., and Remund, J. PYCE: Satellite photovoltaic yield control and evaluation. Proceedings of the 21st European Photovoltaic Solar Energy Conference, 2613–2616, 2006.

Syafaruddin, W., Karatepe, E., and Hiyama, T. Controlling of artificial neural network for fault diagnosis of photovoltaic array. IEEE 16th International Conference on Intelligent System Application to Power Systems, Hersonissos, Greece, 1–6, 2011.

Toodeji, H., Farokhnia, N., and Riahy, G.H. Integration of PV module and STATCOM to extract maximum power from PV. *International Conference on Electrical Power and Energy Conversion System*, Sharjah, United Arab Emirates, 1–6, 2009.

Tsai, H.-L., Tu, C.-S., and Su, Y.-J. Development of generalized photovoltaic model using MATLAB/SIMULINK. Proceedings of the World Congress on Engineering and Computer Science WCECS, San Francisco, 2008.

Twining, E., and Holmes, D.G. Grid current regulation of a three phase voltage source inverter with an LCL input filter. *IEEE Transactions on Power Electronincs*, 18(3): 888–895, May 2003.

Vergura, S., Acciani, G., Amoruso, V., Patrono, G.E., and Vacca F. Descriptive and inferential statistics for supervising and monitoring the operation of PV plants. *IEEE Transactions On Industrial Electronics*, 56: 4456–4463, 2009.

Verma, A., Singh, B., and Sahani, D.T. *Grid Interfaced Photovoltaic Power Generating System with Power Quality Improvement at AC Mains*. Kathmandu, Nepal: IEEE ICSET, 2012.

Verma, A., Singh, B., and Sahani, D.T. Power balanced theory based grid interfaced photovoltaic power generating system with power quality improvement at AC mains. In Proceeding for the IEEE International conference on Power Electronics, Drives and Energy systems, New Delhi, India, 16–19 December 2012.
Vieira, J., and Mota, A. Maximum power point tracker applied in batteries charging with PV panels. *IEEE International Symposium on Industrial Electronics, USIE*, 1: 202–206, 2008.
Villalva, M.G., Gazoli, J.R., and Filho, E.R. Comprehensive approach to modeling & Simulation of Photovoltaic Arrays. *IEEE Transactions on Power Electronics*, 24(5): 1198–1208, 2009.
Wadhwa, C.L. *Electrical Power Systems*. Chennai, India: New Age International, 2009.
Wai, R., and Wang, W. Grid-connected photovoltaic generation system. *IEEE Transactions on Circuits and Systems-I*, 55(3): 953–963, April 2008.
Watson, J.D., Watson, N.R., Santos-Martin, D., Wood, A.R., Lemon, S., and Miller. Impact of solar photovoltaics on the low-voltage distribution network in New Zealand. *IET Generation, Transmission & Distribution*, 10: 1–9, 2015.
Wheeler, H.A. Simple inductance formulas for radio coils, *Proceedings of the Institute of Radio Engineers*, 16(10): 1398, 22 June 2015.
Wikipedia, Energy storage. 2017. Retrieved from https://en.wikipedia.org/wiki/Energy_storage.
Wikipedia. "Solar micro-inverter," August 2017. Retrieved from https://en.wikipedia.org/wiki/Solar_micro-inverter.
Won, C.Y., Kim, D.H., Kim, S.C., Kim, W.S., and Kim, H.S. A new maximum power point tracker of photovoltaic arrays using fuzzy controller. In Proceedings of 25th Annual IEEE Power Electronics Specialist Conference, Taipei, Taiwan, 396–403, 1994.
Worden, J., and M. Zuercher-Martinson, M. How Inverters Work: What Goes on Inside the Magic Box, SolarPro, Issue 2.3, Apr/May 2009. Retrieved from http://solarprofessional.com/articles/products-equipment/inverters/how-inverters-work#.WZYZDIqQzOQ.
World Scientific and Engineering Academy and Society (WSEAS). www.wseas.us/
www.mnre.gov.in/solar-mission/jnnsm/introduction.
www.renewableenergyworld.com/articles/2016/
works.bepress.com
www.arcmroofing.com
www.en.wikipedia.org
www.greenrhinoenergy.com
www.internationalscienceindex.org
www.intersolar.in
www.ncpre.iitb.ac.in
www.pes.ee.ethz.ch
www.powermin.nic.in
www.scirp.org
www.scribd.com
www.waset.org
Yaduvir, S. *Electro Magnetic Field Theory*. India: Pearson Education, pp. 65, 2011.
Yang, Y., and Blaabjerg, F. Synchronization in single phase grid connected photovoltaic systems under grid faults. 3rd IEEE International Symposium on Power Electronics for Distributed Generation Systems (PEDG), Aalborg, Denmark, 2012.
Yazdani, D., Bakhshai, A., and Jain, P. *Adaptive Notch Filtering Based GridSynchronization Techniques for Converter Interfaced Distributed Generation Systems*. Porto, Portugal: IEEE, 3963–3969, 2009.
Yazdani, D., Bakhshai, A., Joós, G., and Mojiri, M., *A real-time extraction of harmonic and reactive current in a nonlinear load for grid-connected converters. IEEE Transactions on Industrial Electronics*, 56(6): 2185–2189, June 2009.
Yu, G.J., Jung, M.W., Song, J., Cha, I.S., and Hwang, I.H. Maximum power point tracking with temperature compensation of photovoltaic for air conditioning system with fuzzy controller. In Proceedings IEEE Photovoltaic Specialists Conference, Washington, DC, 1429–1432, 1996.

Zainudin, H.N., and Mekhilef, S. Comparison study of maximum power point tracker techniques for PV systems. *Proceedings of the 14th International Middle East Power Systems Conference (MEPCON'10)*, Cairo University, Egypt, 750–775, 2010.
Zainudin, H.N., and Mekhilef, S., Comparison study of maximum power point tracker techniques for PV systems. Proceedings of the 14th International Middle East Power Systems Conference (MEPCON'10), Cairo University, Egypt, 19–21 December 2010.
Zgonena, T. "Photovoltaic DC Arc-Fault Circuit Protection and UL Subject 1699B." Presentation for Solar ABCs Annual Stakeholder Meeting, September 14, 2012. Retrieved from www.solarabcs.org/about/publications/meeting_presentations_minutes/2012/09/pdfs/8-UL1699B_PV_AFCI-Zgonena-14Sept2012.pdf.
Zuercher-Pattinson, M., and Schlesinger, R. "AC Side Surge Protection,"from Protecting Electrical PV Systems from the Effects of Lightning (p. 5), white paper, n.d. Retrieved from www.solectria.com//site/assets/files/1483/solectria_protecting_electrical_pv_systems_white_paper.pdf.
Zweibel, K. *Harnessing Solar Power: The Photovoltaics Challenge*. New York, NY: Plenum Press, 1990.
Zweibel, K. Thin Films: Past, Present, Future. *Progress in Photovoltaics*. 3(5), 279–293, Sept/Oct 1995; revised April 1997.

Index

I

J

K

L

M

Q

R